THE PHILOSOPHY OF EXPERTISE

THE PHILOSOPHY OF

Expertise

EDITED BY

EVAN SELINGER AND ROBERT P. CREASE

COLUMBIA UNIVERSITY PRESS

NEW YORK

Columbia University Press
Publishers Since 1893
New York   Chichester, West Sussex

Library of Congress Cataloging-in-Publication Data
The philosophy of expertise / edited by Evan Selinger and Robert P.
    Crease.
        p. cm.
    Includes bibliographical references and index.
    ISBN 978-0-231-13644-0 (cloth : alk. paper)
    1. Expertise. 2. Authority.   I. Selinger, Evan, 1974–
II. Crease, Robert P.
BD209.P45 2006
001 — dc22

                                        2005036403

∞
Columbia University Press books are printed on permanent and durable acid-free paper.
Printed in the United States of America

CONTENTS

THE PHILOSOPHY OF EXPERTISE

EVAN SELINGER AND ROBERT P. CREASE

THIS COLLECTION OF papers surveys the theoretical dimensions of experts and expertise. While the topic lies at the intersection of core issues involving learning, skill, knowledge, and experience, it has rarely attracted explicit philosophical attention. Philosophers broach the issue indirectly through rubrics such as "authority," "power," "rational debate," and "colonization of the lifeworld," but a clearly defined discussion of the philosophical aspects of expertise is in an early stage of development. Thus Alvin Goldman, in the first article of this collection, characterizes his work on expertise and deference as exploratory, and contrasts it with a definitive approach. This volume is intended as another step in that direction. It is intended to display the key issues and indispensable features that would need to be included in any framework that would address expertise in a philosophically comprehensive way.

The lack of explicit philosophical inquiry is surprising, given how thoroughly experts and expertise permeate society, on many levels and in both the public and private spheres. Economic, scientific, social, and technological decisions are regularly delegated to experts. Politicians, judges, businessmen, and ordinary citizens rely on experts not only in circumstances involving technical dimensions but also routinely, in everyday situations. Many essential features of contemporary life, from automotive repair to medical treatment to cybersecurity measures, are now unthinkable without experts and expertise. Each profession is in effect constituted by expertise, and the need to reproduce, maintain, and supervise it. The dependency on expertise involves different kinds of experts, takes place in a wide range of circumstances, and is due to a variety of reasons.

The nature, scope, and application of expertise appears deceptively simple both to understand and to cope with. The word "expert" comes from the Latin *expertus*, the past participle of *experiri*, to try. According to the *Oxford English Dictionary*, the following are all definitions of expert: "experienced in," "having experience of," "trained by experience or practice," "tried by experience," "to know by experience," "one who has gained skill from experience," "one whose special knowledge of skill causes him or her to be regarded as an authority," and "specialist." An expert knows things by virtue of being experienced in the relevant ways of the world; when one uttered the Latin phrase *expertus sum* in the seventeenth century, one meant, "I have experienced." These definitions and formulations suggest that it should be relatively easy to clarify the characteristics of expertise, at least for practical purposes.

But ongoing and seemingly intractable problems with the practical use of experts and expertise hint at the existence of deep and unresolved issues. Despite the vast and profound reliance of modern society on experts and expertise, their value and use are constantly being challenged. Disagreement exists over who the real experts are, and over the objectivity of certain fields of expertise. Different kinds of experts, ranging from governmental experts, industry experts, public-interest experts, and self-appointed experts, have contested one another's claims and produced contradictions in expert testimony. In light of these contradictions, the American judicial system appears to be in a state of turmoil. While social psychological research on laypersons' reasoning about scientific evidence suggests that jurors may be unable to recognize fundamental problems with the methodology used by experts in their research, expertise plays such a pivotal role in the courtroom that the introduction of expert testimony by the prosecution tends to increase conviction rates, while testimony from a defense expert tends to lessen the likelihood of convictions.

Ironically, even if it were desirable to be skeptical about expert pronouncements, such skepticism could never become radical. The ability to doubt particular expert claims necessitates appealing to an alternative base of knowledge, much of which must also be imparted by experts.

Simply put, the modern world needs experts, and frequently consults them. This has necessitated practical solutions in the different arenas in which experts are employed. The courtroom is one of the most significant of these arenas: Which experts should be admitted into courtrooms, and how should expert testimony be handled in court decisions? The practical value of answering these two questions hardly needs to be mentioned. It is socially beneficial to insure that charlatans and ideologues are screened out and that knowledgeable and unbiased scientists be fully and clearly heard. This was the subject of a landmark and still controversial 1993 decision by the U.S. Supreme Court, *Daubert v. Merrell Dow Pharmaceuticals. Daubert* was an attempt to lay out a practical solution to the difficulties, but questions about its effectiveness remain, and it has had a tendency to produce expensive and time-consuming pretrial hearings, which have been viewed as discouraging the kind of sound gatekeeping that the decision was intended to establish. The ruling continues to generate much discussion and literature, as the problem of expert testimony remains a pressing issue.

Other arenas continue to face difficult and unresolved issues involving the role of experts: legislation, peer review, child rearing and education, and the media. The practical solutions reached here, too, are challenged and subject to ongoing discussion. The vast changes that modern society is undergoing, as well as the seeming failure of practical solutions to the problem of experts and expertise, have led some people to suggest that the boundaries between these and other social groups are vanishing. In the prestigious "Essays on Science and Society" section of the magazine *Science*, Bruno Latour suggests that science and society are so deeply "entangled" that the traditional voices of expertise are finding that they need to take "research" from a diverse array of nontraditional stakeholders seriously.[1] Michel Callon, one of Latour's science and technology studies collaborators, goes so far as to characterize his research as taking part in "the subversion of modern institutional framing by challenging the oppositions that we had come to take for granted, yet that are crucial, such as the distinction between expert and layperson."[2]

The deepest and most controversial issue surrounding experts and expertise, however, concerns the nature and exercise of political rationality itself. The fundamental tension can be posed most simply as follows: Democracy depends not only on an educated citizenry, but also on educated decision making in the myriad judgments that have to be made in the day-to-day operations of government. The United States and other countries attempt to accomplish this in a practical way by incorporating experts into governmental operations through their participation in various agencies, regulatory and review panels, committees, and advisory capacities. Yet the authority so conferred on experts seems to collide with the democratic and antielitist urge to accord equality to all opinions; it also risks elitism, ideology, and partisanship sneaking in under the guise of value-neutral expertise. Indeed, charges that this has in fact happened in how the United States has handled its science policy have become more direct and specific than ever before. Can this fundamental tension ever be resolved? If it cannot be resolved, how can it be best approached? The answers to these questions are central to the theory and practice of democracy.

This volume, the first multiauthored collection to explore the fundamental philosophical issues involving experts and expertise, surveys the key issues which, we believe, are ultimately the key to understanding the practical controversies surrounding issues of experts and expertise. These controversies, in which opposing armies mass on either side of a seemingly firm boundary, provide, we think, a perfect foil for exhibiting the value of philosophical exploration and clarification. For the role of philosophy is to detect, expose, and rework the confusions and ambiguities that make such boundaries seem intractable, thus giving rise to the conflict. It does not dissolve the boundaries, but rather restores the weight to the important differences, showing where room for moving the boundaries exists. A philosophical exploration would articulate and clarify the issues that make the practical solutions continually subject to challenge. It may not immediately produce more compelling practical solutions, and may even seem superficially to interfere with solutions insofar as it can point to further complexities. Nevertheless, addressing the philosophical issues is essential to generating effective practical solutions. One has only to point out that any practical solution necessarily involves taking a stand, however implicitly, on philosophical issues. The *Daubert* decision, for instance, was shaped by philosophical concepts, and such concepts are bound to be heavily implicated in the inevitable elaborations and revisions of this decision. By this avenue alone, philosophical analysis of the legitimacy and consistency of positions is bound to affect pragmatic decision making about expertise.

Furthermore, the philosophical analysis of expertise has important implications for philosophy itself. For example, one of the key assumptions of post-foundational philosophy—which finds expression in American pragmatism from Dewey through Rorty, and in postmodern and postphenomenological movements—is that it is untenable to ground ethical and political normativity in foundational claims that are of a universal and ahistorical kind. If it should it turn out that rational principles of deferring to experts can be provided satisfactorily, it would force profound revisions to this assumption. Contrarily, if rational principles for deferring to experts cannot be generated in a satisfying manner, and expertise can only be examined in contextually specific ways, then any philosophical project that aims to provide general rational criteria for deciding when to defer and when not to defer to experts is probably a waste of time, and alternative frameworks for thinking critically about the subject will be required.

For these reasons, it is timely to pose fundamental questions concerning the justification of expert authority and the nature and scope of expert decision making from within a context that minimizes instrumental biases. In this spirit, fifteen essays have been selected that highlight the following questions:

- To what extent is the reliance on expertise localized in an embodied human subject or distributed in a network of tools and practices?
- What models apply to the reasoning process of experts?
- What is the logic or epistemic structure of an appeal to intellectual authority?
- What is the social character of expertise?
- What is the ideological character of expertise?
- What is the relation, if any, between an "objective" and a "reputational" sense of expertise?
- Are there epistemic openings that might make it possible for laypeople to critique expert advice?
- What forces drive what Shelia Jasanoff calls the new class struggle between experts and nonexperts?
- What kind of apprenticeship is involved in the acquisition of expertise, and how does this affect the answers to the above questions?
- What problems does the reliance on expertise pose for liberal democratic theory, either by the way it fosters inequalities or governmental non-neutrality in debates involving a technological dimension?

• How has the incorporation of science into state bureaucracy reshaped scientific activity and its reliance on experts?

This volume divides the issues raised by these questions into three sections, corresponding to three different perspectives that one can adopt with respect to the question of expertise. One can focus on expertise as a relation between those who "have" the expertise and those who use or "consume" it, or one can focus on one or the other pole of this relation. While these three perspectives inevitably overlap, breaking down the question of expertise in this way has the advantage of allowing key issues to emerge. Approaching expertise as a relation allows one to identify and describe the characteristics and dynamics of this relation. Looking at the expert pole of the relationship illuminates the special kind of knowledge that expertise involves, while looking at the user pole brings to the fore an entirely different set of issues, including the evaluation of expert pronouncements and the lay regulation of experts.

Part 1, "Trusting Experts," contains essays that approach the problem of expertise through the issue of trust and deference, thus as a relation between two parties. These five essays study this relationship and its structures in a range of different contexts. Alvin Goldman, for instance, is concerned with the extent to which a rational approach can be constructed for dealing with the novice-expert problem. Harry Collins and Robert Evans call the issue of deference to experts "the pressing intellectual problem of the age," and point out certain serious limitations to the way that the social studies of science have addressed this deference. Scott Brewer's landmark article analyzes the reasoning process of judges and juries in evaluating the trustworthiness of expert witness testimony. Stephen Turner's essay explores the implications, apparent contradictions, and possible rapprochement between liberal democracy and the reliance on expert knowledge. In defending the existence of moral expertise, Peter Singer's article discusses the different positions that moral philosophers and laypersons find themselves in when trying to formulate a moral decision.

Part 2, "Expertise and Practical Knowledge," is concerned with the phenomenological clarification of expertise, which seeks to discover universal structures of embodied cognition and affect. These essays focus on one pole of the expert-layperson relationship by addressing the structure of the expert's knowledge and of the process of knowledge acquisition. Julia Annas argues that the reason why we find it hard to understand the Greek view of morality is that in modern times, obligations are characterized according to prescriptive rules for doing or refraining from doing some action. By contrast, she suggests that the Greeks viewed moral conduct as practical expertise, and that in order to understand their view, a practical understanding of expertise is required. Hubert Dreyfus, a leading figure in the phenomenological tradition, argues that the practical involvements of living bodies ultimately grounds any knowledge about the world, including expert knowledge, and the essay reprinted here contains his influential if controversial model of skill acquisition. Our own article discusses and critiques that model, pointing out its value in shifting the discussion of expertise to universal structures of embodied cognition and affect, but also indicating its drawbacks, in the form of a lack of hermeneutic sensitivity. Evan Selinger and John Mix contend that in order to develop a meaningful concept that can designate adequately the identity and social value of the special kind of expertise that Harry Collins characterizes as "interactional expertise," both phenomenological and sociological resources need to be consulted. Hélène Mialet's case study of two experts, meanwhile, examines how, on the one hand, expertise is materially distributed in specific tools, practices, and social networks, while on the other hand, it still requires an analysis of subjectivity in order to be correctly understood.

Part 3, "Contesting Expertise," examines issues arising from the social and technical externalization of expertise. These issues tend to center around the other pole of the expert-layperson relationship (though inevitably there is some overlap with other essays) in focusing on the social impact of expertise and its implications and conflicts. Paul Feyerabend's essay outlines his famous claim concerning the ideological construction of an expert of view, and his defense of lay regulation of expertise. Steve Fuller's article argues that expertise is "constitutively social." John Hardwig's essay attempts to develop rational maxims for expertise and an account of the rationality of deferring to the authority of experts. It is a classic formulation of the expert-layperson divide and its implications. Edward Said identifies a dominant reason why contemporary humanities scholarship fails to aid emancipatory politics: the modern university is beholden to a ideology of expertise; the separation of fields, objects, and disciplines is sanctioned to such a degree that isolated interpretative fiefdoms flourish and it becomes institutionally undesirable to produce writing that can address a broader community meaningfully. Don Ihde discusses the difference between literary-aesthetic criticism and science criticism in order to determine why it is difficult for the layperson to criticize expert pronouncements.

Nearly all of these essays adopt a critical rather than a synoptic approach; that is, they aim to weaken or undermine common and deeply held assumptions rather than to build a framework for understanding expertise that will stand on its own. However, this volume will contribute to the construction of that framework because, by virtue of its range of critical approaches and philosophical perspectives, it points out certain features that will be indispensable to that framework. The essays in the first section show that expertise is not a simple property or relation, but rather a multilayered set of interactions between individuals and institutions whose complexity will have to be respected in any adequate account. The essays in the second section reveal the need for a phenomenology that can detail how expertise is acquired and maintained and that is hermeneutically sensitive—that is, that takes into account the cultural and situational embeddedness of experts. The essays in the third section highlight the need to be aware of how experts and those whom they address are socially and culturally positioned with respect to each other; this in turn will help clarify the conditions under which outsiders might question or challenge experts with respect to their judgments.

NOTES

1. Bruno Latour, "From the World of Science to the World of Research?" *Science* 280, issue 5361 (1998): 208–209.

2. Michel Callon, "Researchers in the Wild and the Rise of Technical Democracy," unpublished paper, 2001.

PART I
TRUSTING EXPERTS

THE FIVE ARTICLES in this section represent different approaches to describing the process of trusting and deferring to experts. Alvin Goldman's article, "Experts: Which Ones Should You Trust?" serves as an excellent introduction to the challenges that expertise poses to traditional philosophy. The article reminds us of how ancient the problem is—"it was squarely formulated and addressed by Plato," Goldman points out—and of how essential the issue is in the contemporary world, yet also how unfamiliar the problem is within the framework of contemporary epistemology. Goldman mentions several contemporary philosophers who have broached the problem indirectly, outlines what philosophical realignments might have to be made to address the problem more fully, and reaches some tentative conclusions. He argues, for instance, that even though it seems that by virtue of not being experts, laypeople cannot rationally determine whom amongst competing experts to trust, an analysis of the novice-expert problem can reveal a rational way for dealing with the problem of deference. Goldman's "exploratory" article, as he calls it, outlines the form that a comprehensive philosophical approach to experts and expertise might take.

Harry Collins and Robert Evans's contribution, "The Third Wave of Science Studies: Studies of Expertise and Experience," approaches the same issue from a quite different vantage point: the social studies of science. The authors argue that while their field has demystified expert authority, it has not yet addressed adequately the issue of expertise itself, therein making social studies of science research less socially and politically relevant than it could be. The authors know full well the importance of the issue, writing that "one of the most important contributions of sociology of scientific knowledge has been to make it much harder to make the claim: 'Trust scientists because they have special access to truth.' Our question is: 'If it no longer is clear that scientists and technologists have special access to the truth, why should their advice be specially valued?' This, we think, is the pressing intellectual problem of the age." Collins and Evans note that the price that science and technology studies (STS) practitioners have paid for deconstructing expert authority is that their analyses fail to treat expertise as real and as being more substantive than the judgment of history and the play of competing social attributions. "By emphasizing the ways in which scientific knowledge is like other forms of knowledge, sociologists have become uncertain about how to speak about what makes it different; in the same way they have become unable to distinguish between experts and non-experts." For this reason, Collins and Evans say, STS practitioners have become stuck in several different kinds of problems on the issue, including what the authors call the "expert's regress," which they define by analogy with Collins's concept "experimenter's regress": "Because of the experimenter's regress the class of successful replications of an experiment can be identified only with hindsight; because of the expert's regress the class of experts can only be identified with hindsight." What the "expert's regress" suggests is that sociologists can only identify recognized experts after historical judgment on the matter has been passed. This retrospective dimension limits the value of sociological analysis because "decisions of public concern have to be made . . . before the scientific dust has settled because the pace of politics is faster than the pace of scientific consensus formation." Collins and Evans thus plead for STS theorists to change their research focus and start pursuing "Studies of Experience and Expertise." In this new research program, general claims that demarcate experts from nonexperts should be made, as well as normative pronouncements about how different kinds of experts ought to interact with other kinds of experts as well with laypersons. As things currently stand, Collins and Evans show, STS analyses that primarily deconstruct expertise are too narrowly construed; they do not display enough sensitivity to the fact that from both practical and theoretical points of view, the concept of expertise has become so deeply woven in shared understanding that discussions about ultimate goods and ends would be rendered unintelligible if the appeal to expertise were not an appeal to epistemological superiority.

Scott Brewer's influential and frequently cited article "Scientific Expert Testimony and Intellectual Due Process" attempts to analyze the reasoning process of judges and juries in evaluating the testimony of expert witnesses. He carefully lays out what it is the law wants from expert witnesses, and surveys the assumptions that are buried in the practical alternatives to going about it—devoting an entire section, for instance, to the "philosophy of science" of the *Daubert* decision. Brewer's article is especially insightful in the way it brings to light how the traditional ways of evaluating experts (such as credentials, reputation, and presentation) often lead to epistemologically arbitrary results, which run against an important norm that Brewer calls "intellectual due process." The ultimate aim of Brewer's article is to examine "whether and under what conditions it is possible for scientific experts to convey justified beliefs to nonexpert judges and juries."

Stephen Turner's essay, "What Is the Problem with Experts?" aims to "break down the problem of expertise to its elements in political theory." The standard way of doing so, Turner argues, treats expertise as a privileged kind of knowledge, which runs squarely against "the conditions of rough equality presupposed by democratic accountability." Moreover, noting certain arguments by Feyerabend, Turner notes that a state's interest in expertise would seem to conflict with "the impartiality the liberal state must exhibit in the face of rival opinions in order to ensure the possibility of a genuine, fair and open discussion." Focusing on these two problems, Turner carries out a searching exploration of the implications, apparent contradictions, and possible peaceful coexistence between reliance on expertise and liberal democracy.

Finally, Peter Singer argues in "Moral Experts" against the frequently advanced view that there is no such thing as moral expertise—that moral philosophers possess no special information nor special reasoning faculty and are not moral experts. But in the absence of a perfect and undisputed moral code, Singer points out, morally good individuals must think through for themselves the question of what ought to be done. This often difficult and time-consuming process requires gathering information, assessing it, incorporating it into one's moral views, and guarding against bias. Someone having more familiarity with moral concepts and arguments, and more time to gather and reflect on information, is more likely to reach a "soundly based decision." This, Singer concludes, entails not only the possibility of moral expertise, but also that moral philosophers have advantages in achieving it over others.

These essays address expertise as a dynamic relationship involving trust and deference, discuss the structures that can be identified and described in that relationship, and illustrate a range of contexts in which that relationship functions.

1. EXPERTS

WHICH ONES SHOULD YOU TRUST?

ALVIN I. GOLDMAN

1. EXPERTISE AND TESTIMONY

Mainstream epistemology is a highly theoretical and abstract enterprise. Traditional epistemologists rarely present their deliberations as critical to the practical problems of life, unless one supposes—as Hume, for example, did not—that skeptical worries should trouble us in our everyday affairs. But some issues in epistemology are both theoretically interesting and practically quite pressing. That holds of the problem to be discussed here: how laypersons should evaluate the testimony of experts and decide which of two or more rival experts is most credible. It is of practical importance because in a complex, highly specialized world people are constantly confronted with situations in which, as comparative novices (or even ignoramuses), they must turn to putative experts for intellectual guidance or assistance. It is of theoretical interest because the appropriate epistemic considerations are far from transparent; and it is not clear how far the problems lead to insurmountable skeptical quandaries. This paper does not argue for flat-out skepticism in this domain; nor, on the other hand, does it purport to resolve all pressures in the direction of skepticism. It is an exploratory paper, which tries to identify problems and examine some possible solutions, not to establish those solutions definitively.

The present topic departs from traditional epistemology and philosophy of science in another respect as well. These fields typically consider the prospects for knowledge acquisition in "ideal" situations. For example, epistemic agents are often examined who have unlimited logical competence and no significant limits on their investigational resources. In the present problem, by contrast, we focus on agents with stipulated epistemic constraints and ask what they might attain while subject to those constraints.

Although the problem of assessing experts is non-traditional in some respects, it is by no means a new problem. It was squarely formulated and addressed by Plato in some of his early dialogues, especially the *Charmides*. In this dialogue Socrates asks whether a man is able to examine another man who claims to know something to see whether he does or not; Socrates wonders whether a man can distinguish someone who pretends to be a doctor from someone who really and truly is one (*Charmides* 170d–e). Plato's term for posing the problem is *techne*, often translated as "knowledge" but perhaps better translated as "expertise" (see Gentzler 1995, LaBarge 1997).[1]

In the recent literature the novice/expert problem is formulated in stark terms by John Hardwig (1985, 1991). When a layperson relies on an expert, that reliance, says Hardwig, is necessarily *blind*.[2] Hardwig is intent on denying full-fledged skepticism; he holds that the receiver of testimony can acquire "knowledge" from a source. But by characterizing the receiver's knowledge as "blind", Hardwig seems to give us a skepticism of sorts. The term "blind" seems to imply that a layperson (or a scientist in a different field) cannot be *rationally justified* in trusting an expert. So his approach would leave us with testimonial skepticism concerning rational justification, if not knowledge.

There are other approaches to the epistemology of testimony that lurk in Hardwig's neighborhood. The authors I have in mind do not explicitly urge any form of skepticism about testimonial belief; like Hardwig, they wish to expel the specter of skepticism from the domain of testimony. Nonetheless, their solution to the problem of testimonial justification appeals to a minimum of *reasons* that a hearer might have in trusting the assertions of a source. Let me explain who and what I mean.

The view in question is represented by Tyler Burge (1993) and Richard Foley (1994), who hold that the bare assertion of a claim by a speaker gives a hearer prima facie reason to accept it, quite independently of anything the hearer might know or justifiably believe about the speaker's abilities, circumstances, or opportunities to have acquired the claimed piece of knowledge. Nor does it depend on empirically acquired evidence by the hearer, for example, evidence that speakers generally make claims only when they are in a position to know whereof they speak. Burge, for example, endorses the following Acceptance Principle: "A person is entitled to accept as true something that is presented as true and that is intelligible to him, unless there are stronger reasons not to do so" (1993: 467). He insists that this principle is not an empirical one; the "justificational force of the entitlement described by this justification is not constituted or enhanced by sense experiences or perceptual beliefs" (1993: 469). Similarly, although Foley does not stress the a priori status of such principles, he agrees that it is reasonable of people to grant *fundamental* authority to the opinions of others, where this means that it is "reasonable for us to be influenced by others even when we have no special information indicating that they are reliable" (1994: 55). Fundamental authority is contrasted with *derivative* authority, where the latter is generated from the hearer's *reasons for thinking* that the source's "information, abilities, or circumstances put [him] in an especially good position" to make an accurate claim (1994: 55). So, on Foley's view, a hearer need not have such reasons about a source to get prima facie grounds for trusting that source. Moreover, a person does not need to acquire empirical reasons for thinking that people generally make claims about a subject only when they are in a position to know about that subject. Foley grants people a fundamental (though prima facie) epistemic right to trust others even in the absence of any such empirical evidence.[3] It is in this sense that Burge's and Foley's views seem to license "blind" trust.

I think that Burge, Foley, and others are driven to these sorts of views in part by the apparent hopelessness of reductionist or inductivist alternatives. Neither adults nor children, it appears, have enough evidence from their personal perceptions and memories to make cogent inductive inferences to the reliability of testimony (cf. Coady 1992). So Burge, Foley, Coady and others propose their "fundamental" principles of testimonial trustworthiness to stem the potential tide of testimonial skepticism. I am not altogether convinced that this move is necessary. A case might be made that children are in a position to get good inductive evidence that people usually make claims about things they are in a position to know about.

A young child's earliest evidence of factual reports is from face-to-face speech. The child usually sees what the speaker is talking about and sees that the speaker also sees what she is talking about, e.g., the furry cat, the toy under the piano, and so forth. Indeed, according to one account of cognitive development (Baron-Cohen 1995), there is a special module or mechanism, the "eye-direction detector", that attends to other people's eyes, detects their direction of gaze, and interprets them as "seeing" whatever is in the line of sight.[4] Since seeing commonly gives rise to knowing, the young child can determine a certain range of phenomena within the ken of speakers. Since the earliest utterances the child encounters are presumably about these *speaker-known* objects or events, the child might easily conclude that speakers usually make assertions about things within their ken. Of course, the child later encounters many utterances where it is unclear to the child whether the matters reported are, or ever were, within the speaker's ken. Nonetheless, a child's early experience is of speakers who talk about what they apparently know about, and this may well be a decisive body of empirical evidence available to the child.

I don't want to press this suggestion very hard.[5] I shall not myself be offering a full-scale theory about the justification of testimonial belief. In particular, I do *not* mean to be advancing a sustained defense of the reductionist or inductivist position. Of greater concern to me is the recognition that a hearer's evidence about a source's reliability or unreliability can often *bolster* or *defeat* the hearer's justifiedness in accepting testimony from that source. This can be illustrated with two examples.

As you pass someone on the street, he assertively utters a sophisticated mathematical proposition, which you understand but have never previously assessed for plausibility. Are you justified in accepting it from this stranger? Surely it depends partly on whether the speaker turns out to be a mathematics professor of your acquaintance or, say, a nine-year-old child. You have prior evidence for thinking that the former is in a position to know such a proposition, whereas the latter is not. Whether or not there is an a priori principle of default entitlement of the sort endorsed by Burge and Foley, your empirical evidence about the identity of the speaker is clearly relevant. I do not claim that Burge and Foley (etc.) cannot handle these cases. They might say that your recognition that the speaker is a math professor *bolsters* your *overall* entitlement to accept the proposition (though not your prima facie entitlement); recognizing that it is a child *defeats* your prima facie entitlement to accept the proposition. My point is, however, that your evidence about the properties of the speaker is crucial evidence for your overall entitlement to accept the speaker's assertion. A similar point holds in the following example. As you relax behind the wheel of your parked car, with your eyes closed, you hear someone nearby describing the make and color of the passing cars. Plausibly, you have prima facie justification in accepting those descriptions as true, whether this prima facie entitlement has an a priori or inductivist basis. But if you then open your eyes and discover that the speaker is himself blindfolded and not even looking in the direction of the passing traffic, this prima facie justification is certainly defeated. So what you empirically determine about a speaker can make a massive difference to your overall justifiedness in accepting his utterances.

The same obviously holds about two putative experts, who make conflicting claims about a given subject-matter. Which claim you should accept (if either) can certainly be massively affected by your empirical discoveries about their respective abilities and opportunities to know the truth of the matter (and to speak sincerely about it). Indeed, in this kind of case, default principles of the sort advanced by Burge and Foley are of no help whatever. Although a hearer may be prima facie entitled to believe each of the speakers, he cannot be entitled *all things considered* to believe both of them; for the propositions they assert, we are supposing, are incompatible (and transparently incompatible to the hearer). So the hearer's all-things-considered justifiedness vis-à-vis their claims will depend on what he empirically learns about each speaker, or about the opinions of other speakers. In the rest of this paper I shall investigate the kinds of empirical evidence that a novice hearer might have or be able to obtain for believing one putative expert rather than her rival. I do not believe that we need to settle the "foundational" issues in the general theory of testimony before addressing this issue. This is the working assumption, at any rate, on which I shall proceed.[6]

2. The Novice/Expert Problem Versus the Expert/Expert Problem

There are, of course, degrees of both expertise and novicehood. Some novices might not be so much less knowledgeable than some experts. Moreover, a novice might in principle be able to turn himself into an expert, by improving his epistemic position vis-à-vis the target subject-matter, e.g., by acquiring more formal training in the field. This is not a scenario to be considered in this paper, however. I assume that some sorts of limiting factors—whether they be time, cost, ability, or what have you—will keep our novices from becoming experts, at least prior to the time by which they need to make their judgment. So the question is: Can novices, while remaining novices, make justified judgments about the relative credibility of rival experts? When and how is this possible?

There is a significant difference between the novice/expert problem and another type of problem, the expert/expert problem. The latter problem is one in which experts seek to appraise the authority or credibility of other experts. Philip Kitcher (1993) addresses this problem in analyzing how scientists ascribe authority to their peers. A crucial segment of such authority ascription involves what Kitcher calls "calibration" (1993: 314–22). In *direct* calibration a scientist uses his own opinions about the subject-matter in question to evaluate a target scientist's degree of authority. In *indirect* calibration, he uses the opinions of still other scientists, whose opinions he has previously evaluated by direct calibration, to evaluate the target's authority. So here too he starts from his own opinions about the subject-matter in question.

By contrast, in what I am calling the novice/expert problem (more specifically, the novice/2-expert problem), the novice is not in a position to evaluate the target experts by using his own opinion; at least he does not think he is in such a position. The novice either has no opinions in the target domain, or does not have enough confidence in his opinions in this domain to use them in adjudicating or evaluating the disagreement between the rival experts. He thinks of the domain as properly requiring a certain expertise, and he does not view himself as possessing this expertise. Thus, he cannot use opinions of his own in the domain of expertise—call it the *E-domain*—to choose between conflicting experts' judgments or reports.

We can clarify the nature of the novice/expert problem by comparing it to the analogous listener/eyewitness problem. (Indeed, if we use the term "expert" loosely, the latter problem may just be a species of the novice/expert problem.) Two putative eyewitnesses claim to have witnessed a certain crime. A listener—for example, a juror—did not himself witness the crime, and has no prior beliefs about who committed it or how it was committed. In other words, he has no personal knowledge of the event. He wants to learn what transpired by listening to the testimonies of the eyewitnesses. The question is how he should adjudicate between their testimonies if and when they conflict. In this case, the E-domain is the domain of propositions concerning the actions and circumstances involved in the crime. This E-domain is what the listener (the "novice") has no prior opinions about, or no opinions to which he feels he can legitimately appeal. (He regards his opinions, if any, as mere speculation, hunch, or what have you.)

It may be possible, at least in principle, for a listener to make a reasonable assessment of which eyewitness is more credible, even without having or appealing to prior opinions of his own concerning the E-domain. For example, he might obtain evidence from others as to whether each putative witness was really present at the crime scene, or, alternatively, known to be elsewhere at the time of the crime. Second, the listener could learn of tests of each witness's visual acuity, which would bear on the accuracy or reliability of their reports. So in this kind of case, the credibility of a putative "expert's" report can be checked by such methods as independent verification of whether he had the opportunity and ability to see what he claims to have seen. Are analogous methods available to someone who seeks to assess the credibility of a "cognitive" expert as opposed to an eyewitness expert?

Before addressing this question, we should say more about the nature of expertise and the sorts of experts we are concerned with here. Some kinds of experts are unusually accomplished at certain skills, including violinists, billiards players, textile designers, and so forth. These are not the kinds of experts with which epistemology is most naturally concerned. For epistemological purposes we shall mainly focus on cognitive or intellectual experts: people who have (or claim to have) a superior quantity or level of knowledge in some domain and an ability to generate new knowledge in answer to questions within the domain. Admittedly, there are elements of skill or know-how in intellectual matters too, so the boundary between skill expertise and cognitive expertise is not a sharp one. Nonetheless, I shall try to work on only one side of this rough divide, the intellectual side.

How shall we define expertise in the cognitive sense? What distinguishes an expert from a layperson, in a given cognitive domain? I'll begin by specifying an objective sense of expertise, what it is to *be* an expert, not what it is to have a reputation for expertise. Once the objective sense is specified, the reputational sense readily follows: a reputational expert is someone widely believed to be an expert (in the objective sense), whether or not he really is one.

Turning to objective expertise, then, I first propose that cognitive expertise be defined in "veritistic" (truth-linked) terms. As a first pass, experts in a given domain (the E-domain) have more beliefs (or high degrees of belief) in true propositions and/or fewer beliefs in false propositions within that domain than most people do (or better: than the vast majority of people do). According to this proposal, expertise is largely a comparative matter. However, I do not think it is wholly comparative. If the vast majority of people are full of false beliefs in a domain and Jones exceeds them slightly by not succumbing to a few falsehoods that are widely shared, that still does not make him an "expert" (from a God's-eye point of view). To qualify as a cognitive expert, a person must possess a substantial body of truths in the target domain. Being an expert is not simply a matter of veritistic superiority to most of the community. Some non-comparative threshold of veritistic attainment must be reached, though there is great vagueness in setting this threshold.

Expertise is not all a matter of possessing accurate information. It includes a capacity or disposition to deploy or exploit this fund of information to form beliefs in true answers to new questions that may be posed in the domain. This arises from some set of skills or techniques that constitute part of what it is to be an expert. An expert has the (cognitive) know-how, when presented with a new question in the domain, to go to the right sectors of his information-bank and perform appropriate operations on this information; or to deploy some external apparatus or data-banks to disclose relevant material. So expertise features a propensity element as well as an element of actual attainment.

A third possible feature of expertise may require a little modification in what we said earlier. To discuss this feature, let us distinguish the *primary* and *secondary* questions in a domain. Primary questions are the principal questions of interest to the researchers or students of the subject-matter. Secondary questions concern the existing evidence or arguments that bear on the primary questions, and the assessments of the evidence made by prominent researchers. In general, an expert in a field is someone who has (comparatively) extensive knowledge (in the weak sense of knowledge, i.e., true belief) of the state of the evidence, and knowledge of the opinions and reactions to that evidence by prominent workers in the field. In the central sense of "expert" (a strong sense), an expert is someone with an unusually extensive body of knowledge on both primary and secondary questions in the domain. However, there may also be a weak sense of "expert", in which it includes someone who merely has extensive knowledge on the secondary questions in the domain. Consider two people with strongly divergent views on the primary questions in the domain, so that one of them is largely right and the other is largely wrong. By the original, strong criterion, the one who is largely wrong would not qualify as an expert. People might disagree with this as the final word on the matter. They might hold that anyone with a thorough knowledge of the existing evidence and the differing views held by the workers in the field deserves to be called an expert. I concede this by acknowledging the weak sense of "expert".

Applying what has been said above, we can say that an expert (in the strong sense) in domain D is someone who possesses an extensive fund of knowledge (true belief) and a set of skills or methods for apt and successful deployment of this knowledge to new questions in the domain. Anyone purporting to be a (cognitive) expert in a given domain will claim to have such a fund and set of methods, and will claim to have true answers to the question(s) under dispute because he has applied his fund and his methods to the question(s). The task for the layperson who is consulting putative experts, and who hopes thereby to learn a true answer to the target question, is to decide who has superior expertise, or who has better deployed his expertise to the question at hand. The novice/2-experts problem is whether a layperson can *justifiably* choose one putative expert as more credible or trustworthy than the other with respect to the question at hand, and what might be the epistemic basis for such a choice?[7]

3. ARGUMENT-BASED EVIDENCE

To address these issues, I shall begin by listing five possible sources of evidence that a novice might have, in a novice/2-experts situation, for trusting one putative expert more than another. I'll then explore the prospects for utilizing such sources, depending on their availability and the novice's exact circumstance. The five sources I shall discuss are:

(A) Arguments presented by the contending experts to support their own views and critique their rivals' views.
(B) Agreement from additional putative experts on one side or other of the subject in question.
(C) Appraisals by "meta-experts" of the experts' expertise (including appraisals reflected in formal credentials earned by the experts).
(D) Evidence of the experts' interests and biases vis-à-vis the question at issue.
(E) Evidence of the experts' past "track-records".

In the remainder of the paper, I shall examine these five possible sources, beginning, in this section, with source (A).[8]

There are two types of communications that a novice, N, might receive from his two experts, E_1 and E_2.[9] First, each expert might baldly state her view (conclusion), without supporting it with any evidence or argument whatever. More commonly, an expert may give detailed support to her view in some public or professional context, but this detailed defense might only appear in a restricted venue (e.g., a professional conference or journal) that does not reach N's attention. So N might not encounter the two experts' defenses, or might encounter only very truncated versions of them. For example, N might hear about the experts' views and their support from a second-hand account in the popular press that does not go into many details. At the opposite end of the communicational spectrum, the two experts might engage in a full-scale debate that N witnesses (or reads a detailed reconstruction of). Each expert might there present fairly developed arguments in support of her view and against that of her opponent. Clearly, only when N somehow encounters the experts' evidence or arguments can he have evidence of type (A). So let us consider this scenario.

We may initially suppose that if N can gain (greater) justification for believing one expert's view as compared with the other by means of their arguments, the novice must at least understand the evidence cited in the experts' arguments. For some domains of expertise and some novices, however, even a mere grasp of the evidence may be out of reach. These are cases where N is an "ignoramus" vis-à-vis the E-domain. This is not the universal plight of novices. Sometimes they can understand the evidence (in some measure) but aren't in a position, from personal knowledge, to give it any credence. Assessing an expert's evidence may be especially difficult when it is disputed by an opposing expert.

Not every statement that appears in an expert's argument need be epistemically inaccessible to the novice. Let us distinguish here between *esoteric* and *exoteric* statements within an expert's discourse. Esoteric statements belong to the relevant sphere of expertise, and their truth-values are inaccessible to N— in terms of his personal knowledge, at any rate. Exoteric statements are outside the domain of expertise; their truth-values may be accessible to N—either at the time of their assertion or later.[10] I presume that esoteric statements comprise a hefty portion of the premises and "lemmas" in an expert's argument. That's what makes it difficult for a novice to become justified in believing any expert's view on the basis of arguments per se. Not only are novices commonly unable to assess the truth-values of the esoteric propositions, but they also are ill-placed to assess the support relations between the cited evidence and the proffered conclusion. Of course, the proponent expert will claim that the support relation is strong between her evidence and the conclusion she defends; but her opponent will commonly dispute this. The novice will be ill-placed to assess which expert is in the right.

At this point I wish to distinguish *direct* and *indirect argumentative justification*. In direct argumentative justification, a hearer becomes justified in believing an argument's conclusion by becoming justified in believing the argument's premises and their (strong) support relation to the conclusion. If a speaker's endorsement of an argument helps bring it about that the hearer has such justificational status vis-à-vis its premises and support relation, then the hearer may acquire "direct" justification for the conclusion via that speaker's argument.[11] As we have said, however, it is difficult for an expert's argument to produce direct justification in the hearer in the novice/2-expert situation. Precisely because many of these matters are esoteric, N will have a hard time adjudicating between E_1's and E_2's claims, and will therefore have a hard time becoming justified vis-à-vis either of their conclusions. He will even have a hard time becoming justified in trusting one conclusion *more* than the other.

The idea of indirect argumentative justification arises from the idea that one speaker in a debate may demonstrate dialectical superiority over the other, and this dialectical superiority might be a plausible *indicator*[12] for N of greater expertise, even if it doesn't render N directly justified in believing the superior speaker's conclusion. By dialectical superiority, I do not mean merely greater debating skill. Here is an example of what I do mean.

Whenever expert E_2 offers evidence for her conclusion, expert E_1 presents an ostensible rebuttal or defeater of that evidence. On the other hand, when E_1 offers evidence for her conclusion, E_2 never manages to offer a rebuttal or defeater to E_1's evidence. Now N is not in a position to assess the truth-value of

E_2's defeaters against E_2, nor to evaluate the truth-value or strength of support that E_1's (undefeated) evidence gives to E_1's conclusion. For these reasons, E_1's evidence (or arguments) are not directly justificatory for N. Nonetheless, in "formal" dialectical terms, E_1 seems to be doing better in the dispute. Furthermore, I suggest, this dialectical superiority may reasonably be taken as an indicator of E_1's having superior expertise on the question at issue. It is a (nonconclusive) indicator that E_1 has a superior fund of information in the domain, or a superior method for manipulating her information, or both.

Additional signs of superior expertise may come from other aspects of the debate, though these are far more tenuous. For example, the comparative quickness and smoothness with which E_1 responds to E_2's evidence may suggest that E_1 is already well familiar with E_2's "points" and has already thought out counterarguments. If E_2's responsiveness to E_1's arguments displays less quickness and smoothness, that may suggest that E_1's prior mastery of the relevant information and support considerations exceeds that of E_2. Of course, quickness and smoothness are problematic indicators of informational mastery. Skilled debaters and well-coached witnesses can appear better-informed because of their stylistic polish, which is not a true indicator of superior expertise. This makes the proper use of indirect argumentative justification a very delicate matter.[13]

To clarify the direct/indirect distinction being drawn here, consider two different things a hearer might say to articulate these different bases of justification. In the case of direct argumentative justifiedness, he might say: "In light of this expert's argument, that is, in light of the truth of its premises and the support they confer on the conclusion (both of which are epistemically accessible to me), I am now justified in believing the conclusion." In indirect argumentative justifiedness, the hearer might say: "In light of the way this expert has argued—her argumentative *performance*, as it were—I can infer that she has more expertise than her opponent; so I am justified in inferring that her conclusion is probably the correct one."

Here is another way to explain the direct/indirect distinction. Indirect argumentative justification essentially involves an *inference to the best explanation*, an inference that N might make from the performances of the two speakers to their respective levels of expertise. From their performances, N makes an inference as to which expert has superior expertise in the target domain. Then he makes an inference from greater expertise to a higher probability of endorsing a true conclusion. Whereas *in*direct argumentative justification essentially involves inference to the best explanation, direct argumentative justification need involve no such inference. Of course, it *might* involve such inference; but if so, the topic of the explanatory inference will only concern the objects, systems, or states of affairs under dispute, not the relative expertise of the contending experts. By contrast, in indirect argumentative justifiedness, it is precisely the experts' relative expertise that constitutes the target of the inference to the best explanation.

Hardwig (1985) makes much of the fact that in the novice/expert situation, the novice lacks the expert's reasons for believing her conclusion. This is correct. Usually, a novice (1) lacks all or some of the premises from which an expert reasons to her conclusion, (2) is in an inferior position to assess the support relation between the expert's premises and conclusions, and (3) is ignorant of many or most of the defeaters (and "defeater-defeaters") that might bear on an expert's arguments. However, although novice N may lack (all or some of) an expert's reasons R for believing a conclusion p, N *might* have reasons R^* for believing *that* the expert has good reasons for believing p; and N might have reasons R^* for believing that one expert has *better* reasons for believing her conclusion than her opponent has for hers. Indirect argumentative justification is one means by which N might acquire reasons R^* without sharing (all or any) of either experts' reasons R.[14] It is this possibility to which Hardwig gives short shrift. I don't say that a novice in a novice/2-expert situation invariably has such reasons R^*; nor do I say that it is easy for a novice to acquire such reasons. But it does seem to be possible.

4. AGREEMENT FROM OTHER EXPERTS: THE QUESTION OF NUMBERS

An additional possible strategy for the novice is to appeal to further experts. This brings us to categories (B) and (C) on our list. Category (B) invites N to consider whether other experts agree with E_1 or with E_2. What proportion of these experts agree with E_1 and what proportion with E_2? In other words, to the extent that it is feasible, N should consult the numbers, or degree of consensus, among all relevant (putative) experts. Won't N be fully justified in trusting E_1 over E_2 if almost all other experts on the subject agree with E_1, or if even a preponderance of the other experts agree with E_1?

Another possible source of evidence, cited under category (C), also appeals to other experts but in a slightly different vein. Under category (C), N should seek evidence about the two rival experts' relative degrees of expertise by consulting third parties' assessments of their expertise. If "meta-experts" give E_1 higher "ratings" or "scores" than E_2, shouldn't N rely more on E_1 than E_2? Credentials can be viewed as a special case of this same process. Academic degrees, professional accreditations, work experience, and so forth (all from specific institutions with distinct reputations) reflect certifications by other experts of E_1's and E_2's demonstrated training or competence. The relative strengths or weights of these indicators might be utilized by N to distill appropriate levels of trust for E_1 and E_2 respectively.[15]

I treat ratings and credentials as signaling "agreement" by other experts because I assume that established authorities certify trainees as competent when they are satisfied that the latter demonstrate (1) a mastery of the same methods that the certifiers deem fundamental to the field, and (2) knowledge of (or belief in) propositions that certifiers deem to be fundamental facts or laws of the discipline. In this fashion, ratings and conferred credentials ultimately rest on basic agreement with the meta-experts and certifying authorities.

When it comes to evaluating specific experts, there is precedent in the American legal system for inquiring into the degree to which other experts agree with those being evaluated.[16] But precedented or not, just how good is this appeal to consensus? If a putative expert's opinion is joined by the consensual opinions of other putative experts, how much warrant does that give a hearer for trusting the original opinion? How much evidential worth does consensus or agreement deserve in the doxastic decision-making of a hearer?

If one holds that a person's opinion deserves prima facie credence, despite the absence of any evidence of their reliability on the subject, then numbers would seem to be very weighty, at least in the absence of additional evidence. Each new testifier or opinion-holder on one side of the issue should add weight to that side. So a novice who is otherwise in the dark about the reliability of the various opinion-holders would seem driven to agree with the more numerous body of experts. Is that right?

Here are two examples that pose doubts for "using the numbers" to judge the relative credibility of opposing positions. First is the case of a guru with slavish followers. Whatever the guru believes is slavishly believed by his followers. They fix their opinions wholly and exclusively on the basis of their leader's views. Intellectually speaking, they are merely his clones. Or consider a group of followers who are not led by a single leader but by a small elite of opinion-makers. When the opinion-makers agree, the mass of followers concur in their opinion. Shouldn't a novice consider this kind of scenario as a possibility? Perhaps (putative) expert E_1 belongs to a doctrinal community whose members devoutly and uncritically agree with the opinions of some single leader or leadership cabal. Should the numerosity of the community make their opinion more credible than that of a less numerous group of experts? Another example, which also challenges the probity of greater numbers, is the example of rumors. Rumors are stories that are widely circulated and accepted though few of the believers have access to the rumored facts. If someone hears a rumor from one source, is that source's credibility enhanced when the same rumor is repeated by a second, third, and fourth source? Presumably not, especially if the hearer knows (or justifiably believes) that these sources are all uncritical recipients of the same rumor.

It will be objected that additional rumor spreaders do not add credibility to an initial rumor monger because the additional ones have no established reliability. The hearer has no reason to think that any of their opinions is worthy of trust. Furthermore, the rumor case doesn't seem to involve "expert" opinions at all and thereby contrasts with the original case. In the original case the hearer has at least some prior reason to think that each new speaker who concurs with one of the original pair has *some* credibility (reliability). Under that scenario, don't additional concurring experts increase the total believability of the one with whom they agree?

It appears, then, that greater numbers should add further credibility, at least when each added opinion-holder has positive initial credibility. This view is certainly presupposed by some approaches to the subject. In the Lehrer-Wagner (1981) model, for example, each new person to whom a subject assigns "respect" or "weight" will provide an extra vector that should push the subject in the direction of that individual's opinion.[17] Unfortunately, this approach has a problem. If two or more opinion-holders are totally *non-independent* of one another, and if the subject knows or is justified in believing this, then the subject's opinion should not be swayed—even a little—by more than one of these opinion-holders. As in the case of a guru and his blind followers, a follower's opinion does not provide any additional grounds for accepting the guru's view (and a second follower does not provide additional grounds for accepting a first follower's view) even if all followers are precisely as reliable as the guru himself (or as one another)—which followers must be, of course, if they believe exactly the same things as the guru (and one another) on the topics in question. Let me demonstrate this through a Bayesian analysis.

Under a simple Bayesian approach, an agent who receives new evidence should update his degree of belief in a hypothesis H by conditioning on that evidence. This means that he should use the ratio (or quotient) of two likelihoods: the likelihood of the evidence occurring if H is true and the likelihood of the evidence occurring if H is false. In the present case the evidence in question is the belief in H on the part of one or more putative experts. More precisely, we are interested in comparing (A) the result of conditioning on the evidence of a single putative expert's belief with (B) the result of conditioning on the evidence of concurring beliefs by two putative experts. Call the two putative experts X and Y, and let X(H) be X's believing H and Y(H) be Y's believing H. What we wish to compare, then, is the magnitude of the likelihood quotient expressed in (1) with the magnitude of the likelihood quotient expressed in (2).

$$(1) \quad \frac{P(X(H) / H)}{P(X(H) / \sim H)}$$

$$(2) \quad \frac{P(X(H) \ \& \ Y(H) / H)}{P(X(H) \ \& \ Y(H) / \sim H)}$$

The principle we are interested in is the principle that the likelihood ratio given in (2) is always larger than the likelihood ratio given in (1), so that an agent who learns that X and Y both believe H will always have grounds for a larger upward revision of his degree of belief in H than if he learns only that X believes H. At least this is so when X and Y are each somewhat credible (reliable). More precisely, such comparative revisions are in order if the agent is *justified* in believing these things in the different scenarios. I am going to show that such comparative revisions are not always in order. Sometimes (2) is not larger than (1); so the agent—if he knows or justifiably believes this—is not justified in making a larger upward revision from the evidence of two concurring believers than from one believer.

First let us note that according to the probability calculus, (2) is equivalent to (3).

$$(3) \quad \frac{P(X(H) \,/\, H) \; P(Y(H) \,/\, X(H) \,\&\, H)}{P(X(H) \,/\, {\sim}H) \; P(Y(H) \,/\, X(H) \,\&\, {\sim}H)}$$

While looking at (3), return to the case of blind followers. If Y is a blind follower of X, then anything believed by X (including H) will also be believed by Y. And this will hold whether or not H is true. So,

(4) $P(Y(H) \,/\, X(H) \,\&\, H) = 1$,

and

(5) $P(Y(H) \,/\, X(H) \,\&\, {\sim}H) = 1$.

Substituting these two values into expression (3), (3) reduces to (1). Thus, in the case of a blind follower, (2) (which is equivalent to (3)) is the same as (1), and no larger revision is warranted in the two-concurring-believers case than in the single-believer case.

Suppose that the second concurring believer, Y, is not a *blind* follower of X. Suppose he would sometimes agree with X but not in all circumstances. Under that scenario, does the addition of Y's concurring belief always provide the agent (who possesses this information) with more grounds for believing H? Again the answer is no. The appropriate question is whether Y is more likely to believe H when X believes H and H is true than when X believes H and H is false. If Y is just as likely to follow X's opinion whether H is true or false, then Y's concurring belief adds nothing to the agent's evidential grounds for H (driven by the likelihood quotient). Let us see why this is so.

If Y is just as likely to follow X's opinion when H is false as when it's true, then (6) holds:

(6) $P(Y(H) \,/\, X(H) \,\&\, H) = P(Y(H) \,/\, X(H) \,\&\, {\sim}H)$

But if (6) holds, then (3) again reduces to (1), because the right-hand sides of both numerator and denominator in (3) are equal and cancel each other out. Since (3) reduces to (1), the agent still gets no extra evidential boost from Y's agreement with X concerning H. Here it is not required that Y is certain to follow X's opinion; the likelihood of his following X might only be 0.80, or 0.40, or whatever. As long as Y is just as likely to follow X's opinion when H is true as when it's false, we get the same result.

Let us describe this last case by saying that Y is a *non-discriminating reflector* of X (with respect to H). When Y is a non-discriminating reflector of X, Y's opinion has no extra evidential worth for the agent above and beyond X's opinion. What is necessary for the novice to get an extra evidential boost from Y's belief in H is that he (the novice) be justified in believing (6'):

$$(6')\ P(Y(H)\ /\ X(H)\ \&\ H) > P(Y(H)\ /\ X(H)\ \&\ {\sim}H)$$

If (6') is satisfied, then Y's belief is at least partly *conditionally independent* of X's belief. Full conditional independence is a situation in which any dependency between X and Y's beliefs is accounted for by the dependency of each upon H. Although full conditional independence is not required to boost N's evidence, *partial* conditional independence is required.[18]

We may now identify the trouble with the (unqualified) numbers principle. The trouble is that a novice cannot automatically count on his putative experts being (even partially) conditionally independent of one another. He cannot automatically count on the truth of (6'). Y may be a non-discriminating reflector of X, or X may be a non-discriminating reflector of Y, or both may be non-discriminating reflectors of some third party or parties. The same point applies no matter how many additional putative experts share an initial expert's opinion. If they are all non-discriminating reflectors of someone whose opinion has already been taken into account, they add no further weight to the novice's evidence.

What type of evidence can the novice have to justify his acceptance of (or high level of credence in) (6')? N can have reason to believe that Y's *route* to belief in H was such that even in possible cases where X fails to recognize H's falsity (and hence believes it), Y *would* recognize its falsity. There are two types of causal routes to Y's belief of the right sort. First, Y's route to belief in H might entirely *bypass* X's route. This would be exemplified by cases in which X and Y are causally independent eyewitnesses of the occurrence or non-occurrence of H; or by cases in which X and Y base their respective beliefs on independent experiments that bear on H. In the eyewitness scenario X might falsely believe H through misperception of the actual event, whereas Y might perceive the event correctly and avoid belief in H. A second possible route to Y's belief in H might go *partly through X* but not involve uncritical reflection of X's belief. For example, Y might listen to X's reasons for believing H, consider a variety of possible defeaters of these reasons that X never considered, but finally rebut the cogency of these defeaters and concur in accepting H. In either of these scenarios Y's partly "autonomous" causal route made him poised to avoid belief in H even though X believes it (possibly falsely). If N has reason to think that

Y used one of these more-or-less autonomous causal routes to belief, rather than a causal route that guarantees agreement with X, then N has reason to accept (6′). In this fashion, N would have good reason to rate Y's belief as increasing his evidence for H even after taking account of X's belief.

Presumably, novices could well be in such an epistemic situation vis-à-vis a group of concurring (putative) experts. Certainly in the case of concurring *scientists*, where a novice might have reason to expect them to be critical of one another's viewpoints, a presumption of partial independence might well be in order. If so, a novice might be warranted in giving greater evidential weight to larger numbers of concurring opinion-holders. According to some theories of scientific opinion formation, however, this warrant could not be sustained. Consider the view that scientists' beliefs are produced entirely by negotiation with other scientists, and in no way reflect reality (or Nature). This view is apparently held by some social constructionists about science, e.g., Bruno Latour and Steve Woolgar (1979/1986); at least this is Kitcher's (1993: 165–66) interpretation of their view.[19] Now if the social constructionists are right, so interpreted, then nobody (at least nobody knowledgeable of this fact) would be warranted in believing anything like (6′). There would never be reason to think that any scientist is more likely to believe a scientific hypothesis H when it's true (and some other scientist believes it) than when it's false (and the other scientist believes it). Since causal routes to scientific belief never reflect "real" facts—they only reflect the opinions, interests, and so forth of the community of scientists—(6′) will never be true. Anybody who accepts or inclines toward the indicated social-constructionist thesis would never be justified in believing (6′).[20]

Setting such extreme views aside, won't a novice normally have reason to expect that different putative experts will have some causal independence or autonomy from one another in their routes to belief? If so, then if a novice is also justified in believing that each putative expert has some slight level of reliability (greater than chance), then won't he be justified in using the numbers of concurring experts to tilt toward one of two initial rivals as opposed to the other? This conclusion might be right when *all* or *almost all* supplementary experts agree with one of the two initial rivals. But this is rarely the case. Vastly more common are scenarios in which the numbers are more evenly balanced, though not exactly equal. What can a novice conclude in those circumstances? Can he legitimately let the greater numbers decide the issue?

This would be unwarranted, especially if we continue to apply the Bayesian approach. The appropriate change in the novice's belief in H should be based on two sets of concurring opinions (one in favor of H and one against it), and it should depend on *how reliable* the members of each set are and on *how (conditionally) independent* of one another they are. If the members of the smaller group are more reliable and more (conditionally) independent of one another than the members of the larger group, that might imply that the evidential weight of the smaller group exceeds that of the larger one. More precisely, it depends on what the novice is *justified* in believing about these matters. Since the novice's justifiedness on these matters may be very weak, there will be many situations in which he has no distinct or robust justification for going by the relative numbers of like-minded opinion-holders.

This conclusion seems perfectly in order. Here is an example that, by my own lights, sits well with this conclusion. If scientific creationists are more numerous than evolutionary scientists, that would not incline me to say that a novice is warranted in putting more credence in the views of the former than in the views of the latter (on the core issues on which they disagree). At least I am not so inclined on the assumption that the novice has roughly comparable information as most philosophers currently have about the methods of belief formation by evolutionists and creationists respectively.[21] Certainly the numbers do not *necessarily* outweigh considerations of individual reliability and mutual conditional independence. The latter factors seem more probative, in the present case, than the weight of sheer numbers.[22]

5. EVIDENCE FROM INTERESTS AND BIASES

I turn now to the fourth source of possible evidence on our original list: evidence of distorting interests and biases that might lie behind a putative expert's claims. If N has excellent evidence for such bias in one expert and no evidence for such bias in her rival, and if N has no other basis for preferential trust, then N is justified in placing greater trust in the unbiased expert. This proposal comes directly from common sense and experience. If two people give contradictory reports, and exactly one of them has a reason to lie, the relative credibility of the latter is seriously compromised.

Lying, of course, is not the only way that interests and biases can reduce an expert's trustworthiness. Interests and biases can exert more subtle distorting influences on experts' opinions, so that their opinions are less likely to be accurate even if sincere. Someone who is regularly hired as an expert witness for the defense in certain types of civil suits has an economic interest in delivering strong testimony in any current trial, because her reputation as a defense witness depends on her present performance.

As a test of expert performance in situations of conflict of interest, consider the results of a study published in the *Journal of American Medical Association* (Friedberg et al., 1999). The study explored the relationship between published research reports on new oncology drugs that had been sponsored by pharmaceutical companies versus those that had been sponsored by nonprofit organizations. It found a statistically significant relationship between the funding source and the qualitative conclusions in the reports. Unfavorable conclusions were reached by 38% of nonprofit-sponsored studies but by only 5% of pharmaceutical company–sponsored studies.

From a practical point of view, information bearing on an expert's interests is often one of the more accessible pieces of relevant information that a novice can glean about an expert. Of course, it often transpires that *both* members of a pair of testifying experts have interests that compromise their credibility. But when there is a non-negligible difference on this dimension, it is certainly legitimate information for a novice to employ.

Pecuniary interests are familiar types of potential distorters of an individual's claims or opinions. Of greater significance, partly because of its greater opacity to the novice, is a bias that might infect a whole discipline, sub-discipline, or research group. If all or most members of a given field are infected by the same bias, the novice will have a difficult time telling the real worth of corroborating testimony from other experts and meta-experts. This makes the numbers game, discussed in the previous section, even trickier for the novice to negotiate.

One class of biases emphasized by feminist epistemologists involves the exclusion or underrepresentation of certain viewpoints or standpoints within a discipline or expert community. This might result in the failure of a community to gather or appreciate the significance of certain types of relevant evidence. A second type of community-wide bias arises from the economics or politics of a sub-discipline, or research community. To advance its funding prospects, practitioners might habitually exaggerate the probativeness of the evidence that allegedly supports their findings, especially to outsiders. In competition with neighboring sciences and research enterprises for both resources and recognition, a given research community might apply comparatively lax standards in reporting its results. Novices will have a difficult time detecting this, or weighing the merit of such an allegation by rival experts outside the field.[23]

6. Using Past Track Records

The final category in our list may provide the novice's best source of evidence for making credibility choices. This is the use of putative experts' past track records of cognitive success to assess the likelihoods of their having correct answers to the current question. But how can a novice assess past track records? There are several theoretical problems here, harking back to matters discussed earlier.

First, doesn't using past track records amount to using the method of (direct) "calibration" to assess a candidate expert's expertise? Using a past track record means looking at the candidate's past success rate for previous questions in the E-domain to which she offered answers. But in our earlier discussion (section 2), I said that it's in the nature of a novice that he has no opinions, or no confidence in his own opinions, about matters falling within the E-domain. So how can the novice have any (usable) beliefs about past answers in the E-domain by which to assess the candidate's expertise? In other words, how can a novice, *qua* novice, have any opinions at all about past track records of candidate experts?

A possible response to this problem is to revisit the distinction between *esoteric* and *exoteric* statements. Perhaps not every statement in the E-domain is esoteric. There may also be a body of exoteric statements in the E-domain, and they are the statements for which a novice might assess a candidate's expertise. But does this really make sense? If a statement is an exoteric statement, i.e., one that is epistemically accessible to novices, then why should it even be included in the E-domain? One would have thought that the E-domain is precisely the domain of propositions accessible only to experts.

The solution to the problem begins by sharpening our esoteric/exoteric distinction. It is natural to think that statements are categorically either esoteric or exoteric, but that is a mistake. A given (timeless) statement is esoteric or exoteric only *relative* to an epistemic standpoint or position. It might be esoteric relative to one epistemic position but exoteric relative to a different position. For example, consider the statement, "There will be an eclipse of the sun on April 22, 2130, in Santa Fe, New Mexico." Relative to the present epistemic standpoint, i.e., the standpoint of people living in the year 2000, this is an esoteric statement. Ordinary people in the year 2000 will not be able to answer this question correctly, except by guessing. On the other hand, on the very day in question, April 22, 2130, ordinary people on the street in Santa Fe, New Mexico will easily be able to answer the question correctly. In that different epistemic position, the question will be an exoteric one, not an esoteric one.[24] You won't need specialized training or knowledge to determine the answer to the question. In this way, the epistemic status of a statement can change from one time to another.

There is a significant application of this simple fact to the expert/novice problem. A novice might easily be able to determine the truth-value of a statement after it has become exoteric. He might be able to tell *then* that it is indeed true. Moreover, he might learn that at an earlier time, when the statement was esoteric for the likes of him, another individual managed to believe it and say that it is (or would be) true. Furthermore, the same individual might repeatedly display the capacity to assert statements that are esoteric at the time of assertion but become exoteric later, and she might repeatedly turn out to have been right, as determined under the subsequently exoteric circumstances. When this transpires, novices can infer that this unusual knower must possess some special manner of knowing—some distinctive expertise—that is not available to them. They presumably will not know exactly what this distinctive manner of knowing involves, but presumably it involves some proprietary fund of information and some methodology for deploying that information. In this fashion, a novice can verify somebody else's expertise in a certain domain by verifying their impressive track record within that domain. And this can be done without the novice himself somehow being transformed into an expert.

The astronomical example is just one of many, which are easily proliferated. If an automobile, an air-conditioning system, or an organic system is suffering some malfunction or impairment, untrained people will often be unable to specify any true proposition of the form, "If you apply treatment X to system Y, the system will return to proper functioning." However, there may be people who can repeatedly specify true propositions precisely of this sort.[25] Moreover, that these propositions are true can be verified by novices, because novices might be able to "watch" the treatment being applied to the malfunctioning system and see that the system returns to proper functioning (faster than untreated systems do). Although the truth of the proposition is an exoteric matter once the treatment works, it was an esoteric matter before the treatment was applied and produced its result. In such a case the expert has knowledge, and can be determined to have had knowledge, at a time when it was esoteric.[26]

It should be emphasized that many questions to which experts provide answers, at times when they are esoteric, are not merely yes/no questions that might be answered correctly by lucky guesses. Many of them are questions that admit of innumerable possible answers, sometimes indefinitely many answers. Simplifying for purposes of illustration, we might say that when a patient with an ailment sees a doctor, he is asking her the question, "Which medicine, among the tens of thousands of available medicines, will cure or alleviate this ailment?" Such a question is unlikely to be answered correctly by mere guesswork. Similarly, when rocket scientists were first trying to land a spaceship on the moon, there were indefinitely many possible answers to the question, "Which series of steps will succeed in landing this (or some) spaceship on the moon?" Choosing a correct answer from among the infinite list of possible answers is unlikely to be a lucky guess. It is feats like this, often involving technological applications, that rightly persuade novices that the people who get the correct answers have a special fund of information and a special methodology for deploying it that jointly yield a superior capacity to get right answers. In this fashion, novices can indeed determine that others are experts in a domain in which they themselves are not.

Of course, this provides no algorithm by which novices can resolve all their two-expert problems. Only occasionally will a novice know, or be able to determine, the track records of the putative experts that dispute an issue before him. A juror in a civil trial has no opportunity to run out and obtain track record information about rival expert witnesses who testify before him. Nonetheless, the fact that novices can verify track records and use them to test a candidate's claims to expertise, at least in principle and in some cases, goes some distance toward dispelling utter skepticism for the novice/2-expert situation. Moreover, the possibility of "directly" determining the expertise of a few experts makes it possible to draw plausible inferences about a much wider class of candidate experts. If certain individuals are shown, by the methods presented above, to have substantial expertise, and if those individuals train others, then it is a plausible inference that the trainees will themselves have comparable funds of information and methodologies, of the same sort that yielded cognitive success for the original experts.[27] Furthermore, to the extent that the verified experts are then consulted as "meta-experts" about the expertise of others (even if they didn't train or credential them), the latter can again be inferred to have comparable expertise. Thus, some of the earlier skepticism engendered by the novice/2-expert problem might be mitigated once the foundation of expert verification provided in this section has been established.

7. CONCLUSION

My story's ending is decidedly mixed, a cause for neither elation nor gloom. Skeptical clouds loom over many a novice's epistemic horizons when confronted with rival experts bearing competing messages. There are a few silver linings, however. Establishing experts' track-records is not beyond the pale of possibility, or even feasibility. This in turn can bolster the credibility of a wider class of experts, thereby laying the foundation for a legitimate use of numbers when trying to choose between experts. There is no denying, however, that the epistemic situations facing novices are often daunting. There are interesting theoretical questions in the analysis of such situations, and they pose interesting practical challenges for "applied" social epistemology. What kinds of education, for example, could substantially improve the ability of novices to appraise expertise, and what kinds of communicational intermediaries might help make the novice-expert relationship more one of justified credence than blind trust.[28]

NOTES

1. Thanks to Scott LaBarge for calling Plato's treatment of this subject to my attention.

2. In his 1991 paper, Hardwig at first says that trust must be "at least partially blind" (p. 693). He then proceeds to talk about knowledge resting on trust and therefore being blind (pp. 693, 699) without using the qualifier "partially".

3. However, there is some question whether Foley can consistently call the epistemic right he posits a "fundamental" one, since he also says that it rests on (A) my justified *self-trust*, and (B) the *similarity* of others to me—presumably the *evidence* I have of their similarity to me (see pp. 63–64). Another question for Foley is how the fundamentality thesis fits with his view that in cases of conflict I have more reason (prima facie) to trust myself than to trust someone else (see p. 66). If my justified trust in others is really fundamental, why does it take a backseat to self-trust?

4. Moreover, according to Baron-Cohen, there is a separate module called the "shared attention mechanism", which seeks to determine when another person is attending to the same object as the self is attending to.

5. For one thing, it may be argued that babies' interpretations of what people say is, in the first instance, constrained by the assumption that the contents concern matters within the speakers' perceptual ken. This is not an empirical finding, it might be argued, but an a priori posit that is used to fix speakers' meanings.

6. Some theorists of testimony, Burge included, maintain that a hearer's justificational status vis-à-vis a claim received from a source depends partly on the justificational status of the source's own belief in that claim. This is a *transpersonal, preservationist,* or *transmissional* conception of justifiedness, under which a recipient is not justified in believing p unless the speaker has a justification and entitlement that he *transmits* to the hearer. For purposes of this paper, however, I shall not consider this transmissional conception of justification. First, Burge himself recognizes that there is such a thing as the recipient's "proprietary" justification for believing an interlocutor's claim, justification localized "in" the recipient, which isn't affected by the source's justification (1993: 485–486). I think it is appropriate to concentrate on this "proprietary" justification (of the recipient) for present purposes. When a hearer is trying to "choose" between the conflicting claims of rival speakers, he cannot appeal to any inaccessible justification lodged in the heads of the speakers. He can only appeal to his *own* justificational resources. (Of course, these might include things *said* by the two speakers by way of defense of their contentions, things which also are relevant to *their own* justifications.) For other types of (plausible) objections to Burge's preservationism about testimony, see Bezuidenhout (1998).

7. In posing the question of justifiedness, I mean to stay as neutral as possible between different approaches to the concept of justifiedness, e.g., between internalist versus externalist approaches to justifiedness. Notice, moreover, that I am not merely asking whether and how the novice can justifiably decide to accept one (candidate) expert's view *outright*, but whether and how he can justifiably decide to give *greater* credence to one than to the other.

8. I do not mean to be committed to the exhaustiveness of this list. The list just includes some salient categories.

9. In what follows I shall for brevity speak about two experts, but I shall normally mean two *putative* experts, because from the novice's epistemic perspective it is problematic whether each, or either, of the self-proclaimed experts really is one.

10. It might be helpful to distinguish *semantically* esoteric statements and *epistemically* esoteric statements. (Thanks to Carol Caraway for this suggestion.) Semantically esoteric statements are ones that a novice cannot assess because he does not even *understand* them; typically, they utilize a technical vocabulary he has not mastered. Epistemically esoteric statements are statements the novice understands but still cannot assess for truth-value.

11. By "direct" justification I do not, of course, mean anything having to do with the basicness of the conclusion in question, in the foundationalist sense of basicness. The distinction I am after is entirely different, as will shortly emerge.

12. Edward Craig (1990: 135) similarly speaks of "indicator properties" as what an inquirer seeks to identify in an informant as a guide to his/her truth-telling ability.

13. Scott Brewer (1998) discusses many of the same issues about novices and experts canvassed here. He treats the present topic under the heading of novices' using experts' "demeanor" to assess their expertise. Demeanor is an especially untrustworthy guide, he points out, where there is a lucrative "market" for demeanor itself—where demeanor is "traded" at high prices (1998: 1622). This practice was prominent in the days of the sophists and is a robust business in adversarial legal systems.

14. Of course, in indirect argumentative justification the novice must at least *hear* some of the expert's premises—or intermediate steps between "ultimate" premises and conclusion. But the novice will not share the expert's *justifiedness* in believing those premises.

15. These items fall under Kitcher's category of "unearned authority" (1993: 315).

16. Appealing to other experts to validate or underwrite a putative expert's opinion—or, more precisely, the *basis* for his opinion—has a precedent in the legal system's procedures for deciding the admissibility of scientific expert testimony. Under the governing test for admitting or excluding such testimony that was applicable from 1923 to 1993, the scientific principle (or methodology) on which a proffered piece of testimony is based must have "gained general acceptance in the particular field in which it belongs". (*Frye v. United States,* 292 F. 1013 D.C. Cir. (1923)). In other words, appeal was made to the scientific community's opinion to decide whether the basis of an expert's testimony is sound enough to allow that testimony into court. This test has been superseded as the uniquely appropriate test in a more recent decision of the Supreme Court (*Daubert v. Merrell Dow Pharmaceuticals,* 509 U.S. 579 (1993)); but the latter decision also appeals to the opinions of other experts. It recommends that judges use a combination of four criteria (none of them necessary or sufficient) in deciding whether proffered scientific expert testimony is admissible. One criterion is the old general acceptance criterion and another is whether the proffered evidence has been subjected to peer review and publication. Peer review, obviously, also introduces the opinions of other experts. Of course, the admissibility of a piece of expert testimony is not the same question as how heavily a hearer—e.g., a juror—should trust such testimony if he hears it. But the two are closely intertwined, since courts make admissibility decisions on the assumption that jurors are likely to be influenced by any expert testimony they hear. Courts do not wish to admit scientific evidence unless it is quite trustworthy. Thus, the idea of ultimately going to the opinions of other experts to assess the trustworthiness of a given expert's proffered testimony is certainly a well-precedented procedure for trying to validate an expert's trustworthiness.

17. Lehrer and Wagner say (p. 20) that one should assign somebody else a positive weight if one does not regard his opinion as "worthless" on the topic in question—i.e., if one regards him as better than a random device. So it looks as if every clone of a leader should be given positive weight—arguably, the same weight as the leader himself, since their beliefs always coincide—as long as the leader receives positive weight. In the Lehrer-Wagner model, then, each clone will exert a positive force over one's own revisions of opinion just as a leader's opinion will exert such force; and the more clones there are, the more force in the direction of their collective opinion will be exerted.

18. I am indebted here to Richard Jeffrey (1992: 109–10). He points out that it is only conditional independence that is relevant in these kinds of cases, not "simple independence" defined by the condition: $P(Y(H) / X(H)) = P(Y(H))$. If X and Y are even slightly reliable independent sources of information about H, they won't satisfy this latter condition.

19. I myself interpret Latour and Woolgar as holding a more radical view, viz., that there is no reality that could causally interact, even indirectly, with scientists' beliefs.

20. This is equally so under the more radical view that there are no truths at all (of a scientific sort) about reality or Nature.

21. More specifically, I am assuming that believers in creation science have greater (conditional) dependence on the opinion leaders of their general viewpoint than do believers in evolutionary theory.

22. John Pollock (in a personal communication) suggests a way to bolster support for the use of "the numbers". He says that if one can argue that $P(X(H) / Y(H) \& H) = P(X(H) / H)$, then one can cumulate testimony on each side of an issue by counting experts. He further suggests that, in the absence of countervailing evidence, we should believe that $P(X(H) / Y(H) \& H) = P(X(H) / H)$. He proposes a general principle of probabilistic reasoning, which he calls "the principle of nonclassical direct inference", to the effect that we are defeasibly justified in regarding additional factors about which we know nothing to be irrelevant to the probabilities. In Pollock (2000) (also see Pollock 1990) he formulates the idea as follows. If factor C is irrelevant (presumably he means *probabilistically* irrelevant) to the causal relation between properties B and A, then conjoining C to B should not affect the probability of something's being A. Thus, if we have no reason to think that C is relevant, we can assume defeasibly that $P(Ax / Bx \& Cx) = P(Ax / Bx)$. This principle can be applied, he suggests, to the case of a concurring (putative) expert. But, I ask, is it generally reasonable for us—or for a novice—to assume that the opinion of one expert is probabilistically irrelevant to another expert's holding the same view? I would argue in the negative. Even if neither expert directly influences the opinion of the other, it is extremely common for two people who work in the same intellectual domain to be influenced, directly or indirectly, by some common third expert or group of experts. Interdependence of this sort is widespread, and could be justifiably believed by novices. Thus, probabilistic irrelevance of the sort Pollock postulates as the default case is highly questionable.

23. In a devastating critique of the mental health profession, Robyn Dawes (1994) shows that the real expertise of such professionals is, scientifically, very much in doubt, despite the high level of credentialism in that professional community.

24. In the present discussion only *epistemic* esotericness, not *semantic* esotericness, is in question (see note 10).

25. They can not only recognize such propositions as true when others offer them; they can also produce such propositions on their own when asked the question, "What can be done to repair this system?"

26. I have discussed such cases in earlier writings: Goldman 1991 and Goldman 1999 (p. 269).

27. Of course, some experts may be better than others at transmitting their expertise. Some may devote more effort to it, be more skilled at it, or exercise stricter standards in credentialing their trainees. This is why good information about training programs is certainly relevant to judgments of expertise.

28. For helpful comments on earlier drafts, I am indebted to Holly Smith, Don Fallis, Peter Graham, Patrick Rysiew, Alison Wylie, and numerous participants at the 2000 Rutgers Epistemology Conference, the philosophy of social science roundtable in St. Louis, and my 2000 NEH Summer Seminar on "Philosophical Foundations of Social Epistemology".

REFERENCES

Baron-Cohen, Simon (1995). *Mindblindness.* Cambridge, MA: MIT Press.

Bezuidenhout, Anne (1998). "Is Verbal Communication a Purely Preservative Process?" *Philosophical Review* 107: 261–88.

Brewer, Scott (1998). "Scientific Expert Testimony and Intellectual Due Process," *Yale Law Journal* 107: 1535–681.

Burge, Tyler (1993). "Content Preservation," *Philosophical Review* 102: 457–88.

Coady, C. A. J. (1992). *Testimony.* Oxford: Clarendon Press.

Craig, Edward (1990). *Knowledge and the State of Nature—An Essay in Conceptual Synthesis.* Oxford: Clarendon Press.

Dawes, Robyn (1994). *House of Cards: Psychology and Psychotherapy Built on Myth.* New York: Free Press.

Foley, Richard (1994). "Egoism in Epistemology," in F. Schmitt, ed, *Socializing Epistemology.* Lanham, MD: Rowman & Littlefield.

Friedberg, Mark et al. (1999). "Evaluation of Conflict of Interest in Economic Analyses of New Drugs Used in Oncology," *Journal of the American Medical Association* 282: 1453–57.

Gentzler, J. (1995). "How to Discriminate between Experts and Frauds: Some Problems for Socratic Peirastic," *History of Philosophy Quarterly* 3:227–46.

Goldman, Alvin (1991). "Epistemic Paternalism: Communication Control in Law and Society," *Journal of Philosophy* 88: 113–31.

Goldman, Alvin (1999). *Knowledge in a Social World.* Oxford: Clarendon Press.

Hardwig, John (1985). "Epistemic Dependence," *Journal of Philosophy* 82: 335–49.

Hardwig, John (1991). "The Role of Trust in Knowledge," *Journal of Philosophy* 88: 693–708.

Jeffrey, Richard (1992). *Probability and the Art of Judgment.* New York: Cambridge University Press.

Kitcher, Philip (1993). *The Advancement of Science.* New York: Oxford University Press.

LaBarge, Scott (1997). "Socrates and the Recognition of Experts," in M. McPherran, ed., *Wisdom, Ignorance and Virtue: New Essays in Socratic Studies.* Edmonton: Academic Printing and Publishing.

Latour, Bruno and Woolgar, Steve (1979/1986). *Laboratory Life: The Construction of Scientific Facts.* Princeton: Princeton University Press.

Lehrer, Keith and Wagner, Carl (1981). *Rational Consensus in Science and Society.* Dordrecht: Reidel.

Pollock, John (1990). *Nomic Probability and the Foundations of Induction.* New York: Oxford University Press.

Pollock, John (2000). "A Theory of Rational Action." Unpublished manuscript, University of Arizona.

2. THE THIRD WAVE OF SCIENCE STUDIES

STUDIES OF EXPERTISE AND EXPERIENCE

H. M. COLLINS AND ROBERT EVANS

lay' man one of the laity; a non-professional person; someone who is not an expert.[1]

THE PROBLEM OF LEGITIMACY AND THE PROBLEM OF EXTENSION

Technical decision-making in the public domain is where the pigeons of much recent social science are coming home to roost. The problem can be stated quite simply: Should the political legitimacy of technical decisions in the public domain be maximized by referring them to the widest democratic processes, or should such decisions be based on the best expert advice? The first choice risks technological paralysis: the second invites popular opposition.

By 'technical decision-making' we mean decision-making at those points where science and technology intersect with the political domain because the issues are of visible relevance to the public: should you eat British beef, prefer nuclear power to coal-fired power stations, want a quarry in your village, accept the safety of anti-misting kerosene as an airplane fuel, vote for politicians who believe in human cloning, support the Kyoto agreement, and so forth. These are areas where both the public and the scientific and technical community have contributions to make to what might once have been thought to be purely technical issues.

Like many others, what we want to do is consider how to make good decisions in the right way. But our particular concern is to find a rationale which is not inconsistent with the last three decades of work in science studies. Our initial claim is that though many others working within the science studies tradition have studied the problem, and contributed valuably to the debate about technical decision-making, they have not solved it in a way that is completely intellectually satisfying. For us to claim to have solved anything would be to give a hostage to fortune, but we think we can indicate, firstly, the reasons why there may be grounds for both academic and political discomfort and, secondly, a direction in which the work might go.

This paper is not about social relations between scientists and society. For example, it is not about whether scientists are trustworthy, or whether they behave in a way that inspires trust in the public, or whether the institutions through which their advice and influence are mediated inspire trust. At least, in so far as the paper is about these things, it is only indirectly about them. What it is about is the reason for using the advice of scientists and technologists in virtue of the things they do *as* scientists and technologists, rather than as individuals or as members of certain institutions. In other words, it is about the value of scientists' and technologists' knowledge and experience as compared with others' knowledge and experience. The dominant and fruitful trend of science studies research in the last decades has been to replace epistemological questions with social questions, but we return to a rather old-fashioned approach, asking about the grounds of knowledge. What is different here, as compared with the debates about the grounds of knowledge that took place before the 'sociological turn' in science studies, is that we try to shift the focus of the epistemology-like discussion from *truth* to *expertise and experience*. We think we need to start pursuing 'SEE'—Studies of Expertise and Experience.

One of the most important contributions of the sociology of scientific knowledge (SSK) has been to make it much harder to make the claim: 'Trust scientists because they have special access to the truth'. Our question is: 'If it is no longer clear that scientists and technologists have special access to the truth, why should their advice be specially valued?' This, we think, is the pressing intellectual problem of the age.[2] Since our answer turns on expertise instead of truth, we will have to treat expertise in the same way as truth was once treated—as something more than the judgement of history, or the outcome of the play of competing attributions. We will have to treat expertise as 'real', and develop a 'normative theory of expertise'.[3]

To those who share our feelings of political and academic unease with the existing situation in science studies, we want to suggest that the problem lies with the tension described in our first paragraph: the tension between what we will call 'the Problem of Legitimacy' and 'the Problem of Extension'. Though science studies has resolved the Problem of Legitimacy by showing that the basis of technical decision-making can and should be widened beyond the core of certified experts, it has failed to solve the Problem of Extension: 'How far should participation in technical decision-making extend?' In other words, science studies has shown that there is more to scientific and technical expertise than is encompassed in the work of formally accredited scientists and technologists, but it has not told us how much more.

To save misunderstanding, let us admit immediately that the practical politics of technical decision-making still most often turn on the Problem of Legitimacy; the most pressing work is usually to try to curtail the tendency for experts with formal qualifications to make *ex-cathedra* judgements curtained with secrecy. Nevertheless, our problem is not this one. Our problem is academic: it is to find a clear rationale for the expansion of expertise. But a satisfying justification for expansion has to show, in a natural way, where the limits are. Perhaps this is not today's practical problem, but with no clear limits to the widening of the base of decision-making it might be tomorrow's. It is just possible, of course, that setting a limit on the extension of expertise will soothe the fears of those who resist any widening of participation, on the grounds that it will open the floodgates of unreason. It is just possible, then, that this exercise will help with today's practical problems, even though we approach the matter with a different aim in view.

PAINTING WAVES WITH A BROAD BRUSH

This is in some ways a polemical paper, and we proceed in a direct manner. We start by sketching idealized models of what we call 'three waves' of science studies. Violence is often done when one compresses the work of many authors and thinkers into a few simple formulae, as the ludicrous accounts of SSK associated with the 'science wars' show us. 'Ask not for the meaning but the use', Wittgenstein tells us; but here we are setting out meanings with a somewhat cavalier attitude to use. So we apologize to all the contributors to these movements whose work we caricature, and hope the violence is not too great; fortunately, the project depends not on historical or scholarly accuracy, but on sketching the broad sweep of ideas.[4]

If what we paint with a broad brush is not totally unreasonable, then it shows that the First Wave of Science Studies had no Problem of Extension, and was unaware of the Problem of Legitimacy. It shows why the Second Wave of Science Studies was good for solving the Problem of Legitimacy that it inherited from Wave One, but replaced it with the Problem of Extension. We propose that the Third Wave of Science Studies (and we might only be labelling a movement that already exists in embryonic form) should accept the Second Wave's solution to the Problem of Legitimacy, but still draw a boundary around the body of 'technically-qualified-by-experience' contributors to technical decision-making.

To show that our argument is more than a programmatic gesture, we will indicate one way to start to build a normative theory of expertise, and what it would mean for technical decision-making. There are, no doubt, many other ways to go about such an exercise, but to focus attention on the goal by providing an example of one approach to it is at least a start.

LANGUAGE AND PRESENTATION

Though we are going to talk about widening participation in technical decision-making, we will abandon the oxymoron 'lay expertise'.[5] As we see at the head of this paper, the dictionary definition of 'layman' includes the sentiment 'someone who is not an expert', and this makes it all too easy to over-interpret the term 'lay expertise'. If those who are not experts can have expertise, what special reference does expertise have? It might seem that anyone can be an expert. We say that those referred to by some other analysts as 'lay experts' are just plain 'experts'—albeit their expertise has not been recognized by certification; crucially, they are not spread throughout the population, but are found in small specialist groups. Instead of using the oxymoron, we will refer to members of the public who have special technical expertise in virtue of experience that is not recognized by degrees or other certificates as 'experience-based experts'.

Since all humans have enormous expertise in language speaking and every other accomplishment that requires an understanding of social context, the term 'experience-based expertise', if it is to do any work in this context, has to be used to refer to specialist abilities. To use the term to mean something more general would strip it of its power to solve the Problem of Extension.[6]

The nature of the exercise means that we need to move swiftly into our arguments and therefore, in the main, we discuss earlier work which bears upon them in an Appendix (pp. 77–95). The main body of the paper still contains some references, acknowledgements, and discussions of previous work (mostly in the Notes), but the Appendix deals more fully, if not exhaustively, with the existing literature. It shows where we agree and disagree, in a substantive way, with certain others who have looked at the same problems, and where our approach differs markedly from work which may, at first glance, look similar. And the Appendix shows, if it needed showing, that the idea of a *normative theory of decision-making* has been discussed by a long line of distinguished scholars. It shows, then, that in so far as we are trying to do something new, our aim is modest—to try to discover a systematic rationale for a *normative theory of expertise* that is compatible with SSK, and that contributes to the normative theory of decision-making that others have essayed. The Appendix follows the major section headings of the main text, in so far as that is possible.

Three Waves of Social Studies of Science

The First Wave of Science Studies

To simplify outrageously, let us say that there was once what seemed to many to be a golden age before 'the expertise problem' raised its head. In the 1950s and 1960s, social analysts generally aimed at understanding, explaining and effectively reinforcing the success of the sciences, rather than questioning their basis.[7] In those days, for social scientists and public alike, a good scientific training was seen to put a person in a position to speak with authority and decisiveness in their own field, and often in other fields too. Because the sciences were thought of as esoteric as well as authoritative, it was inconceivable that decision-making in matters that involved science and technology could travel in any other direction than from the top down. This wave of 'positivism' began to run into shallow academic waters in the late 1960s with Thomas Kuhn's book and all that followed. By the end of the 1970s, *as an academic movement*, it had crashed on to the shore.[8]

THE SECOND WAVE OF SCIENCE STUDIES

The following wave of science studies, which has run from the early 1970s, and continues to run today, is often referred to as 'social constructivism', although it has many labels and many variants. One important variant is the sociology of scientific knowledge (SSK). What has been shown under Wave Two is that it is necessary to draw on 'extra-scientific factors' to bring about the closure of scientific and technical debates—scientific method, experiments, observations, and theories are not enough. With science reconceptualized as a social activity, science studies has directed attention to the uses of scientific knowledge in social institutions such as courts of law, schools, and policy processes such as public inquiries. The emphasis on the 'social construction' of science has meant, however, that when expertise is discussed, the focus is often on the attribution of the label 'expert', and on the way the locus of legitimated expertise is made to move between institutions.

By emphasizing the ways in which scientific knowledge is like other forms of knowledge, sociologists have become uncertain about how to speak about what makes it different; in much the same way, they have become unable to distinguish between experts and non-experts. Sociologists have become so successful at dissolving dichotomies and classes that they no longer dare to construct them. We believe, however, that sociologists of knowledge should not be afraid of *their* expertise, and must be ready to claim their place as experts in the field of knowledge itself.

Sociologists of knowledge must be ready to *build* categories having to do with knowledge; we must be ready, then, to develop a 'knowledge science' using knowledge and expertise as *analysts' categories*. SEE, the Third Wave of Science Studies, is one approach.

DOWNSTREAM TO UPSTREAM

An important strand in our argument is to indicate the compatibility of a normative theory of expertise with what has been achieved in Wave Two. The relationship between Wave One and Wave Two is not the same as the relationship between Wave Two and Wave Three. Wave Two replaced Wave One with much richer descriptions of science, based on careful observation and a relativist methodology (or even philosophy). Wave Two showed that Wave One was intellectually bankrupt. Wave Three, however, does not show that Wave Two is intellectually bankrupt. In this strange sea, Wave Two continues to roll on, even as Wave Three builds up.[9] Wave Three is one of the ways in which Wave Two can be applied to a set of problems that Wave Two alone cannot handle in an intellectually coherent way. Wave Three involves finding a special rationale for science and technology even while we accept the findings of Wave Two—that science and technology are much more ordinary than we once thought. The aim of this paper, one might say, is to hammer a piton into the ice wall of relativism with enough delicacy not to shatter the whole edifice (the destruction that so many critics believe is the only solution).

To be willing to find a rationale for a special place for science and technology, now that so much has been deconstructed under the Second Wave of Science Studies, means *reconstructing* knowledge. As we have said, the Third Wave of Science Studies must emphasize the rôle of expertise as an *analyst's category* as well as an *actor's category*, and this will allow *prescriptive*, rather than merely *descriptive*, statements about the rôle of expertise in the public sphere.

The shift to a prescriptive theory of expertise, as commentators have pointed out, seems incompatible with much that the authors of this paper have previously argued. Commentators have said that it follows from Wave Two analyses, with which the authors have been, and still are, closely associated, that the 'experts' who play a rôle in a debate can be distinguished 'only after the dust has settled, after it becomes clear whose claims became convincing in the ongoing course of things'; and that 'it does not matter who defines the expert, actor or analyst; judgements of who the experts are always lay downstream'.[10] We might label this the 'expert's regress', by analogy with the 'experimenter's regress'. Because of the experimenter's regress, the class of successful replications of an experiment can be identified only with hindsight; because of the expert's regress, the class of experts can be identified only with hindsight. The trouble is that the expert's regress gives no more positive help with the problem of technical decision-making in the public domain than the experimenter's regress gives positive help with settling scientific controversies. But decisions of public concern have to be made according to a timetable established within the political sphere, not the scientific or technical sphere; the decisions have to be made *before* the scientific dust has settled, because the pace of politics is faster than the pace of scientific consensus formation. Political decision-makers are, therefore, continually forced to define classes of expert *before* the dust has settled—before the judgements of history have been made. In defining classes of expert actors in the political sphere, they are making history rather than reflecting on it. What we are arguing is that sociologists of scientific knowledge, *per se*, might also have a duty to make history as well as reflect on it; they have a rôle to play in making history in virtue of *their* area of expertise—'knowledge'.[11]

The dilemma is not a new one, and has been present within Wave Two all along, though largely unremarked. One of the current authors (HMC) has discussed it in the course of his work on artificial intelligence (AI). Here, rather than reflecting upon the way the controversy about AI unfolded, he found himself taking an active part in the controversy, using his *knowledge about knowledge* to contribute to the debate. He referred to this activity, in contrast with more reflective science studies, as 'knowledge science'.[12] We might say that in knowledge science one works to affect the flow of the river of history, rather than examining its turns and eddies. In the same way, what we are doing here is 'upstream work' rather than 'downstream work'.[13]

Doing upstream work without abandoning the insights of Wave Two may involve a degree of compartmentalization of activity, but compartmentalization can often be avoided only on pain of paralysis.[14] It is also worth noting that, for better or worse, in Wave-Two work involving tacit knowledge and its consequences for replication of experiments and so forth, experimental and other skills have always been used as an 'upstream' category—something real and fixed that can be transferred from one person to another, or can fail to be transferred. Indeed that idea *has already* been used upstream in an attempt to smooth and steer the course of science.[15]

Let us try one more way of putting the matter. Wave Two deals with the problem: 'How is scientific consensus formed?' Some form of relativism in respect of the outcome of that consensus is vital if the answer is not to risk circularity. Wave Three deals with the problem: 'How do you make decisions based on scientific knowledge before there is an absolute scientific consensus?' Wave Three does not replace Wave Two because the problems are different. For Wave Three, something in addition to relativism is needed. One way to approach the problem of Wave Three is to look at the way science is granted legitimacy in the political, legal, or other spheres, and much existing writing in science studies which deals with science in the public domain has approached the problem in this way. But what we are trying to understand is why science *should* be granted legitimacy because of the kind of knowledge it is. In the case of this question, referring back to the way legitimacy is granted is what carries the risk of circularity. We are asking the same kind of question about what makes science special as the sociologist Robert Merton (and any number of philosophers and political theorists) asked in the aftermath of World War II. But we are trying to answer it in the aftermath of the demise of Wave One.

We understand, of course, that any arguments we put forward will merely enter the stream as another ripple rather than divert it wholesale, and we realize that any ideas that are found useful in the paper will themselves be used as devices within continuing debates about the boundaries of expertise and the like. Realizing this, however, is not a reason to give up—the course of the stream might be at least slightly changed by such an intervention.[16]

CORE-SETS, CORE-GROUPS, AND THEIR SETTINGS

We now build up our approach to Wave Three in yet another way, using diagrams to enrich the exposition. Though our problem is about sciences and technologies in the public domain, we will start with the esoteric sciences. No doubt other approaches are possible, for example treating 'public-domain sciences and technologies' (those which directly affect, rather than merely being of interest to, the public), as essentially different to the esoteric sciences; but we have chosen a different analytic strategy. Starting with the esoteric sciences has the advantage that they are familiar to the authors, and that they are the traditional 'hard case' starting point for more general studies of science. We find that it is possible to work outward from our esoteric starting point in a coherent manner, ending up with the public-domain sciences which are our target.

A core-*set* has been defined as being made up of those scientists deeply involved in experimentation or theorization which is directly relevant to a scientific controversy or debate. A core-set is often quite small—perhaps a dozen scientists, or half-a-dozen groups. A core-*group* is the much more solidaristic group of scientists which emerges after a controversy has been settled for all practical purposes.[17] If the science is esoteric, then only the members of the core-set or core-group (hereafter 'core-scientists') can legitimately contribute to the formation of the consensus, or develop the science thereafter.[18] It is not always easy to define the boundaries of a core-set, because disputes within core-sets often involve the 'boundary-work' of trying to define people in or out—that is, defining them as legitimate or illegitimate commentators.[19] Nevertheless, if one takes a really esoteric scientific controversy such as that over the detection of gravitational waves, or the detection of solar neutrinos, or the likelihood that binary neutron stars will collapse into black holes just before they spiral into each other, or if one takes the sciences that follow after them, then members of Western society know, without having to agonize, that anyone who is not a recognized physicist with a great deal of equipment or special theoretical knowledge will not be, and *should not be*, counted as a member of the set of decision-makers in respect of the *scientific knowledge* itself. Were members to take a different view of this matter, they would no longer participate in Western scientific society as the term is used here. There are those who would not accept that scientists have any special rights even in these esoteric matters, but here we must simply state our starting point that, as members of the scientific community broadly conceived, and contributors to Western scientific society, 'we think they do'. This is a reference to our culture, not a reference to the way political legitimacy is granted in our society. Should any politicians ever want to dismantle the right of the scientific community to settle esoteric issues within science, we would want to fight them.

As was pointed out to us, under Wave Two much intellectual leverage was gained precisely by denying the kind of sentiment that is taken to be self-evident in the last paragraph and those that follow.[20] We were asked whether we would be so happy to restrict judgement to such a small circle were the issue to be the importance of Tracy Emin's unmade bed, notoriously displayed by the Tate Gallery in London as an *avante-garde* work of art. It was said:

> No-one without the training and exposure to appropriate gallery-going is . . . 'competent' [to make a judgement]. So, can one derive the conclusion that only they *should* judge art?

And the implication is certainly correct that we all feel we have something to say on the matter of 'the bed'. It was suggested that in respect of 'the bed', one could reach the conclusion that only a narrow range of people are competent to judge only if one makes realist-type assumptions about the quality of works of art. It was suggested that the same must go for the argument when applied to science.

These comments are correct in that it is necessary to make an assumption of a kind that is untypical of Wave Two if we want to restrict the sphere of judgement in science. But this seems to be inevitable if we want to do upstream work. That is why we have set out our assumptions as clearly as possible here, and we do not think we can do better than say 'this is the kind of society we like—one where we do consider that scientists with experience of an esoteric specialism are the best people to make judgements about what should count as truth within that specialism'. When placed alongside the terrible experiences of humans and humanity at the hands of others, what follows from the 'post-modernist' approach to knowledge is not that it is impossible to make judgements between plural realities, but that sometimes one just has to make judgements without timeless intellectual justification.[21]

SCIENCE AND ART

Later on, we will consider kinds of expertise that are different to those found in the esoteric sciences. At this point, however, it is worth noting something interesting in the comments made to us. It was suggested that restricting the circle of judges, in the case of esoteric science, to the core-set, is equivalent to restricting the circle of judges, in the case of 'the bed', to those with training and exposure to appropriate gallery-going. The appropriate group of judges, it was said, is not artists in general, nor even artists of the type who display their unmade beds (and the like), but art critics. In language we will explain more fully later in the paper, this is a class of experts with 'interactional expertise' rather than 'contributory expertise'. It may be that this is one of the ways in which science and art are different. The end-point of art, after all, is to be experienced, and that is why it is reasonable to suppose that those with special viewing, or experiencing, expertise—critics—rather than those with special creative expertise—artists—would be the best judges.[22] Science, on the other hand, is less obviously directed at consumers, and it is less clear that the audience has so much in the way of interpretative rights where science is concerned.[23] This might indicate that in the case of science, those who actually do it (who have contributory expertise) might have more relative merit as judges of scientific value than critics (those with interactional expertise), as compared with the case of art. So though we will cleave to our claim about the self-evidence of the nature of judgement in esoteric science, we would not want to be drawn to generalize the claim in respect of art or other cultural endeavours, and this analysis shows us one reason why one might not be a good analogy for the other.[24]

POLITICS IS INTRINSIC TO ESOTERIC SCIENCE, NOT EXTRINSIC

In setting out our view about who has the best claim to judge truth in the esoteric sciences, we have made a prescriptive statement about expertise! Since, as we have intimated, it is hard to get an 'ought' from an 'is', our prescriptive statement is based on a clearly stated preference about a certain form-of-life and what follows from it. We have argued that this preference and its corollaries do not necessarily hold in the case of art. We will also argue that they do not hold in the case of public-domain science. Thus, given common agreement on the self-evident nature of the case for esoteric sciences, we have already established that to understand the importance of contributions to technical decision-making by different elements of society, it is necessary to know what kind of science we are dealing with. Our analytic strategy is to proceed by drawing out the systematic differences between esoteric sciences and other sciences, starting with the core-set as found in the esoteric sphere.

One can represent a core-set as the bull's-eye of a 'target diagram' with two or more rings surrounding it.

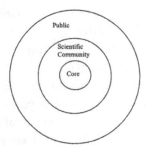

Figure 1. Core Scientists in the Wider Setting

In Figure 1, the 'bull's eye' is the core-set, the first ring out represents the scientific community who have no special knowledge or experience of the esoterica which concern the core-set, while the third ring represents the general public. Other rings might be used to represent the media and/or scientific funders and policy-makers, but we do not need them for the purposes of this analysis.

It might be thought that the stress on the scientific pre-eminence and exclusiveness of core-scientists in esoteric sciences flies in the face of the whole Wave Two analysis. According to the sociology of scientific knowledge, politics is never absent from the centre. Sometimes this will be the politics of the scientific community, but sometimes 'big-P' politics will play a significant rôle. This claim remains valid even for us, self-proclaimed, members of Western scientific society; indeed, one of the authors of this very paper was one of the first to describe the issue in print. How is this position compatible with the prescriptive statement we have just made, that in the case of esoteric sciences: 'anyone who is not a recognized physicist with a great deal of equipment or special theoretical knowledge will not be, and *should not be*, counted as a member of the decision-making group?'

The answer is that in these cases it is the esoteric decision-making group alone that disposes of the political influences that bear upon it. Thus, in what is probably the clearest example of this genre which involved 'big-P' politics, Steven Shapin showed that in 19th-century Edinburgh, scientists studying the brain observed features that were homologous with their position in local Edinburgh politics, and that the core-set were influenced by such considerations in reaching their conclusions.[25] And yet it would be quite wrong to say that because the phrenology debate was influenced by Edinburgh politics it would have been right for the brain scientists, and the public they served, to have consulted local Edinburgh politicians in order to form their opinions on brain structure. Such a view would, quite properly, be counted as encouraging 'bias', and would be incompatible with the 'form-of-life' of Western science. Anyone who held such a view would, by that fact, prefer to inhabit a different social and conceptual space to the authors of this paper. What Shapin's and similar studies show is that politics of this sort may influence science, but not that it is a *legitimate input* to scientific decision-making. Setting aside Lysenkoism and the like—still seen as pathologies by members of Western scientific society—one would never set out to design scientific or political institutions to enhance the influence of 'big-P' politics on the content of such an esoteric science: one would do quite the opposite.[26] We might say that the SSK studies show that politics is 'intrinsic' to science, but they do not license 'extrinsic' political influence.[27]

Thus, while SSK-type studies reveal various influences on the formation of views within science, they do not legitimate it any more than the revelation of similar influences in the justice system would legitimate their enhancement. In justice, as in esoteric science, one always tries to minimize external influence. Later, we will re-examine the question for less esoteric sciences and find it more complicated, but at least we have shown how to *break into* the Problem of Extension, in spite of what has been learned under Wave Two. More exactly, we have discovered that those of us who think of ourselves as living in Western scientific society have always lived with a partial solution to the Problem of Extension, even while we were emphasizing the non-expert influences on expert decisions. The 'circle was squared' because under Wave Two our analyses were descriptive not prescriptive: the hidden preferences were preserved because we never discussed them. We might say, in respect of our search for a boundary between legitimate and illegitimate inputs into esoteric scientific knowledge-making, that even those of us who have been practising SSK without compromise, have been 'speaking prose all along'—it has been the prose of the form-of-life of Western science. This is to reiterate the point already made about the degree of largely unremarked compartmentalization already found in Wave Two.[28]

BEYOND THE CORE

According to the sociology of scientific knowledge, 'distance lends enchantment'.[29] Core-scientists are continually exposed, in case of dispute, to the counter-arguments of their fellows and, as a result, are slow to reach complete certainty about any conclusion. In general, it is those in the next ring out in Figure 1—the non-specialists in the scientific community—who, in the short term, reach the greatest certainty about matters scientific.

These outsiders reach certainty more easily than core-scientists because they learn of goings-on in the core of the science only through digested sources, such as conversations with their colleagues, scientific journals, the scientific media, and the broadcast media. Inevitably, such sources condense and simplify—that is their job. Only exposure to the lived history of the core-set can reveal the richness of a dispute and its potential for being re-opened. For those at the heart of matters, scientific disputes are seen to linger on long after the wider community takes matters to be settled.[30]

There is a second reason why debate closes down in the wider community before it closes in the core-set. The consumers, as opposed to the producers, of scientific knowledge have no use for small uncertainties. Decisions about action generally involve binary choices—'we will fund cold-fusion research, or we won't'; 'we will impose a carbon tax, or we won't'. When a decision is made to act, it can 'read back' on the scientific debate at its core and make any remaining doubts harder to sustain. (Though there are circumstances in which exposure to the public opens controversy, rather than closes it down.)

Figure 2. Core Elongated in Time

We can represent some of these processes by modifying the target diagram. The diagram can be stretched horizontally, and the left-to-right dimension used to represent the passage of time. The vertical dimension will be used to represent uncertainty. Thus, in Figure 2, the processes that take place in a core-set are represented by a narrowing triangle; as time goes on uncertainty decreases, though never quite reaching an apex of certainty.

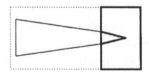

Figure 3. Apex of Certainty Constructed by Wider Scientific Community

The next figure, Figure 3, reintroduces the wider scientific community—the rectangle. Now we see that an apex of certainty has been added to the core-set, but it has been added, not by the core-set's deliberations, but by those in the next area out—the wider scientific community. The apex of certainty is shown, therefore, as belonging to this group. The line representing the wider scientific community changes from dashes to solid only as it begins to play its part in the perception of the outcome of the science. This is the process that has been labelled 'distance lends enchantment'.

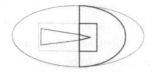

Figure 4. Apex of Certainty Visible to Public

Figure 4 introduces the general public. In esoteric sciences which are controversial, the public merely watches as disputes play out, but when the science becomes popularized, the apex of certainty becomes public property. The next generation of scientists are also introduced to these certainties by textbook writers who collapse the time dimension of the science they write about. It is only the apex of certainty that is visible to these new generations, and all the preceding years of experimentation and argument disappear into it.[31]

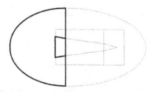

Figure 5. Science Becomes Visible before it Becomes Certain

The processes represented in Figures 1 to 4 show how the nature and the history of science are turned into a mythology as a result of exposure beyond the core-set. Exposing and exploring the details of the process has been a major part of Wave Two. The potential for damage to science occurs when sciences find themselves exposed to the public early on, before consensus has developed within the core-set. Under these circumstances, expectations developed from exposure to the view represented in Figure 4 are applied to sciences at a time when core-set debates are too unsettled (too wide in the vertical dimension), to give rise to a robust apex of certainty. This situation is represented in Figure 5: the dotted areas to the right of the solid vertical line are still in the future; the public sees the left-hand end of the core-set and expects, or at least is generally believed by policy-makers to expect, the same kind of outcomes as they have previously seen at the right-hand end.[32] But now they find that the scientists, who previously revealed a relatively united and robust front, argue with each other with different sides having rough parity; they change their minds, and are no longer a source of confidence. It is easy to understand why scientists prefer to keep their work private until they have reached something closer to unity.

We might look at this situation in the following way: in the 1970s, sociologists began to study scientific controversies as 'breaching experiments' which opened up the hard, formal, though mythical, shell of science, exposing the soft social inside filled with seeds of everyday thought.[33] When the left-hand end of a core-set is publicly exposed, it too is a kind of breaching experiment, but one visible to all; everyone gets to see the soft flesh of the scientific fruit and the familiar passions and arguments that constitute it. Both kinds of breaching experiment show that scientists rely on ordinary reasoning to bring their technical arguments to a conclusion, and this closes the gap between science and the rest of us. Suddenly, the conclusions formally wrought by science alone are the property of everyone, and each has a right to contribute their opinion along with that of the no-longer-so-special scientists. This is where Wave Two, just like the approaches of Ulrich Beck and Anthony Giddens (see Appendix), struggles with the question of how to weight the opinions of the myriad potential contributors.

THE THIRD WAVE OF SCIENCE STUDIES?

The Third Wave of Science Studies, SEE, turns, as we have said, on a normative theory of expertise. The aim is to approach the question of who should and who should not be contributing to decision-making in virtue of their expertise. Crucially, rights based on expertise must be understood one way, while rights accruing to other 'stakeholders', who do not have any special technical expertise, must be understood another way. Stakeholder rights are not denied, but they play a different role to the rights emerging from expertise. In a rather old-fashioned way, reminiscent of Wave One, Wave Three separates the scientific and technical input to decision-making from the political input. This is not an attempt to go back to Wave One, because Wave Three takes into account all that has been learned during Wave Two and, as we stress, Wave Two runs on as strongly as before; we are trying, under Wave Three, to reconstruct knowledge, not rediscover it. Thus, under Wave One, political rights made almost no contribution to technical decision-making, being almost entirely overwhelmed by top-down expertise; under Wave Three, expert and political rights can be seen to be much more balanced because of the new understanding of contested science that emerged from Wave Two. To represent this feature of Wave Three, we cut the diagram in half horizontally, reserving the bottom half for political and stakeholder rights, and the top half for scientific and technical debate. Scientists and technologists appear twice in this diagram. They appear in the bottom section of the diagram because they have rights as citizens and stakeholders; they appear in the top half in virtue of the rights that grow out of their specialist expertise.

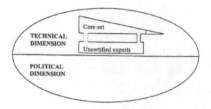

Figure 6. Uncertified Experts and the Core

In Figure 6, the top half of the oval contains the core-set. We have simplified the Figure in one important respect compared to the previous diagrams. *The wider scientific community no longer plays any special part in the decision-making process.* Henceforward, in our treatment, the wider scientific community is indistinguishable from the citizenry in general.[34] This, we would argue, is more than an analytic convenience: the wider scientific community *should* be seen as indistinguishable from the citizenry as a whole; the idea that scientists have special authority purely in virtue of their scientific qualifications and training has often been misleading and damaging. Scientists, as scientists, have nothing special to offer toward technical decision-making in the public domain where the specialisms are not their own; therefore scientists as a group are found in the bottom half of the diagram. In making this clear, Wave Three differs markedly from Wave One.

Within the top half of the oval, Figure 6 shows a small rectangle representing experience-based experts. The rectangle of experience-based experts feeds into the core-set. This is another way in which Wave Three departs from Wave Two. Wave Three, as we have said, distinguishes between two kinds of citizens' rights in technical decision-making. There are those from the bottom half of the oval, which we have already mentioned. And there are those in the top half, which accrue in virtue of the existence of *pockets of expertise* among the citizenry, and which are properly described as being within the technical rather than the political domain. Under Wave Two, it has been easy to confuse these types of expertise with rights accruing within the political sphere. Wave One located all expertise within the scientific community; Wave Two, reacting to this incorrect picture, made it hard to distinguish between scientific expertise and political rights; Wave Three is intended to re-establish the distinction, but with the dividing line set in a different place within the population. This difference in approach is summed up in Figure 7.

Figure 7 shows the location of expertise as conceptualized under the three idealized waves of science studies. Crucially, under Wave One, the dividing line was horizontal, separating the certified scientific community from the laity; under Wave Three it is vertical, separating specialist experts, whether certified or not, from non-specialists, whether certified or not.

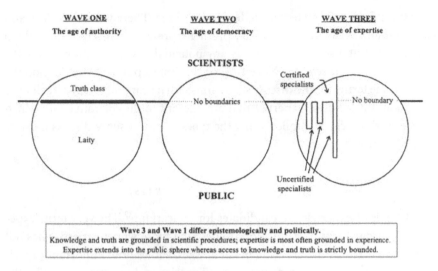

Figure 7. Three Waves of Science Studies

THE NATURE OF EXPERTISE

We now begin to develop a classification of expertise which will help us understand what is in the rectangular box that lies within the technical sphere (Figure 6), and the odd-shaped pockets of expertise found among the lay public (Figure 7). The classification will show how the rights that accrue from expertise differ from more diffuse political rights. The object is to develop a discourse of expertise which will help to put citizens' expertise in proper perspective alongside scientists' expertise.

To carry out this task, it is necessary to recognize and categorize, not only different types of expertise, but also different types of science. Much excellent work has been done under Wave Two by 'deconstructing dichotomies', dissolving boundaries, and the like, but like it or not, the world is made of distinctions and boundaries. One of the styles of Wave Two argument is to concentrate on boundary problems. It is shown, and it can often be easily shown, that the boundary between entity 'A' and entity 'B' is unclear, and it is often argued that this means that A and B are not really separate things at all. Interesting studies of the way actors create and patrol boundaries can then be carried out.[35] Some writers, however, have gone further, and taken the fuzziness of many boundaries as the empirical counterpart of a philosophical prescription: 'Dissolve all dichotomies'. Here we approach from the other direction. We intend to point to differences by starting at the extreme points of our continuum—we will take 'ideal types' of this kind or that kind of expertise as our initial examples, and worry about the boundary problems later. In this way it is possible to begin to think about how different kinds of expertise combine in social life, and how they should combine in technical decision-making. There will be no clean and easy solution, because the boundary problems present themselves, not only to the analyst, but also to the actors as any potential new institution enters the arena of political discourse. Nevertheless, the first step must be to develop the appropriate terms for the discourse; we must learn a language which facilitates talk about the kinds of expertise that are relevant to the dilemma with which the discussion began. It follows that the types of expertise we discuss must be treated as 'real'.[36]

EXPERIENCE AND EXPERTISE

What kinds of expertise are candidates for reification?[37] The very term 'experience-based experts' that we adopt to describe those whose expertise has not been recognized in the granting of certificates, shows how important experience is to our exercise in demarcation. Experience, however, cannot be the defining criterion of expertise. It may be necessary to have experience in order to have experience-based expertise, but it is not sufficient. One might, for example, have huge experience of lying in bed in the morning, but this does not make one an expert at it (except in an amusing ironic sense). Why not? Because it is taken for granted that anyone could master it immediately without practice, so nothing in the way of 'skill' has been gained through the experience.

More difficult, one might have huge experience at drawing up astrological charts, but one would not want to say that this gives one the kind of expertise that enables one to contribute to technical decision-making in the public domain. Why not? Here, unlike lying in bed, an esoteric skill has been mastered which could not be mimicked by just anyone—at least not to the extent that it could pass among skilled practitioners of astrological charts. Astrology is, rather, disqualified by its content. It is hard to say much about which kinds of expertise are excluded in this way, but we can say something.[38]

Stephen Turner divides expertise up into five kinds, according to the way they obtain legitimacy from their consumers.[39] For Turner, the first kind of expertise (Type I) is like that of physics, which has gained a kind of universal authority across society in virtue of what everyone believes to be its efficacy. Type II expertise has been granted legitimacy only among a restricted group or sect of adherents; Turner gives theology as his example, and we might put astrological expertise in the same category. Type III experts, such as new kinds of health or psychological 'therapist', create their own adherents, or groups of followers. Type IV and Type V experts have their adherents created for them by professional agencies which set themselves up to promote a new kind of expert, or, like government departments, become specialist consumers of new kinds of expertise.

Our concern, in this paper, is very largely with Type I expertise. But what *argument* might we provide to justify stopping at Type I expertise? This remains an unresolved problem for upstream work in SSK. The best we can do is note that the adherents of all the kinds of expertise we value positively, were they to have what we will call 'contributory expertise' (see below), could make a reasonable claim to be members of the core-set relevant to any particular technical decision. That is to say, their expertise would be continuous with the core-set's expertise, rather than discontinuous with it; astrology and theology are discontinuous with those of radiation ecology, whereas the expertise of sheep farmers is not.[40]

In drawing a boundary around legitimate contributors to decision-making, then, two kinds of judgement are made in logical sequence. The first judgement is about what fields of experience are relevant. We might decide, for example, that astronomy is relevant to some question and that astrology is not, in spite of the claims of its adherents. But we have almost nothing to say here about this choice except the groping remarks just made about continuity and discontinuity, and a reference back to the form-of-life of Western science. Our views on which fields are legitimate and which are not are certainly not fixed for all time, and they may change as the flux of history brings one field out from the cold and pushes another into it.[41] Nevertheless, this choice has to be made ahead of the choice of who is an expert within a field. The point is clarified in Figure 8.

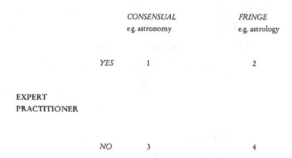

Figure 8. Expertise in Consensual and Fringe Sciences

In box 4 of Figure 8 are inexpert and inexperienced practitioners of fringe fields, whereas box 2 contains expert astrologers, remote-viewers, and forth. In box 3 are found inexpert practitioners of fields which are acceptable to the broad consensus of Western science—people who are just bad at science. In box 1 are the experts in consensual fields. We have, as indicated, very little to say about the horizontal dimension in this table except that it is there—our concern is with the vertical dimension of the left-hand side of the table. The burden of this paper is that there are people whom qualified and credentialled scientists would want to consign to box 3, who ought to belong in box 1. Our licence for saying this is the expertise, drawn from SSK, in the matter of the nature of science and knowledge. In particular, Wave Two studies of the essential craft content of science have shown that it is more difficult to separate the credentialled scientist from the experienced practitioner than was once thought: when we move toward experience as a criterion of expertise the boundary around science softens, and the set of activities known as 'science' merges into expertise in general. But box 3, as we have stressed, still contains only small subsets of the population at large.[42]

Choice between fields and expertise within a field is, then, orthogonal. Disagreement is on yet another orthogonal dimension (in Figure 8, one can visualize it coming up out of the page). Thus, as the core-set studies show over and over again, experts, all of whom belong in box 1, disagree. It merely complicates matters a little that they sometimes express their disdain for other experts with whom they disagree by saying that they do not belong in box 1 at all, but should be relegated to box 2. It is, of course, no surprise that officially appointed radiation scientists might disagree with the views of Cumbrian sheep farmers, even though their knowledge of sheep is continuous with the sheep farmers' knowledge, and it is no surprise that they might express their disagreement by saying that these sheep farmers are not experts. We (that is, sociologists of scientific knowledge) claim the right to disagree about this last judgement.

To sum up once more, what we are dealing with is types of expertise that are actually or potentially continuous with what Turner calls Type I expertise. The classification that we need lies, then, *within* the envelopes of the categories of expertise discussed by Turner, and mostly within the envelope of his Type I. Our classification is, perforce, of a kind which is quite different to his.[43]

THREE TYPES OF EXPERTISE

There are dozens of ways of classifying competence and expertise.[44] Classifying competence is the basis of much educational theory, psychology of intelligence, sociology of employment, and so forth. It also forms the foundation of the study of artificial intelligence and expert systems. One of the authors of this paper has himself classified expertise in several different ways.[45] Here we choose our starting point for our classification on the basis of familiarity. We thought it might be persuasive to begin with something that many potential readers know about from first-hand experience. We start, then, with ourselves and our practices as sociologists of scientific knowledge.

One way in which the group of analysts who practise SSK have to confront the concept of expertise is in the problem they themselves face in trying to gain a cultural foothold in the areas of those sciences they want to analyse. Typically, SSK fieldworkers enter scientific fields which they do not know, and try to learn enough about them to do sociological analyses. Rarely, however, do they reach the level of expertise of a full-blown participant. In the case of the esoteric sciences, the fieldworker hardly ever participates in the science itself. Thus, to begin with, by reflecting on certain sociologists' fieldwork experiences, we can distinguish three levels of expertise:

1) *No Expertise:* That is the degree of expertise with which the fieldworker sets out; it is insufficient to conduct a sociological analysis or do quasi-participatory fieldwork.

2) *Interactional Expertise:* This means enough expertise to interact interestingly with participants and carry out a sociological analysis.

3) *Contributory Expertise:* This means enough expertise to contribute to the science of the field being analysed.

Since reflecting on the practice of sociologists tells us that there is a difference between these three states, we have made another important step: we have begun to understand expertise as an analyst's category, as well as an actor's category. In this case, expertise is an analyst's category in a very direct way: it is a category which analysts use to think about themselves, and this is why we think it is a persuasive starting point. Since we already use this language to describe ourselves (speaking prose all along!), there should be less obstacle to using it to describe other actors.[46]

Of course, these three categories are ideal types and, as with most such classifications, there will be boundary problems. For example, the attainment of category 2 is hotly disputed by 'science warriors', who frequently claim that sociologists do not have enough scientific expertise to carry out their sociological analyses, and have failed to escape category 1. We could, if we wanted, give an attributory account of our own experience. What does count as having enough expertise to do fieldwork, or even to contribute to a science? The trick, however, is not to become paralysed by these problems, but to proceed with an imperfect set of classifications, just as other experts proceed. Instead of worrying about the imperfections of our science, we should note that the very fact that we, as sociologists of science, work hard to acquire a level of expertise that enables us to defend ourselves against the charge of being insufficiently expert shows that *we do not* act as though it is all a matter of attribution. As empirical researchers in the sociology of knowledge, we act as though there is something in the nature of expertise that can be acquired if we work at it. Furthermore, at least some of us have found that we have been unable to acquire sufficient expertise to analyse certain scientific fields to our satisfaction, and this too is a salutary experience.[47] Again, just occasionally, we do manage to acquire enough expertise in some field we study to begin to contribute.[48] Thus, though there are boundary and definitional problems, they do not have to be fatal.

Having accepted that to categorize expertise makes sense in spite of the boundary problems, the task is to begin to work out what these types of expertise mean and how they fit together. For example, having interactional expertise does not give one contributory expertise, but one might think the former was a necessary condition for the latter. But it may not be! We will work out some of these differences by referring to what is fast becoming the paradigm study of so-called 'lay expertise'—Brian Wynne's study of the relationship between scientists and sheep farmers after the radioactive fallout from the Chernobyl disaster contaminated the Cumbrian fells.[49]

Wynne found that the sheep farmers knew a great deal about the ecology of sheep, and about their behaviour (and that of rainwater) on the fells, that was relevant to the discussion of how the sheep (and the fells) should be treated so as to minimize the impact of the contamination. Since the Windscale-Sellafield plant was built soon after World War II, the farmers in the locality had long experience of the ecology of sheep exposed to (radioactive) waste. The farmers have all the characteristics of core-group experts in terms of experience in the ecology of hill sheep on (mildly radioactive) grassland, even though they had no formal qualifications. In our terms, the farmers had contributory expertise which in some respects exceeded that of scientists working for the relevant government department.

The scientists, however, were reluctant to take any advice from the farmers. Now, for the farmers to have contributed to the science they would *not* have had to engage in a symmetrical conversation with the scientific experts—all that would have been necessary was for the scientific experts to try to learn from the farmers. This seemingly trivial point helps us to understand what expertise is, but also points out where the social location of change needs to be. The normative point that follows is that the body of expertise that should have emerged in respect of Cumbrian sheep was a combination of the separate contributory expertise possessed by the scientists and the farmers. The scientists' expertise was not at risk of being displaced by that of the farmers; it was, or should have been, added to by that of the farmers. Should the situation have been symmetrical, it might have been an arbitrary matter whether the farmers' expertise was absorbed by the scientists or the scientists' expertise was absorbed by the farmers, but it was not symmetrical. To produce the optimum outcome, the scientists needed to have the *interactional expertise* to absorb the expertise of the farmers. Unfortunately, they seemed reluctant either to develop or to use such expertise.[50] Here we begin to see how our theory of the interrelationship of types of expertise might gain in richness and practical relevance. We have two theses:

Thesis 1: *Only one set of experts need have interactional competence in the expertise of another set of experts for a combination of contributory expertises to take place.*

Thesis 2: *In such a case, only the party with the interactional expertise can take responsibility for combining the expertises.*
Thesis 2 says something about the social responsibility of different parties, but there could also be a 'Thesis 3' that says something more prescriptive still about how these different parties should interact:

Thesis 3: *In such circumstances the party without the interactional expertise in respect of the other party should be represented by someone with enough interactional expertise to make sure the combination is done with integrity.*

In effect, Thesis 3 is suggesting that the Cumbrian farmers might well have had more success in their dealings with the scientists from the UK Ministry of Agriculture, Fisheries and Food (MAFF) and from British Nuclear Fuels Ltd (BNFL), if their concerns were mediated by a Greenpeace scientist, a Brian Wynne, or the like. Clearly such an individual would need to be briefed by the farmers about what the certified scientists were doing wrong, but such a person may have been able to phrase the problem in ways more familiar to the scientists, making it more credible (or less resistible). This problem was recognized by AIDS treatment activists in the USA, who found that they had to learn the language of science if they were to represent the interests of the wider community within the clinical trials process.[51]

REFERRED EXPERTISE

Sometimes expertise in one field can be applied in another. A third category of expertise which seems useful is 'Referred Expertise'; it is, as it were, expertise 'at one remove'. Consider the managers and leaders of large scientific projects. In general they will not possess contributory expertise in respect of the many fields of science they must coordinate. In the field study of one of the authors, this became a bone of contention. As one expert scientist put it:

> . . . What I found disappointing was that after two years the project manager still didn't really know what it meant to do interferometric detection of gravitational waves.

Whereas a manager saw it this way:

> Once you professionalize, the guys who are very good in the lab where you control everything, no longer have their arms around it all. Other people can work very well in that environment. They interface with the experts who built the electronics and understand what they need to of that; they interface with the computer people and do very well at that; and some people can work in this broader environment technically. Some people make the mistake of saying that as soon as you are in this broader environment

it's a management problem; it's not a management problem! The technical part is actually more technical and more sophisticated.

If we stay with the terminology we have developed so far, we would have to say that in respect of the science they are managing, the managers have only interactional expertise (and, one would hope, discrimination, and also the ability to translate: see below). But in so far as they have contributory expertise, it is expertise in management rather than in science. Does this mean that their scientific expertise is no greater than that of, say, the visiting sociologist? The answer has to be 'no', or our theory would be reduced to the absurd (as well as disagreeing head-on with the last sentiment expressed in the above quotation).[52]

The resolution seems to be that to manage a scientific project at a technical level requires, not contributory expertise in the sciences in question, but *experience of* contributory expertise in some related science.[53] In other words, the managers must know, from their own experience, what it is to have contributory expertise; this puts them in a position to understand what is involved in making a contribution to the fields of the scientists they are leading at one remove. As one might have 'referred pain' in a leg as a result of back injury, this is 'referred expertise'. It would be quite reasonable to expect that managers of scientific projects with referred contributory expertise would manage much better (and with much more authority and legitimacy) than those without it.[54]

TRANSLATION

There are at least two other kinds of ability that go into the making of technical judgements of the kind we are discussing: the first is the ability to translate. For groups of experts to talk to each other, translation may be necessary. Some people have a special ability to take on the position of the 'other', and to alternate between different social worlds and translate between them.[55] The translator will have to have at least interactional expertise in both areas.

Thesis 4: *A necessary but not sufficient condition of translation is the achievement of interactional expertise in each of the fields between which translation is to be accomplished.*

If the translator has one or more bodies of contributory expertise, so much the better, but contributory expertise is not a necessary condition for translation. Returning to Thesis 3, it seems important that those who represent one group to another group must be able to translate.

If the ability to translate consists of more than having multiple interactional expertises, what is the extra bit? Presumably it has to do with the skills of the journalist, the teacher, the novelist, the playwright, and so forth, skills notoriously hard to explain—as qualitative sociologists know to their cost.[56]

DISCRIMINATION

The second requisite is the ability to discriminate. Social actors can sometimes make judgements between knowledge-claims based on something other than their scientific knowledge. Judgements of this sort can be made on the basis of actors' social knowledge: does the author of a view come from within the right social networks, and has he or she the appropriate experience to make their claim credible? Such things as the personal demeanour of the expert might be the crucial inputs to these judgements, rather as one might judge a politician. The questions are: Does the author of a claim seem to have integrity? Is the author of a claim known to have made unreliable claims in the past?[57] There are also secondary features of a claim itself that can be judged with only minimal scientific understanding: Is a claim internally consistent or inconsistent, or consistent with other claims made by the same person? Does the claim seem so self-serving as to give concern?

To make the notion of discrimination do any work, it is once more necessary to distinguish between specialists and the population as a whole. Most members of a society, just by being members of that society, are able to discriminate between what counts as science and what counts as non-science. This is the ubiquitous judgement on which we rely when we dismiss astrology and the like as potential contributors to the *scientific* element in technical decision-making. Most members of our society have sufficient judgement to know that the social and cognitive networks of, say, astrologers do not overlap with the social and cognitive networks of scientists with (Turner's) 'Type I' expertise.[58] This kind of discriminatory ability comes with participatory expertise in the matter of living in society!

To see how this works, consider cold fusion. Most reasonably literate members of this society 'know' that cold fusion has been tried and found wanting; though there was a time when cold fusion was contiguous with science as we know it, its cognitive and social networks no longer overlap. This knowledge has nothing to do with scientific competence. On the contrary, it is vital to ignore scientific credentials. Thus Martin Fleischman, the co-founder of the cold fusion field, is immensely well-qualified and has both interactional and contributory expertise in cold fusion, yet still believes in it. What people in Western populations have in common is what they have heard about cold fusion in the broadcast media. Their consensual view emerges from the making of social judgements about *who* ought to be agreed with, not *scientific* judgements about *what* ought to be believed. The crucial judgement is to 'know' when the mainstream community of scientists has reached a level of social consensus that, *for all practical purposes*, cannot be gainsaid, in spite of the determined opposition of a group of experienced scientists who know far more about the science than the person making the judgement. This ability is gained through membership of what the *Guardian* newspaper calls the 'chattering classes'. Note that this is not the sort of judgement we would expect even an immaculately qualified scientist from 'another planet' where the *Guardian* does not circulate, to be able to make.[59] What the members of the chattering classes have is what we might call 'discrimination'.

But contrast this with the more locally informed kind of discrimination of the Cumbrian sheep farmers. The farmers had contributory expertise about the ecology of their farmland, but they could also do a special kind of discrimination. The dispute was between the local community, MAFF and the Sellafield authorities (BNFL), but extensive dealings had already taken place between the parties over the years; as a result, the interaction was tense. Through experience, the farmers had developed discrimination in respect of the pronouncements of (in particular) the Sellafield authorities: they found the authorities more questionable than they otherwise would, and more questionable than they would seem to an outsider with less experience of this particular social and geographical location. (In the Appendix we will discuss another case described by Brian Wynne where there is ambivalence between *local* and *ubiquitous* discrimination.)

The Lack of Expertise of the Wider Scientific Community

As we have mentioned, one big mistake that has been made in the past is to exaggerate the importance of the referred expertise of the wider community of scientists. At the very outset, when we discussed Wave One, we noted that in the 1950s scientists were often attributed with authority to speak on subjects outside their narrow area of specialization. The Second Wave has shown how dangerous it is to take this kind of referred expertise at face value, since the pronouncements of the wider scientific community are nearly always based on simplified and retrospectively constructed accounts of the scientific process. Quite simply, scientists' supposed referred expertise about fields of science distant from their own is nearly always based on mythologies about science, rather than on science itself. That is why we have stressed the continuity between the wider scientific community and the public in all but specialist areas, and represented this point in the Wave Three diagram within Figure 7. Organizations such as the UK Committee on the Public Understanding of Science (COPUS), in its first incarnation, and many self-appointed scientific spokespersons, by making science as a whole the focus of their campaigns, have oversold it; it is the work of *specialists*, not generalists—not the whole scientific community—that should be the focus of campaigns to raise the status of science. Of course, the former type of campaign treats science as a world view—competing with religion and the like—and therefore is accompanied with the thrill of zealotry, or what might be called 'scientific fundamentalism'; the latter type of campaign would be a comparatively mundane enterprise, stressing experience and professionalism rather than priestly virtues.

Examining Wave Three more closely, as it is represented in Figure 7, the same point can be seen from a different angle, as it were. What differentiates core-scientists from their fellow scientists on the other side of the vertical line in the top part of the diagram? It is not credentials! The core comprises those who have actually done relevant experiments, or who have developed or worked with theories relevant to the issue in question.

To express this in more general terms, the core-scientists' special position, apart from their possession of specialist equipment, arises from their long experience and integration into the specialist social group of which such expertise is the collective property. Core-set members do not possess extra formal qualifications, and they have not undergone special periods of formal training over and above what they needed to qualify as certified scientists in the first place. It is not more certification that qualifies them for membership of the core. In terms of formal criteria they are indistinguishable from the rest of the scientific community; the difference between the core-set and the others is informal.[60] This informality—the fact that membership of the most esoteric groups is based on experience—gives us licence to dissolve the boundary between the certified experts and experience-based experts; in sum, the demarcation lines run vertically in Wave Three (Figure 7) because they demarcate the set of all experts, certified or uncertified, with relevant experience, from those without it; in Wave One (Figure 7) they run horizontally, demarcating all of those with scientific qualifications from those without them.[61]

To sum up, in the vertically partitioned world of Wave Three, the right to contribute technically to a technical decision is to be assessed by examining expertise. The appropriate balance of contributory expertise, interactional expertise and referred expertise has still to be worked out, and so has the rôle of discrimination and translation. In this world, certification as a scientist has little or no importance. The rôle of expertise and the rôle of democratic rights are separate.

CASE STUDIES

INCREASING INTERACTION: CUMBRIAN SHEEP
AND AIDS TREATMENT IN SAN FRANCISCO

Let us now see how this categorization of expertise helps us understand some of the well-known case studies. We start by discussing the Cumbrian sheep farmers once more. The sheep farmers had contributory expertise that was complementary to that of the MAFF scientists. They developed it through their long collective experience in the ecology of the fells and the sheep that live on them. They failed to make much impact with this expertise because they lacked interactional expertise; they needed help, either from generous-minded ministry scientists, or from intermediaries with interactional expertise and the ability to translate. The sheep farmers also had a special level of local discrimination in respect of the BNFL scientists; this had developed out of long experience.

Thus, what Brian Wynne's study shows is not what it has often been taken to show—that scientific expertise is to be found among the public—but that, in this particular case, there were not one but two sets of specialists, each with something to contribute. The sheep farmers were a small group in possession of a body of knowledge as esoteric as that of any group of qualified scientists. The sheep farmers were not 'lay' anything—they were not people who were not experts—they were experts who were not certified as such. To repeat, in Wynne's study can be seen the working out of the interactions, not of experts and the public, but of two groups of experts.

Now, it is also true that the sheep farmers had some rights in the matter in virtue of their ownership of the sheep—but this can be distinguished from the matter of their technical expertise with a thought experiment. Imagine that just prior to the Chernobyl explosion a group of London financiers had got together to buy the Cumbrian farms as their private weekend resort, employing the farmers as managers so as to preserve the existing ecology: the financiers, not the farmers, would then be the owners of the sheep, yet all the expertise would remain with the farmers. 'Overnight', much in the way of stakeholder rights would have been transferred from the farmers to the financiers, yet this would not make them members of the core-set; it remains the case, however, that the farmers should have been included in the core-set. Thus it is easy to see the difference between political rights and expertise.

		PHASE	
		Political	Technical
	Politics	Extrinsic	Intrinsic
NATURE	Rights	Stakeholder	Meritocratic
OF	Representation	By Survey	By Action
	Delegation	By Proxy	Impossible

Figure 9. Political and Technical Contributions

To clarify the point further, Figure 9 sets out some indicators of the difference between the political and the technical contributions to technical decision-making; the first two rows in the Figure have already been discussed, the second two rows are newly introduced.

The first row in Figure 9 tells us that the way politics enters the two 'phases' of the decision-making process is different.[62] In the political phase, the politics is readily visible and is treated as an extrinsic feature of the scientific decision. But, as in the Edinburgh phrenology case, politics enters the technical phase intrinsically—it is amalgamated into the science in such a way that its effect is usually hidden unless picked out in studies such as Steven Shapin's.

The second row repeats what we have just said: contributions to technical decision-making are made by right by stakeholders in the political phase (the landowners), but by merit in the technical phase (the experienced farmers and the scientists).

The third row points out that the contribution of stakeholders could be represented by something like an opinion survey or a vote, whereas technical contributions have to respond continually to unanticipated developments in the live science and technology, so that the expertise has to be carried in the person of the contributor.

The fourth row is a corollary of the last point, that stakeholders could appoint a proxy (for example, a solicitor) to represent their interests, whereas no proxy can exercise skills on someone else's behalf.

Returning to events post-Chernobyl, it is not clear whether the Cumbrian sheep farmers' advice ever actually entered the core of the post-Chernobyl discussion, but we can be unabashedly prescriptive and say that it *should* have become part of it.

We have argued that one of the reasons that the sheep farmers made less impact than they might have done was their lack of interactional expertise. The AIDS-treatment controversy in the San Francisco gay community is an example where the non-certified experts succeeded in gaining an entree to the scientific core.[63] But they did not manage this until they gained interactional expertise—that is, until after they learned the language of the relevant science. Their case is represented in Figure 6. The dashed-line nexus between the pocket of experience-based expertise and the core is where the AIDS activists might have made their contribution, but they did not make it until later—as represented by the solid-line nexus.[64] One outcome of our analysis, we hope, will be to encourage the movement of such a nexus to the left—that is, to encourage the involvement of experience-based experts earlier in the game—possibly by encouraging such groups to look for spokespersons with interactional expertise in the science in question, or to encourage the growth of intermediary groups to speak for the scientific knowledge of the uncertified, not as campaigners, nor as experts themselves, but as translators. There is, of course, a certain *naïveté* about this suggestion, but unless all hope of unbiased action is to be abandoned (and why, then, are we academics?), it is our duty to be naïve from time to time.[65]

DECREASING INTERACTION:
CRASHING FUEL FLASKS AND AIRCRAFT

One of the characteristics of the analyses of the relationship between experts and the public under Wave Two is that they all push in the same direction: increased participation by the public to solve the Problem of Legitimacy. One cannot but feel a little uncomfortable when every treatment has the same political recipe, because it makes it all too easy to imagine that the prime motivation is political rather than analytical. A reassuring feature of the Wave Three approach, which puts expertise at the centre of the analysis, is that there are cases which push in the opposite direction—cases where, according to our analysis, participation by the public should have been *decreased*, because their expertise was insufficient to make a contribution.

On 17 July 1984, in Leicestershire, England, the British Central Electricity Generating Board (CEGB) decided to demonstrate the safety of their method of transporting, by train, spent nuclear fuel around the country. They crashed a train travelling at high speed into a nuclear fuel flask. On 1 December 1984, at Edwards Air Force Base, in California, NASA and the FAA deliberately crashed a remotely controlled Boeing 720, carrying 75 dummy passengers and a full load of 'anti-misting kerosene' (AMK) into the ground; AMK, as opposed to ordinary jet-fuel, was supposed to reduce the likelihood of the catastrophic life-taking fires that usually follow otherwise survivable aircraft crashes.

In both of these cases, the public was brought into the heart of the scientific process by being given grandstand seats at the demonstrations—some directly, by being at the scene, and many more indirectly, through what they saw on television.[66] What the public saw was that the flask did survive the spectacular crash with its integrity unscathed, whereas the aircraft was almost completely destroyed by fire. In the case of the flask, the audience was invited by the late Sir Walter Marshall, then Chairman of the CEGB, to draw the conclusion that the flasks were a safe means of transport. He said on television: 'If they're not convinced by this they won't be convinced by anything'. The conclusion of the audience watching the aircraft crash, bolstered by headlines in all the newspapers, was that AMK was a failure. And yet these TV audiences were in no position to make such judgements—they did not have the necessary expertise. In these cases, giving the impression that the public could judge the meaning of the tests was misleading and seemed designed to serve political ends.[67]

Thus, according to other experts, the train crash could not be taken to imply the safety of the method of transport because of certain special features of the test whose significance was evident only to the expert eye. These included the absence of the railway lines beyond the point of impact, and the removal of the wheels of the wagon on which the flask was placed: the lines could have penetrated the flask had they been there, and the wheels could have dug into the ground, enhancing the impact.[68] Likewise, the plane crash could not be taken to imply the non-safety of AMK, according to the experts, because the crash was more severe than was intended (a steel pylon entered one of the engines), the fire was in any case much less severe than it appeared and some passengers could have escaped it, and there was much unburned fuel left in and around the aircraft which helped to cool the flames in the first instance.

Including inexpert members of the public within the groups judging the meaning of these two crashes meant that debate was cut off prematurely before the appropriate expert analysis, of the kind we have sketched in the last paragraph, had time to make a mark; the public who witnessed the events simply did not have the contributory, or even interactional, expertise to make sensible judgements (though they seemed to have enough 'discrimination' to find Sir Walter Marshall's account unconvincing). In these cases then, the irony is strong: the environmental lobby, who are usually in favour of widening public participation in decision-making, would have preferred the interpretation to be the opposite of the immediate one. In these cases a better interpretation would have been accomplished by narrowing the group of decision-makers to certified experts alone.[69] This group, of course, would not have been limited to the 'official' experts and would have included representatives of environmental and safety-conscious lobbies, but they would have had to be *expert* representatives.[70]

UNDERSTANDING INTERACTION: THE MAGICIANS AND BENVENISTE

It might be thought that the prospect of bringing non-certified experts into scientists' core-sets and core-groups is near to zero, even if the idea makes sense. It might be argued that the professional pride of the scientific community would always prevent a change in this direction, just as it did in the case of the Cumbrian fells. It might be said that the San Francisco gay community, even though they were allowed to enter the core discussions, were allowed to do so only after they had adopted the personae, and perhaps the persuasions, of the scientists.[71]

Fortunately, another kind of case shows us that when the circumstances are appropriate, professionalism is not a barrier to the inclusion of experience-based experts into the very heart of scientific decision-making. Thus, in the case of cold fusion, there was a veritable feeding-frenzy of rejection by members of the scientific community who were not core-set members—suddenly there were experts everywhere. In the case of still more heterodox ideas, such as parapsychology, or Jacques Benveniste's claims about the power of zero-solutions (homeopathy works through the molecular 'memory' of water), stage magicians were brought in to pronounce on the propriety of the science, and their work was admired to such an extent that one of them was even given a prestigious MacArthur 'genius' award.

There are two ways of looking at the stage-magician phenomenon. One would be to account for it as an aberration from proper science that needs to be explained in terms of political expediency. Thus, one might see it as a quick and dirty way for scientists to accomplish rejection of 'fringe phenomena', with the maximum publicity, and without having to do the messy, difficult, and immensely time-consuming work of trying to prove a 'null' (there are no paranormal effects; plain water never has special biological properties). In other words, it is a way for core scientists to get their rejections straight into the public arena without going through the ordinary core-set process. If that is what is happening it is hard to remain neutral in the face of the process; we find ourselves wanting to be prescriptive and say that this is 'wrong'—it is a dereliction of scientific duty.[72] After all, among other things, scientists are there to help us know whether there are paranormal effects or homeopathic effects, but their input should be based on their best scientific efforts; ex-cathedra statements, or dirty tricks, are of no special value, nor should scientists pass their responsibility to outside groups. We noted above that scientific expertise cannot be transferred to a proxy, and the business of electing stage-magicians as science's representatives has to be questioned in this light.

Represented in the language of Figure 6, this case would reveal a strong and wide nexus from core-set to specialist pocket. A too-ready passing over of responsibility could be represented by a nexus so wide that the whole core-set would flow down it, like water down a drain, leaving the entire decision to the uncertified specialists. Revealing too ready a willingness to abandon responsibility as scientists—the moral guardians of a certain way of understanding the world—is, for obvious reasons, a dangerous game for scientists to play.

However, another way of analysing the stage-magician phenomenon—and this is how scientists tend to explain it—is precisely in terms of pockets of specialist expertise. Under this interpretation, in employing stage magicians, scientists are reaching toward specialist but experience-based expertise that has particular application in cases where fraud is suspected. Looked at this way, it seems less like an abandonment of scientific responsibility and more like a very reasonable extension of the core-set into a social group who may be formally unqualified and 'uncertificated', but who still have many of the qualities of core-set members in terms of long and relevant experience.[73]

DIFFERENT TYPES OF SCIENCE AND TECHNOLOGY

So far we have only differentiated between types of expertise; we have not differentiated between types of science. Wave Three, however, needs a categorization of sciences as well as a categorization of expertise. This is because the appropriate way to integrate the public into policy processes depends on the nature of the science and technology. In some cases, the public seems to be an integral part of the knowledge-base that is needed to make policy decisions; in other cases, their potential to contribute is much less clear. Let us start with the most well-worked-out case, that of technologies for wide-scale or mass public use.

INTEGRAL PUBLIC EXPERTISE IN PUBLIC-USE TECHNOLOGIES: CARS, BICYCLES, PERSONAL COMPUTERS

Consider technologies, such as cars, bicycles, computers and computer programs, where end users comprise a large proportion of the public, whose preferences are taken into account in the very design process. In these cases, specialist uncertified expertise is integral to the development of the technology (and thus of the related sciences).

There are at least two kinds of experience-based expertise relevant to such cases. First there is the narrow specialist expertise of computer 'buffs' and the like. Indeed, companies now take advantage of this kind of expertise by nurturing 'lead users' among their customers. In effect, these users acquire contributory expertise, and this is then recognized by the companies, who then consult them as experts.[74] Social groups such as computer hackers are similarly expert, although their intervention is not so welcome, at least, not by the computer companies.

Secondly, there is the much broader category of those whose legitimate contribution to the 'closure' of a technological design grows out of the very fact that they, being the users (or active non-users) of the object, are integral to the establishment of its meaning and success. In effect, these groups have some form of contributory expertise that shapes the future design, form and function of the artefact. This kind of argument has been most forcefully put within what has become known as 'the new sociology of technology'.[75] Were this case to be included in a version of Figure 6, it would show that the 'pocket' of specialists in the top, technical, half of the diagram, would exhaust nearly all the space outside the core.

Even in the case of a technology like this, however, there is still a political dimension. For example, in the case of cars, both drivers and non-drivers have a political say in the design of cars, based on their political preferences—often preferring designs which minimize fuel consumption and tax regimes which discourage pollution. These preferences, pressures and rights are better represented as belonging in the bottom half of the diagram, though in this case there is a serious borderline problem—the rights of the public as public, and the public as car-drivers, are very hard to untangle. In spite of the severe borderline problem in this kind of case, it is important, here as elsewhere, not to generalize from such extremes to science and technology as a whole.

INTEGRAL PUBLIC EXPERTISE IN LOCAL-INTEREST TECHNOLOGIES: PLANNING

In planning processes, 'local' knowledge often seems to confer special expertise on certain social groups. Like the car users discussed in the previous paragraph, the users of a locality seem to merit special involvement with the technical experts in the planning process. Local people can be seen as a large pocket of experience-based expertise when the issue within the core is local planning. However, thinking critically about expertise helps to disentangle the force of this localness.

In planning, local knowledge is a kind of expertise because local people can be said to have long experience of the local environment. But this expertise has to be used carefully, because local experience, when it is not combined with other kinds of experience, is partial, and this will frame contributions in a particular way. Thus, in the case of mineral extraction or waste disposal, the local population will tend to have a disproportionate understanding of the *disadvantages* of any development: they will know exactly how such developments will harm the local environment. But they may not have any special knowledge, or even any knowledge at all, of how developments will *advantage* the population of the larger regions within which the locality is embedded—the county, the nation, and so forth—and the users of the product. It is likely to be planning specialists who understand these things.

So far we have said nothing about local political interests, only local expertise. And one can see that local expertise is likely to favour the locality, even before the politics enters the equation. It is tempting to say that any attempt to separate the expertise and the politics is doomed to failure. In practice, this may be so, but the two phases are still easy to separate analytically. Thus, in the case of mineral extraction and waste disposal, local expertise will almost certainly militate against location of new plants in the local area, as such plants are almost certain to damage the local environment and increase public health risks. The political interests are more ambivalent, however, and likely to split along class lines. Thus, the building of a new quarry is likely to have an adverse effect on property prices in the locality, but a positive effect on employment, wages, and the profits of small shops. So the expertise and the political interests of the higher social classes are likely to be congruent, while the expertise and political interests of those who work for a weekly wage are likely to pull in opposite directions. Thus, even in local decision-making, it is still possible and useful to separate the political considerations from the technical considerations.[76]

ESOTERIC AND CONTROVERSIAL SCIENCES

In public-use technologies and planning, the involvement of the public as experts is 'integral' to the science itself. Now let us return to sciences where this is not so. At a first approach, four kinds of science of this type can be distinguished. To these we will apply the labels 'normal science', 'Golem science', 'historical science' and 'reflexive historical science'.

In *normal science* there are no major disputes, and the science is as settled as it ever can be. In these cases, scientists can fill the rôle of consultants without problem unless matters are opened up again by exposure to a controversy, such as in a courtroom or larger dispute. In courtrooms and the like, even the most routinized procedures with the longest historical entrenchment can be the subject of heated and detailed analysis. But this ground has been thoroughly studied by others, so we will not cover it again here.[77]

Golem science is science which has the potential to become normal science, but has not yet reached closure to the satisfaction of the core-set. The exposure of the public to Golem science is represented in our Figure 5. For example, in the debate over genetically modified organisms (GMOs), the argument about whether rats' stomach linings are affected by certain kinds of genetically modified potatoes is science of this kind; in the BSE ('mad cow') debate, the question of the strength of the causal link between BSE and Creutzfeld-Jacob disease is science of this kind. In neither of these cases is there any reason to think that the core-set will not reach a consensus eventually, nor is there any reason to want to say that the decision they reach should be influenced by anyone who does not work in a specialist scientific laboratory or medical school. It seems wrong to argue that the outcomes of these decisions should be the prerogative of the political sphere, indeed much of the complaint from the public is that the science has been prematurely passed to politicians who tried to impose a closure to the debates that would reassure the public about the safety of the new technologies when no closure had been reached by the scientists.[78]

This, of course, is *not* to say that the decision about what should be done now about GMOs and BSE can or should be left to certified specialists alone. There are two reasons: firstly, they do not have the answers; and, secondly, they may not have been given questions that correspond with public concerns. For example, their view of what is acceptable in terms of ethics, or risk, may not match the view of the public. Thus, in the case of Golem sciences, it can be seen that the balance of the two spheres of decision-making separated by the horizontal line—the technical and public—is bound to favour the public, as compared to normal science. It should be expected, however, that as time—and it may be many decades—passes, the balance would slowly shift back again as a core consensus is reached.

Historical sciences, on the other hand, are those in which it is not to be expected that there will be any closure in the core-set debate in the foreseeable future. Such sciences have also been understood for a long time, even though new developments in science and technology have brought them much more to the fore in recent decades. Historical sciences deal with unique historical trends rather than repeatable laboratory tests.[79] The question of global warming is a historical question; long-term weather forecasting is a historical science; the ecological effects, as opposed to the effects on single organisms, of GMOs is a historical problem. The reason closure on these matters should not be expected in the foreseeable future is that the whole system in which they are embedded is too complicated to model accurately, and may even be impossible to predict accurately because of the working of chaotic processes.

In *reflexive historical sciences*, the potential for uncertainty becomes even greater, as the long-term outcomes are affected by the actions of humans themselves. For example, the science of global warming, as well as being historical (as just explained), is also reflexive. This means that the input variables will include the outcome of political and ethical debates among humans.[80]

When an environmental decision has to be made, Golem and historical science are in some ways similar and in some ways different. They are similar in as far as the scientific input is equally uncertain; but they are different in that the certainty which Golem science can eventually reach through normal scientific processes, cannot be attained in historical sciences. In reflexive historical sciences it cannot be approached without social or cultural regulation. Thus, in the case of all historical sciences, society needs certified and experience-based expertise in the scientific fields belonging to the problem, as well as political input; while in reflexive historical sciences, politics, policy, regulation and sociology enter in the top half of the diagram—expertises in the sciences of politics, policy, and so forth, are needed, as well as political input in the more ordinary sense.[81]

In the case of historical science, the rôle of political and social interests is, perhaps, especially prominent, as there is no hope of any major increase in scientific input, so the institutions that are designed to meld the expert and the inexpert would have more permanence than they would in the case of Golem science. In the case of reflexive historical science, futures must be based not just on permanent social institutions for the regulation of science, but on the development and maintenance of new social institutions for the regulation of social life. In this way, these historical policy sciences are more like the public technologies discussed earlier, as they rely on the participation of the lay public (or at least a large portion of it) for their success.[82] It can be seen, then, that, even when the science is esoteric and controversial, thinking critically about the nature of expertise makes it possible to understand how and when different types of decision-making processes are needed.

CONCLUSION

We have argued that, although science studies has made an enormous contribution to our understanding of the relationship between science and society, there is more to do. Wave Two of Science Studies has shown us the many ways in which science cannot solve technical problems in the public domain. In particular, the speed of political decision-making is faster than the speed of scientific consensus formation. As a result of this emphasis, Wave Two's predominant motif has been the need to legitimate technical decisions—to solve the *Problem of Legitimacy.* Decisions will have no legitimacy if they continue to follow the intellectually unsupportable, top-down, authoritarian model of Wave One. Nevertheless, it would be disastrous to solve the Problem of Legitimacy by dissolving the distinction between expertise and democracy. To do this would be to create a new *Problem of Extension.* We argue that expertise *should* feed into the decision-making process, but not in the old Wave One way; solving the Problem of Extension without re-erecting the Problem of Legitimacy depends on recognizing and using new kinds of expertise emerging from non-professional sources. We argue that to do this we need a Third Wave of Science Studies, with the ability to develop a normative theory of expertise. Wave Two has been enormously successful, and continues to be enormously successful, in deconstructing knowledge; without abandoning Wave Two, we now need to reconstruct knowledge and develop Studies of Expertise and Experience—SEE.

We use a series of diagrams to explicate the way science studies has contributed to our understanding of the science/society relationship, and how it might do so in the future. The diagrams indicate that decision-making rights that emerge from the political sphere, and those that grow out of expertise, should not be confounded. We resurrect the old distinction between the political sphere and the sphere of expertise, but in our model the boundary is found in a new place. This boundary is no longer between the class of professional accredited experts and the rest; it is between groups of specialists and the rest. This follows from distinctions that scientists make themselves: in any specialism it is easy to distinguish between a core group of experts and scientists in general, yet the core holds no special professional qualifications. We find that to make these classifications work well we have to distinguish between esoteric sciences, on the one hand, and public technologies, such as cars and computers, on the other.

We go on to indicate, first, that it is possible to have a normative theory of expertise without abandoning the insights or the programme of Wave Two. We begin to show what the components of such a theory might include. We show that we can classify scientific expertise into *interactive expertise* and *contributory expertise.* We show that these ideas emerge from sociologists' own practice, and this offers one persuasive way into a normative theory. We develop some thesis-like propositions using this classification of expertise. We also introduce the ideas of *referred expertise, translation* and *discrimination*. In discussing discrimination, we distinguish between ubiquitous and specialist knowledge that has been gathered as a result of local experience. Using these ideas, we argue that scientists as a class have no special contribution to make to technical decision-making in the public domain, and that if there are to be public defences of science, they should concentrate on scientists as specialists, rather than as generalists.

We briefly re-analyse a series of case studies to show how our new categories work. In particular, we show that Brian Wynne's well-known study of the Cumbrian sheep farmers should not be understood as a defence of 'lay expertise', but as the interaction of two communities of experts, one without certificates. We argue that institutions are needed that can translate the knowledge of such pockets of experience-based expertise so as to make it less easy for certified scientists to resist their advice. Such bodies of experts already exist, but tend to be associated with campaigning organizations.

We re-describe the success of the AIDS activists studied by Steven Epstein, using our new term, 'interactional expertise'. We redescribe Harry Collins's study of crash demonstrations, showing that our theory sometimes leads to the conclusion that there should be less public involvement in technical decision-making. That our theory sometimes indicates more and sometimes less public participation seems to us a strength, as compared with the monotony (in the mathematical sense) of other theories that look at the same area of concern. We show that scientists' use of stage magicians to settle certain disputes reveals that in some circumstances professional scientists are happy to absorb pockets of uncertified expertise.

Finally, we argue that this kind of analysis has a dimension that relates to types of science. We distinguish *normal science, Golem science, historical science,* and *reflexive historical science,* each of which has different implications for our futures. What we have tried to do is to provide a language and some concepts for debating these issues. Each different case of public-domain science will need its own combination of expertise. The sheep farmers were a particularly clear case of the failure to utilize a pocket of experience-based expertise, but the same analysis will not always apply.

The romantic and reckless extension of expertise has many well-known dangers—the public can be wrong.[83] Let us give some examples. When scientific advisers concluded that the battery additive AD-X2, launched in the mid-1940s, had no significant effect, there was an intense lobbying campaign, supported by both industrial and individual users. This campaign eventually led to the Director of the US National Bureau of Standards, Dr Allen Austin, being fired. He was subsequently reinstated following protests from the scientific community, and the battery additive was finally withdrawn from sale in the mid-1960s.[84] More recently, Greenpeace, probably feeding on public acclaim for its actions, blocked the disposal of the 'Brent Spar' oil platform, only to have to admit later that its scientific assessment was incorrect.[85] Similarly, citizen groups, who campaigned in support of Laetrile, a purported cure for cancer that was labelled a hoax by the FDA, seem to have been fooled.[86] More controversially, citizen groups in the USA continue to lobby for creationist science to be taught in schools, while, in the UK, at the time of writing, vaccination levels for measles are falling as a result of an alleged link between the triple measles, mumps and rubella (MMR) vaccine and childhood autism which seems to find virtually no support among the scientific community. These observations merely indicate the kind of work and analysis that has to be done before 'the public' as a whole is attributed with expertise.

The job, as we have indicated, is to start to think about how different kinds of expertise should be combined to make decisions in different kinds of science and in different kinds of cultural enterprise. The job is to contribute to the debate by deploying the science studies community's specialist *contributory expertise* in the matter of the nature of knowledge and expertise. To do this is to embark on SEE, and to act as knowledge scientists. One obvious next step is to find ways to think about how to weight contributory expertise, interactive expertise and referred expertise, along with translation and discrimination, when judgements about a variety of public-domain technologies are made. This has the feel of a classic problem, and we would guess that better scholars than ourselves will discover that the distinction and its consequences have already been discussed in the Greek city state,[87] in the post-1945 debate about the relationship between politicians, civil servants, industrialists, managers, scientists and other producers of culture,[88] and where critics and artists have confronted each other.

Though this is in many ways a programmatic paper, it is meant to do real work in changing the way we look upon the enterprise of science studies and the way it handles questions to do with the relationship between science and society. We argue a case, but also show how the work of building a corresponding structure, theoretical, empirical and institutional, could be carried forward. This is a pressing problem if we are to navigate our way between the Scylla of public disillusion and the Charybdis of technical paralysis.

APPENDIX

This paper draws on a range of existing empirical and theoretical work. This Appendix discusses some of this background, but makes no claim to be exhaustive. Instead, the aim of the Appendix is to show how our analysis shares certain concerns already present in STS, though we reformulate the old problems and approach from a different starting point. As noted in the main text, the structure of the Appendix mirrors that of the main paper and the main subject headings in the paper re-appear here. Items of literature are often relevant to more than one heading, so the arrangement under headings is to some extent arbitrary.

THE PROBLEM OF LEGITIMACY
AND THE PROBLEM OF EXTENSION

However, we are not the first to have noticed that there is a problem with the way in which science interacts with the wider society. For example, in 1977, Edward Lawless discussed 45 cases of controversial science in the USA that occurred between 1948 and 1973, and listed many more.[89] In the UK, the BSE crisis, documented in the Phillips Report, and public opposition to GM ingredients in food, are two well-known examples.[90] Recent policy documents, such as the 'Science and Society' report produced by the House of Lords Science and Technology Committee, the European Union White Paper on Governance and the Loka Institute's citizen panels set up to consider 'telecommunications and democracy' and 'genetically engineered foods', all show that it is widely recognized that there is a problem to which some response is needed.[91]

Almost invariably, the call has been for greater dialogue between science and the public, and for increased participation in decision-making about science and technology. For example, the Report of the House of Lords Science and Technology Committee recommended:

> That direct dialogue with the public should move from being an optional add-on to science-based policy-making and to the activities of research organisations and learned institutions, and should become a normal and integral part of the process.[92]

Studies have shown that suspicion within the wider society does not manifest itself in respect of every area of science and technology. For example, mobile phones, replacement hip joints and microwave ovens are not perceived as problematic by the public. In 2000, a review of science communication and public attitudes to science in Britain showed that:

> 84% of people think that scientists and engineers make a valuable contribution to society, and three quarters think that science and engineering are good careers, and that science, engineering and technology will provide more opportunities for the next generation.[93]

In the USA, surveys that address the topic of science and technology in general, as opposed to their specific applications in various fields, also show a broad support for science and technology. For example, the most recent edition of the National Science Foundation's *Science and Engineering Indicators* reports that: 'In general, Americans express highly favorable attitudes toward science and technology'.[94] Perhaps surprisingly, a recent *Eurobarometer* survey dealing with European attitudes to biotechnology found that, even in cases where scepticism might be expected to be very strong, respondents were still more likely to agree than disagree with the statement that technologies such as telecommunications, information technology, space exploration and biotechnology, will improve life over the next 20 years.[95] This support was not uniform, however, suggesting that where the public do have concerns about science and technology they seem to be about specific aspects or applications. Thus genetic engineering scores lower than communications and information technology—but nuclear energy alone was more distrusted than trusted.

In the same way, suspicion does not extend to all scientists. Although not enjoying the same level of public support as some professions, opinion poll evidence for the UK routinely shows that scientists are amongst the most trusted sources of information in the public domain. For example, a Mori poll, 'Trust in Scientists', conducted in March 2001 for the British Medical Association, found that 65% of the sample would 'generally trust scientists to tell the truth'.[96] The result has to be put into perspective in that 89% of doctors were trusted in spite of the well publicized cases in which doctors have been seen to be less than honest in their dealings with patients.[97] Teachers were trusted by 86% of the population, professors, judges and the clergy by 78%, and television news readers by 75%.[98] All these, then, were trusted more than scientists. On the other hand, scientists were trusted more than civil servants (43%), trade union officials (39%), business leaders (27%), government ministers (20%), journalists (18%) and politicians generally (17%).[99]

Similar sentiments can also be found in the Advisory Report on the Regulation of Biotechnology, in which the section summarizing the findings of the consultation with the People's Panel notes that:

Government advisory groups are a trusted mechanism for decision making, but membership should be broadly based. Scientists and healthcare professionals are seen as particularly important contributors to decision-making. There is a widespread demand for as much information as possible and again Government Advisory Groups, scientists, healthcare professionals and consumer and environmental groups would be trusted to provide this information. But the media, retailers and industry were not trusted.[100]

In formulating our definition of the problem, therefore, it is specific episodes of science in context that we use to illustrate our arguments.

One indication of where things might be going wrong can also be found in the same survey data. It turns out that the trust in scientists expressed by these members of the public is sensitive to the wording of the questionnaire, and falls substantially when the scientist is associated with government or industry. What seems to happen is that the distrust of the scientist's organization outweighs the more positive evaluation of science in general, leading to an outcome that is not dissimilar to that for government or industry in general.[101] As a result, a scientist working for the government or industry is seen as much less trustworthy than one without this affiliation. This is what Brian Wynne has referred to as 'scientific body language'.[102]

These survey results are backed up by a range of more qualitative studies. For example, in 1998, Anne Kerr, Sarah Cunningham-Burley and Amanda Amos published the results of a series of focus group discussions on the public perception of, and reaction to, the new genetics. They found that genetic science was frequently interpreted in the context of a wider understanding of the nature of scientific work, noting that discussion in the focus groups included such topics as:

> ... competition and cooperation among scientists; sources for funding, especially the relationship between the new genetics, pharmaceutical companies, and government; and the relationship between geneticists and the media.[103]

Thus participants in the focus groups did not evaluate science in general, but science in practice, and seemed well aware of the scientists' need for publicity, publications and grant income.

Similar results have been obtained in a wide range of other studies, such as those done at Lancaster University, where the importance of the public perception of scientific institutions has been repeatedly highlighted. Perhaps the most remarkable of these, and certainly the most prescient, is the 1996 study that predicted the controversy over GM foods in the UK a year or two before it actually happened. In the follow-up study (published as *Wising Up*),[104] the same concerns continued to dominate the discussion of GM foods, but the more positive evaluation of mobile phones highlights the fact that the concerns being expressed were not simply general anti-science attitudes, but responses to specific characteristics of the GM food industry. Similar themes have also been identified in a recent paper published in this journal by Steven Yearley, while three overview collections of public understanding of science research, one under the auspices of COPUS, and the others edited by Alan Irwin and Brian Wynne, and by Ian Hargreaves and Galit Ferguson, also contain references to related findings.[105] Together these highlight several of the more robust findings from the public understanding of science research literature. People are not typically concerned with science in general but with particular concrete instances and applications. What is more, in responding to this application of science, they are often sensitive to the uncertainties surrounding the science itself, and do not distinguish between the science and its sponsor. Indeed, one of the most frequent and striking observations of this research (for example, by Alan Irwin, Alison Dale and Denis Smith) is the way the science effectively 'disappears' from the dispute at a relatively early stage.[106]

Finally, surveys conducted in Britain, such as that by Geoffrey Evans and John Durant, and case studies such as that by Ian Welsh, have found that, amongst those most critical of specific applications of science, there can be groups that have considerable scientific knowledge.[107] Likewise, a recent survey of American public opinion, focusing on life sciences and stem cell research, found a similar picture, with 30% of college graduates and 24% of those who considered themselves 'very informed about science' being somewhat or strongly opposed to research using embryonic stem cells.[108] In other words, more knowledge does not necessarily lead to more support. In some cases, the expertise of these stakeholder groups or their representatives will be substantive, and will constitute a direct challenge to the science.

In other cases, stakeholders will question the ability of institutions and regulations to deliver the standards of performance needed for the scientific advice to be implemented safely, challenging, not the science, but the assumptions on which it is based. In the BSE crisis in the UK, regulations concerning the slaughtering and disposal of animals seem not to have been applied, and this seems to be the cause of concern. Other examples from the STS literature include Brian Wynne's discussion of the ways in which people living around the Sellafield (BNFL) nuclear plant interpret statements about safety in the context of previous, largely negative, experiences. Similar themes can also be found in the focus group discussions concerning air quality monitoring in Sheffield that are analysed by Steven Yearley.[109]

These studies inform much contemporary thinking about the way in which the public responds to science. In particular, the studies have been successful in counteracting the 'deficit' model—the argument that public opposition to science followed from public ignorance of science, and can be 'cured' by removing the 'deficit' in the public's knowledge and understanding.[110] It is now acknowledged that what we refer to as 'The Problem of Legitimacy' is much more complex than the deficit model would imply.

Another way to approach the Problem of Legitimacy is from the direction of theoretical developments in social science.[111] Ulrich Beck, for example, argues that modernity is undermining its own institutions and that science, in particular, is increasingly the cause of, and not the solution to, societal problems.[112] At the centre of this development is the recognition that 'invisible' risks, such as those created by radiation, pollution and environmental change, are making concern about the uncertainty and contingency that accompany scientific and technical innovation a central feature of contemporary society. Policy-makers have become pre-occupied with avoiding technologies that may ultimately create more problems than they solve. The emergence of ideas such as 'sustainable development' and the 'precautionary principle', which are central to new policy discourses, shows that the problems exemplified by 'the environment' have crossed traditional boundaries to become, simultaneously, social, cultural, economic, ethical and scientific problems.

A second theme, associated with Anthony Giddens, that runs through the reflexive modernization literature, is the problem of identity that results from the undermining of traditional institutions.[113] The critical element of the risk society literature is that the decoupling of individuals from traditional institutions and rôles has politicized identity and lifestyle. In effect, social life becomes an ever-increasing series of individual choices and responsibilities. In the case of science, this change is manifested in the increasing proliferation of expertise and counter-expertise, and hence the need to make choices about who or what is be trusted in this new context. More generally, the change is reflected in the importance of alternative social movements that provide the social spaces within which the 'sub-politics' of individual life are played out and given meaning. One solution is to find ways of incorporating these new social movements and political alliances within the institutions of governance. We emphasize, however, that this approach begs important questions: How should this be accomplished? How much more inclusive should these new institutions be? Who should be included and who excluded? In our terms, this is the 'Problem of Extension'. The attempt to resolve the Problem of Extension takes us well beyond the reflexive modernization literature.

THREE WAVES OF SCIENCE STUDIES

As work by authors such as Ian Welsh demonstrates,[114] there was opposition to the power of science and technology even during the high days of Wave One. This shows how broad is the brush with which we are painting. Nevertheless, thinking about Wave One as a coherent body of thought can be legitimated by referring to writers such as Karl Mannheim, who insisted that sociological analysis should draw back when it encountered natural science. Michael Mulkay summarized the key points in Mannheim's sociology of knowledge as follows:

In the first place, the phenomena of the material world and the relationships between them are seen as being invariant (Mannheim 1936: 116).

Mannheim regularly refers to the natural world, and to the concepts appropriate to its study, as being 'timeless and static'. Valid knowledge about such objective phenomena he maintains can be obtained only by detached, impartial observation, by reliance on sense data and by accurate measurement (Mannheim 1952: 4–16; 1936: 168–9). Because the empirical relationships of the natural world are unchanging and universal, the criteria of truth by which knowledge claims are to be judged are also permanent and uniform (1936: 168). It follows that natural science develops in a relatively straight line, as errors are eliminated and a growing number of truths discerned. In short, scientific knowledge evolves through the gradual accumulation of permanently valid conclusions about a stable physical world.[115]

Robert Merton's sociology of science, with its identification of the norms of scientific activity, also contributes to the understanding of science as different from other kinds of knowledge-generating culture.[116]

We do not mean to imply that the intellectual arguments that underpinned the First Wave of Science Studies have disappeared completely. The old arguments tend to be promoted by philosophers and, more recently, by scientists concerned to resist what they see as an attack on science. The latter tendency is part of what has become known as the 'Science Wars'.[117]

We feel it unnecessary in this paper to provide extensive references to Wave Two of Science Studies, except where those studies could be seen to overlap on the project described in the main paper, in particular in describing the basis of expertise.

THE NATURE OF EXPERTISE

Sheila Jasanoff's studies are an example of Wave Two work which does overlap with our concerns. Her researches into regulatory legal proceedings in the USA are classic examples of the critical insight that the sociological perspective can give.[118] Jasanoff argues that the adversarial nature of the legal proceedings through which US regulatory policies are tested performs a range of useful functions:

At their most effective, legal proceedings have the capacity not only to bring to light the divergent technical understandings of experts but also to disclose their underlying normative and social commitments in ways that permit intelligent evaluation by lay persons . . . . Controversies about risk are perhaps the domain in which courts have made the most impressive contribution to the civic culture of American science and technology. . . . By insisting on their prerogatives in this regard, courts have repeatedly affirmed that the ultimate power to guide technology policy is vested not in experts but in the citzenry.[119]

In this context, the selection of the witnesses and experts who testify, and their ability to demonstrate their expertise under cross-examination, is crucial. Much effort has been put into developing criteria for the selection of expert witnesses and to instituting various forms of quality control within the courts. Recent examples of this process in action are the three Supreme Court rulings issued in the 1990s that effectively encouraged judges to take a more active role in sifting 'expert' testimony, so that juries were only presented with relevant and reliable testimony. This represented a change from previous practice, when such evaluations were typically left to the jury.

These and many other aspects of the American system are described in detail in Jasanoff's publications, and we will not attempt to summarize them here. Instead, we concentrate on the epistemological status of the outcome of this process. As Jasanoff argues, the decision to prohibit or permit something combines scientific content with regulatory power. For example, in order to reach a decision about whether or not a particular chemical or process is hazardous, the court may have to decide whether the 'LD50 test' is appropriate. (This test measures the toxicity of something by establishing the level of exposure at which 50% of the test animals are killed.) Likewise, the courts can argue about whether results from laboratory animals can be generalized to humans. In other words, the regulatory decision cannot be made without attributing credibility to one set of experts and denying it to the other. This implies making a judgement that has traditionally been the preserve of the core-set scientists alone. In the main paper, we separate out the different dimensions of this process. We say that these decisions fall under the 'political phase' of the decision-making process, which deals with the societal response to scientific uncertainty. Decisions about the content of science fall into the 'technical phase'. Unlike some of the more recent attempts to achieve a scientific consensus, or at least minimize controversy in the political phase, by restricting participation to 'approved experts', in our model the scientific decision invariably gets made after the political one.

The legal examples are important because they are one way in which non-scientists become involved in making scientific decisions. Again, this focuses a concentration, as in our Problem of Extension, on the boundary of the decision-making group. During the 'recombinant DNA' debate of the mid-1970s, the Cambridge Experimental Review Board (CERB) pushed these boundaries out further than the courts. The deliberations of the CERB have been analysed from a rhetorical perspective by Craig Waddell.[120] The CERB gave decision-making powers not to judges but to a citizen panel that, in 1976, was asked by the city authorities in Cambridge, Massachusetts, to make recommendations about whether, and under what conditions, research using recombinant DNA techniques would be permitted in the city's universities.[121] The Board comprised 12 members selected from the city's population, and their job was to weigh the evidence presented by both proponents and opponents of the research and make appropriate recommendations. The outcome of the Panel's deliberations was that research was to be permitted, under conditions that broadly mirrored national guidelines, and the whole process was widely seen as a success by all who participated in it.[122] In the context of our paper, the important aspect of the CERB study is that, like the studies of courtrooms, it shows that non-scientists can lend credibility to decisions concerning science and technology.

These are positive arguments for increasing participation, and there are also negative arguments stemming from the failure of more restricted practices. One set is brought out by a case study of the regulation of the chemical 2,4,5,T.[123] The negative arguments focus on the neglect of the assumptions that underpin and frame scientific knowledge claims, and highlight the problems that arise when scientific knowledge is generalized uncritically. One of the successes of Wave Two was to draw attention to the contingency and uncertainty of scientific knowledge whilst also highlighting alternative knowledge(s) that can (or even ought) to complement or replace it.

The case of the regulation of 2,4,5,T (an organophosphate pesticide used by farm workers) is a well-known example of this argument. A Scientific Advisory Committee in the UK concluded that 2,4,5,T was safe to use, subject to the caveat that appropriate precautions were taken. Farm workers, on the other hand, argued that, because the appropriate precautions could not be taken in the day-to-day settings in which the chemical was actually used, then it was not safe. In this case, the embodied experience of the farm workers is advanced as an alternative, contextual knowledge resource that could (and should) have been a legitimate input to the decision-making process, and to which scientists were largely blind.[124] Other settings in which similar ideas receive empirical support include the gendered and culturally specific experiences of science and technology, such as have been discussed by Evelyn Fox Keller, Helen Longino, Sandra Harding and Donna Haraway;[125] the participation of AIDS treatment activists in clinical trials as described by Steven Epstein;[126] the capacity of people with other illnesses or injuries, such as 'miners' lung', CFS and RSI, to contribute to the medical understanding and treatment of their condition, as researched by Hilary Arksey and Michael Bloor;[127] the contribution of community groups to public inquiries and planning processes, as discussed by Arie Rip, Thomas Misa and Johan Schot;[128] and the development of multidisciplinary teams and end-user groups in industry and research, as analysed by Michael Gibbons and his colleagues.[129]

One could say that the tendency to dissolve the boundary between those inside and those outside the community reaches its apogee in 'Actor Network Theory', as first adumbrated by Bruno Latour and Michel Callon. Here even the boundary between human experts and non-human contributors to the resolution of conflict is taken away.[130]

Another contribution to thought about the meaning of 'expert' is provided in another study by Brian Wynne. He describes the experience of apprentices working in the radioactive materials industry. He suggests that the apprentices felt they had no need to contribute to their own safety by trying to understand the science of radioactivity, because they were 'intuitively competent sociologists' and 'vigilant and active seekers of knowledge . . . tacitly and intuitively, positioning themselves, using their knowledge of their social relationships and institutions'.[131] Wynne argues that the apprentices used their social understanding as a basis of trust in their employers. In a later paper, referring to the same group, he says that these apprentices' 'technical ignorance was a function of social intelligence'.[132]

There are two ways of looking at Wynne's contribution. It could be an example of what we have called 'local discrimination'. In this case, the apprentices would be seen as using their hard-won specialist competence in understanding the trustworthiness of their particular employers and their own place within the social networks of trust operating in that particular workplace, to assess the safety of the procedures to which they were exposed. There is, however, the danger that this analysis is vulnerable to the same kind of ambiguous interpretation that we find in the case of the sheep farmers—namely, that specialist expertise that is not recognized with a certificate is confounded with the capabilities of humans in general, in virtue of their 'socialness'.[133] The danger is, in our terminology, that local discrimination and more ubiquitous discrimination are being confused.

Thus, let us assume that the apprentices also hold bank accounts. Would Wynne want to say that they felt no need to understand economics because they were, to use his phrases, 'intuitively competent sociologists' and 'vigilant and active seekers of knowledge . . . tacitly and intuitively, positioning themselves, using their knowledge of their social relationships and institutions . . .' when they paid their cheques into the bank? Would he argue that the apprentices would be using their social understanding as a basis of trust in their bankers, and that their economic ignorance was a function of social intelligence? The answer is that if he did say this, he would be right, but he would be talking of the relations of trust in general that pertain throughout any smoothly functioning human society, rather than a specific locally acquired discriminatory ability. Once more, for discrimination to be a useful concept, we must solve the Problem of Extension in respect of those who can discriminate.

We have discussed Stephen Turner's paper at length in the text, and note here that though he makes a useful classification of expertise, he does not discuss the levels of competence within an expertise that license contribution to a technical decision. This difference in emphasis may arise from difference in concerns between the UK and the USA. The STS literature shows that public participation and opposition often start with the 'neighbours' of a technical problem, and in particular those who are directly affected by it or unable to avoid its (potentially) negative consequences. As a reading of James Petersen makes clear, in the USA these concerns chime with a wider tendency to be sceptical of government and to place a high value on public involvement as a mechanism for ensuring accountablity.[134] For example, the Economic Opportunity Act of 1964 actually makes the participation of socially and economically disadvantaged groups in community development plans into a political right. In the 1970s, this idea was extended to include a wider range of science and technology policy areas.[135] Some consequences of these developments can be seen in the referenda that are a growing feature of the US political landscape, especially in relation to science and technology issues such as airport expansion, nuclear power, and the like, the increasing use of public opinion research by administrative agencies, and in experiments with deliberative forums. The claim is that:

> . . . substantial public input ensures a more thorough and open debate on questions of science and technology policy. This is especially important in that the public has so frequently been excluded from decisions on technical questions. In this context, extraordinary measures may be required to facilitate effective citizen participation to counterbalance the current elite domination of technical policy making.[136]

This quotation encapsulates the tension that motivates our paper. Although two reasons are given for participation, only one of them is supported by the STS literature. The first claim, and the one that is supported by STS research, is that public participation ensures fuller debate, which has the effect of ensuring that more of the available options and assumptions are questioned and tested, and perhaps more importantly, seen to be tested. Thus, as Ian Welsh argues, one important role played by protest groups is to keep doubts alive in the wider community and maintain this questioning of expert advice.[137] The importance attached to this scepticism has, no doubt, been reinforced by a series of technically based controversies,[138] and the observed failure of past expert advice (for example, with regard to nuclear power as a source of safe and abundant source of energy).

The second reason given, which is not supported by STS, is that public participation redresses elite domination of technical decision-making. This is presumed to be a good thing. But is it? We think the answer begs a solution to the Problem of Extension. The issue can be clarified by asking whether or not the same urge for participation is found in other policy areas and, if it is not, would it make sense to advocate it in these contexts. One of the most striking contrary examples is the case of economic policy, where the tendency in most of the major economies has been to move towards independent central banks, effectively giving the power to make key monetary policy decisions to an elite group. In other words, once the targets have been set, and monitoring mechanisms set up, the responsibility for meeting these targets resides solely with the central bank and its advisers. There is not, however, any direct requirement for more public input to these decisions. In the UK, interest rate decisions are made by the Monetary Policy Committee (MPC), whose members are all economists working mainly in the financial or banking sectors, although industry voices are also represented.[139] In the USA, there is little obvious pressure to make the US Federal Reserve replace Alan Greenspan with a more participatory process.[140]

From a European perspective, the equation between public participation and better decisions is less persuasive, and the 'elitist' starting point to debates about the extension of expertise seems unremarkable. This means that the debate about appropriate sources of advice has a different tone. Although participation is encouraged, this is seen as a problem of efficiency rather than democracy. Thus individuals or groups are said to be able to contribute to a consultation process because they have some relevant experience, rather than in the context of a discourse of rights and accountability. The most formal implementation of the European perspective is to be found in the Constructive Technology Assessment (CTA) approach, described by Arie Rip and his colleagues, which seeks to maximize the benefits from the more informal assessments that are triggered by scientific and technical controversies.[141]

The CTA approach is explicitly sociological, and is closely related to the Second Wave of Science Studies. Within this perspective, the emphasis is on the networked nature of knowledge, with the robustness of knowledge claims being related to the amount of work that one has to do in order to challenge them. Controversies are useful, therefore, because they can destabilize existing networks and expose the work that goes into creating new ones. Social learning occurs as different frames, knowledges and sources of expertise are articulated, and the network made more or less robust. This notion of articulation is important. It implies that only certain types of contributions will promote social learning and that, whilst participation should not be restricted to established institutions and actors, only certain types of contribution are to be welcomed. Thus Rip argues:

> The effectiveness [of extraparliamentary dissent] lies in the attention given to knowledge claims in addition to negotiations between interests, and in the broadening of the agenda that occurs by including more parties in the debate. These advantages are relevant for public participation in general, but have to be set against the disadvantage that rules for interaction and

the emergence of consolidation require some boundary. Introducing a new party in the debate may offset the balance of forces; this should only be done when a gain in articulation is to be expected. Concretely, this implies that participation is not a citizen's right *per se*, but has to be earned on the basis of specific claims about the issues in the controversy.[142]

Characterizing the process as one of 'social learning' also has implications for the nature of the outcomes. Consensus is not necessarily the goal, as the process will have worked if all it does is raise awareness of questions and uncertainties. This is particularly important for many of the scientific controversies that occur in policy debates precisely because the existence of controversy itself signifies a lack of consensus. According to the CTA approach, in such circumstances what is needed is a process that will enable the new network of knowledge to be developed in a context in which it is unclear who knows what and what, if anything, needs to be learned. For this process to develop, there needs to be a means of identifying potential participants, processes for orchestrating the interaction between different parties, and a purpose to motivate their interaction. Our paper is aimed mainly at the first of these problems, but we do not deny that the other stages also pose significant difficulties.

For example, one obvious problem faced by any institution dealing with such controversies is that the arguments are not just about science and facts, but about interests and resources. As a result, the forum created for resolving the conflict can become just another resource within it, and is thus used strategically by the participants. Thus, rather than facilitating a Habermas-type discourse between equals, the participatory forum becomes another part of the already contested decision-making process. In this context, Rip cites the case of the 2,4,5,T debate in the USA, where opponents of the use of the herbicide refused to participate in a second 'dispute resolving conference' because they felt the first such conference had co-opted them, and they did not want to be associated with a similar outcome. Similarly, conferences about power in Holland and Austria became so dominated by pro-nuclear groups that the public they wished to persuade actually stayed away![143]

Despite these problems, the need for social learning remains, and the acknowledgement that controversy exists, even if it is badly organized, is still a more useful response than the repression or denial of its legitimacy. These ideas are reflected in the discussions at the recent EU Workshop, 'Democratising Expertise', in which it was accepted that there was a need to involve experts and stakeholders at the earliest possible stage in technical decision-making processes, and to retain their involvement as decisions need to be revisited and re-evaluated in the light of new evidence.[144] The workshop participants also acknowledged that the definition of expertise needs to be broad, and to include theoretical and practical knowledge from across the range of sciences and

stakeholder groups, including the public at large, whilst also emphasizing that
'democratising expertise' is not about majority voting in science.[145] Instead,
there is a need to elaborate principles on the way expertise is developed, used
and communicated, and to develop mechanisms to make expert advice more
widely available so that representatives can take more informed decisions.[146] In
other words, decisions need to be taken by accountable decision-makers, but
the quality and legitimacy of those decisions are enhanced if they are seen to
take the full range of views into account.[147]

The idea of extending decision-making rights outward from the generally
recognized core-set of certified experts has a resonance with the idea of 'maximum objectivity', which Sandra Harding defines as follows:

> A maximally objective science, natural or social, will be one that includes
> a self-conscious and critical examination of the relationship between the
> social experience of its creators and the kinds of cognitive structures fa-
> vored in its inquiry.[148]

There is, however, a difference between our view and that of the advocates of
standpoint science. In our case, participation is predicated on experience-
based expertise. In the case of standpoint science, however, political position
in society is itself taken to legitimate an input to science; there would be a feminist science, a black science, and so forth.[149] These sciences would be discontinuous with each other. In our model, the contributions of women or members of ethnic groups to science would be continuous with it. Women, blacks,
and other groups, would contribute specific experience-based expertise which
could be gained no other way, except through participation as members of
those groups. They would contribute their special kinds of expertise wherever
such expertise was relevant. We would make no claim to legislate in advance
for where such expertise was relevant—that would be a matter to be settled in
each particular case. But such expertises would almost certainly not be univer-
sally relevant—there would be no female or ethnic physics, just distinctive
contributions to areas of science by women and certain ethnic groups wherever this was appropriate.

To us it seems strange that academics, in particular, should want to adhere
to the opposite view to the one expressed above. The ready acceptance of the
idea that science is politicized through and through rules out the possibility of
complaint when we find that certain scientific and technical arguments are
hopelessly biased by their sources. For example, do we never want to say that
the tobacco industry has for years falsified the implications of epidemiological
studies out of a concern for selling more cigarettes? Do we want to say, rather,
that this was just the tobacco industry's point of view and the only fight there
is to be had with them is a political fight, not a scientific fight? Do we want to

say that the estimates for the success of Patriot in shooting down Scuds during the Gulf War were not 'illegitimately affected' by the interests of the parties, only 'affected'?[150] Accepting the arguments for standpoint sciences would imply that such concerns are a category mistake because science is indistinguishable from politics. Oddly, the one group of people who would be most affected in terms of loss of power is academics, because their only source of power is the legitimacy of their arguments, critical or otherwise.

It seems, then, important to retain a notion, even it is an idealized one, of a core-set community in which expertise is used to adjudicate between competing knowledge-claims and to determine the content of knowledge. The wider society still has a role to play in forming a view about the socially acceptable use of such knowledge and what to do while such knowledge remains contested, but this contribution lies in the political sphere. Lay people as lay people, however, have nothing to contribute to the scientific and technical content of debate. Even specific sets of lay people, as demarcated by gender or colour, have a special contribution to make to science and technology only where it can first be shown that their special experience has a bearing on the scientific and technical matters in dispute.

CASE STUDIES

Cases of common illness also give rise to analyses which confuse expertise among the general public with the experience-based expertise of a specific group. For example, although Hilary Arksey makes much of the expertise of the public in general, the 'Repetitive Strain Injury' (RSI) sufferers whom she studied were actually a specialist group. As one sufferer said, and with some justification in our view:

> We're the experts: not the doctors, or the consultants, or the physios. We're the ones who have to live with it [RSI] day in day out. It's us they ought to be asking if they want to find out about RSI.[151]

It may not be the case that the RSI victims hold all the keys to their illness, but they surely hold some of them. Arksey seems to think that this discovery licenses a much more general positive evaluation of skills among the public. She uses this to critique what she sees as the elitism of the suggestion that the general public were not able to evaluate the crash tests seen on television. And she is not alone in making such inferences. In a similar vein, Simon Locke argues that books such as *The Golem* series underestimate the extent to which the public are able to understand the limitations of technology.[152] Jon Turney, echoing Locke's position, agrees that 'it is . . . possible to doubt that lay publics are quite as sociologically naive about scientists and scientific knowledge as Collins and

Pinch's approach suggests'.[153] Similarly, Brian Wynne says that the public are not 'imprisoned by the experts' control of the technical dimension'.[154]

Our response, as has already been indicated, is to examine the specific expertise involved, rather than to make general claims about the developed expertise of the public. We find that the two are very often confused. Thus Locke considers that the public have little to learn about science from treatments such as are found in *The Golem* series, because they are already chock-full of sociological knowledge about science. What a strange argument! There may indeed be pockets of the public that have such knowledge as a result of their experience, but why think that this is true of the public at large in this one area of (highly disputed) academic study? Once more the discovery of pockets of expertise seems to be being romantically extended to the public as a whole.

Perhaps these authors are mistaking the spreading cynicism about science among the public for sociological expertise. Distrust is easy; sophisticated evaluation is difficult. The hard problem is to make the evaluations sophisticated enough to be able to do more than just criticize; the public has also to struggle with the very, very difficult problem of making positive evaluations if the Problem of Extension is to be resolved for them as well as us.

Types of Science and Technology

Our discussion resonates with Silvio Funtowicz and Jerry Ravetz's term 'post-normal science'. In their work, post-normal science is the uncertain and controversial science, such as the sciences of the environment, in which the stakes for decision-makers are very high but the uncertainties in the knowledge are enormous. As a result, in these sciences it is impossible to separate facts from the value commitments, themselves often controversial, that underpinned their production. Such a science has only limited epistemological authority and, therefore, only a weak claim to compel action, so that managing risks and uncertainty, which is a political rather than a scientific matter, becomes increasingly important. This means that there is a need for new, more inclusive, decision-making processes:

> Only a dialogue between all sides, in which scientific expertise takes its place at the table with local and environmental concerns, can achieve creative solutions to such problems, which can then be implemented and enforced. Otherwise, either crude commercial pressures, inept bureaucratic regulations, or counterproductive protests will dominate, to the eventual detriment of all concerned.[155]

Although agreeing that there are cases where science alone clearly cannot provide the answers, we believe that the idea of a 'post-normal science' does not help with the Problem of Extension. The trouble is that the concept of 'post-normal' science conflates different themes from the public understanding of science literature by treating different types of expertise and knowledge as if they were interchangeable. These problems are discussed and investigated empirically by Steven Yearley,[156] who also compares the approach of Funtowicz and Ravetz with that of Wynne.

Yearley starts from a similar point to the one we have argued—that there are certain robust findings that have come out of the public understanding of science tradition, but that more general criteria for applying these findings to new contexts remain elusive. According to Funtowicz and Ravetz, we need to focus on the 'quality control' procedures that warrant knowledge claims. In the case of post-normal science, the proposed resolution is for a process of extended peer review, where non-scientist groups bring in 'extended' facts that may be relevant to the matter. As Yearley points out, there are some problems with this, even as a conceptual system. For example, when only one of 'decision stakes' or 'system uncertainty' is high, the need for extended peer review is less clear (for example, in cosmology and major industrial disasters), so that the theory does not always work in practice. Secondly, where any particular issue should be positioned—that is, where the boundaries between normal science, consultancy and post-normal science lie—is itself potentially subject to controversy, so that the identification of post-normal science itself is part of the problem.[157]

Assuming that post-normal science can be identified, the solution proposed by Funtowicz and Ravetz is that controversy be resolved through reducing either system or stake uncertainty through further research such that professional consultancy becomes appropriate. The problem with such an approach is that the research itself is potentially contestable, and it is not obvious how the membership of this extended peer community would be established and inclusion/exclusion criteria maintained. We would argue that the problem thus lies in the conflation of the technical and political phases of the decision-making processes. More research that might ultimately reduce uncertainty could be an appropriate response, and we would argue that its conduct will be enhanced if it draws on a wide range of expertise in its conduct, but it is unlikely to resolve the political problems in a realistic time-frame. Hence in these cases a separate process, based on different criteria, is needed to resolve the political need for action in the short term. In other words, the technical and political processes need to be conducted in parallel, with priority in the first instance going to the political phase.

In arguing this view, we are thus much closer to the position of Wynne, who typically talks about expertise rather than facts (or extended facts). Wynne's work is based on a distinction between different types of uncertainty: risk (where odds are known), uncertainty (where parameters but not odds are known), ignorance (where not even parameters are known), and indeterminacy (where the way in which systems will be used by others cannot be guaranteed). From Wynne's perspective, which is shared amongst much of the CTA literature, there is the sense that although scientific expertise is partial (in the sense that it rests on cultural assumptions and norms, and so on) its 'gaps' can be 'filled' by others with complementary expertise in the relevant areas. These areas might include: local knowledge about the system (natural or social) in which science is to be applied (sheep farmers, farmworkers, slaughterhouse employees, and the like) and knowledge about the past behaviours of the institutions involved, so as to enable them to make (better) informed judgements about whom to trust and whom not to trust (Sellafield inhabitants, people living around chemical factories, and so on). In this way criteria for inclusion emerge based on participation in particular social/cultural settings.

The regulatory problem for Wynne is to increase the attention given to what we don't know (ignorance and indeterminacy), and to find ways of bringing these inside regulatory programmes (that is, to build commitment to precautionary or anticipatory regulation), without reducing them (as typically happens) to the category of 'risk'. It is this translation, and imposition of a particular model of regulation, that is the problem in Wynne's view, which is sceptical both of public acceptance of experts in the past and of the role played by expert disagreement now. Once the translation of scientist's knowledge of the 'social' into pseudo-science is recognized, the power of lay knowledge as a critique of science is much more powerful than reflexive modernization as promoted by Beck and Giddens suggests.

Our view is not dissimilar from Wynne's, in terms of how we would understand the nature of knowledge and expertise. Where we do differ is in our willingness to be prescriptive about what should follow from this. In particular, Wynne's categorization of uncertainty and knowledge is typically very effective in structuring empirical data. The problem is how to turn these observations into an institutional response. Wynne repeatedly emphasizes the case-by-case and local nature of knowledge, suggesting no easy algorithm to its identification and incorporation in regulatory processes. Yearley, on other hand, is more optimistic, suggesting that focus groups can perform something like the peer review function suggested by Funtowicz and Ravetz, and provide some of the expertise needed to inform the 'broader debate' that Wynne says is needed. Our aim has been to go one step further and to articulate some of the criteria that could be used to institutionalize these responses more effectively.

We have divided what Funtowicz and Ravetz call post-normal science into three phases. Once these kinds of science are separated they do not seem to involve any deep abandonment of the expertise/value distinction, even though they argue for a stronger input from the bottom half of our diagrams. Thus, in the case of Golem science, the argument would be that the public has no particular role to play in developing the scientific consensus, although it may have a legitimate input to policy processes that decide what to do in the absence of scientific consensus. Crucially, however, it may still be the case that among the public there are pockets of expertise that do have a legitimate claim to enter into the core-set, and that these specialist groups should therefore contribute to the developing scientific consensus in a special way.

NOTES

The provenance of this paper is the theoretical work done at Cardiff University in putting together an application for an ESRC Research Centre, the 'Centre for the Study of Expertise and Environmental Policy' (SEEP). This initial work was done in the autumn of 1999, and the bid was submitted on 20 January 2000. Here is the opening paragraph of the submission: 'We face a crisis over the way we make decisions about the environment. We find ourselves caught on the horns of a dilemma: do we maximise the political legitimacy of our decisions by referring them to the widest democratic processes, and risk technical paralysis, or do we base our decisions on the best expert advice and invite popular opposition? This is the crisis that SEEP will address'. A little way below we find: 'Thus, on the academic side we want to create a new way of talking and thinking about *expertise* and *experience* to replace the old discourse about science and truth'. It can be seen that the framework of the argument presented in this paper was already in place at this point. We are grateful to various members of three Cardiff departments—the Schools of City and Regional Planning, Journalism and Media Studies, and Social Sciences—for providing an environment in which the theory could be beaten out. The paper also benefitted from critical comments by members of audiences at Gothenburg University (where a version was presented by Collins in September 1999) and at Cornell University (where a nearly finished draft was presented in November 2001). We are also grateful to Ingemar Bohlin, Martin Kusch, Arie Rip, Steve Yearley, Anne Murcott and members of the Cardiff KES group for comments on earlier versions. We also thank the referees of the first submitted draft for providing us with the opportunity to improve the paper markedly.

1. *The Chambers Dictionary* (Edinburgh, UK: Chambers Harrap Publishers, 1993), 951.

2. For an approach which grows out of political philosophy, see Stephen Turner, 'What is the Problem With Experts?', *Social Studies of Science*, Vol. 31, No. 1 (February 2001), 123–49. Turner argues that it initially seems hard to square the notion of liberal democracy with the idea of elite groups of experts whose knowledge takes them beyond the reach of normal political judgements. He concludes, however, that a rationale exists for expertise to function in modern democratic societies.

3. Many other contemporary social analysts of science and technology have normative commitments but, so far as we know, none has developed a normative theory of expertise.

4. Among the variations to be found are those expressed by one of our referees, who insisted that the discovery of the negotiability of the boundaries of expertise was in no way connected to the idea that anyone should have a say in expert decisions. It seems to us that if there is no defining criterion for expertise, it follows that there is no way of defining people out of the category, and this invites unlimited extension. It also seems to us that many

have read precisely this conclusion into 'Wave Two'. That the referee did not agree merely shows how difficult it is to describe a broad sweep in a way that will take everyone's interpretation into account.

5. The term is used liberally in, for example, Hilary Arksey, *RSI and the Experts: The Construction of Medical Knowledge* (London: UCL Press, 1998).

6. The wider use of the notion of expertise does, of course, do immense work in the debate about artificial intelligence, but precisely because it shows that so much expertise is *restricted* to humans, not machines—that is to say, it *extends* only to the boundary of social beings, and no further: see H. M. Collins, *Artificial Experts: Social Knowledge and Intelligent Machines* (Cambridge, MA: MIT Press, 1990); H. M. Collins and Martin Kusch, *The Shape of Actions: What Humans and Machines Can Do* (Cambridge, MA: MIT Press, 1998). If the term 'lay expertise' has an application, a better one might be to the kind of expertise that is distributed throughout the human race.

7. There were questions concerning the 'social responsibility' of science, but these very problems arose out of science's power; to raise questions of social responsibility is to ignore questions of the foundation of knowledge. That things were not as uniformly simple as our broad brush suggests can be seen in publications such as Anthony Standen, *Science is a Sacred Cow* (London: Sheed & Ward, 1952), and Ian Welsh, *Mobilizing Modernity: The Nuclear Moment* (London: Routledge, 2001).

8. Let us bear in mind that being philosophically 'high and dry' does not mean that positivism does not remain immensely strong in terms of political and economic power, as well as being the predominant driving idea in the tremendously successful natural sciences.

9. We resisted the pun. Both authors continue, unabashed, with their Wave Two–type studies, and so do their colleagues and students.

10. The quotation marks here indicate where we are quoting the words of the referees of an earlier draft of the paper.

11. We understand, of course, that any such contribution is not going to settle the problem 'once and for all' (to quote a critical referee).

12. See the works cited in note 6.

13. The distinction made here is not to be confused with the similar methodological difference between investigating the flow of the river of history while standing in the stream—by studying a contemporaneous science—and by studying it as a historian, after the river has reached its outflow. This is a distinction of methodology, not aims.

14. As has been argued in H. M. Collins and Steven Yearley, 'Epistemological Chicken', in Andrew Pickering (ed.), *Science as Practice and Culture* (Chicago, IL: The University of Chicago Press, 1992), 301–26.

15. As well as the work cited in note 6, see Trevor Pinch, H. M. Collins and Larry Carbone, 'Inside Knowledge: Second Order Measures of Skill', *Sociological Review*, Vol. 44, No. 2 (May 1996), 163–86; H. M. Collins, 'Tacit Knowledge, Trust and the Q of Sapphire', *Social Studies of Science*, Vol. 31, No. 1 (February 2001), 71–85.

16. Collins & Yearley, op. cit. note 14. Let us hasten to add that many Wave Two authors have made valuable 'upstream' contributions, and many of these are discussed in the Appendix and in the main body of the paper. We are simply trying to describe, systematize and set on a firmer foundation, the contribution of the sociology of scientific knowledge to what happens upstream. To give one example, Evelleen Richards has argued that it was part of the

duty of science studies to give positive advice on the conduct of science, and Trevor Pinch, in reviewing her work, referred to it as 'third generation' SSK (though his 'first' and 'second' generations did not coincide with our First and Second Waves): see, for example, Brian Martin, Evelleen Richards and Pam Scott, 'Who's a Captive? Who's a Victim? Response to Collins's Method Talk', *Science, Technology, & Human Values*, Vol. 16, No. 2 (Spring 1991), 252–55; Trevor Pinch, 'Generations of SSK' (Review of Richards, *Vitamin C and Cancer, &* Sapp, *Where the Truth Lies*), *Social Studies of Science*, Vol. 23, No. 2 (May 1993), 363–73. Richards and Collins disagreed about whether her work was SSK, and Collins would still say it was not—it was knowledge science. This is the kind of distinction that we are trying to resolve here.

17. H. M. Collins, *Changing Order: Replication and Induction in Scientific Practice* (Beverly Hills, CA & London: Sage, 1st edn, 1985; Chicago, IL: The University of Chicago Press, rev. 2nd edn, 1992), passim; H. M. Collins, 'The Meaning of Data: Open and Closed Evidential Cultures in the Search for Gravitational Waves', *American Journal of Sociology*, Vol. 104, No. 2 (September 1998), 293–337.

18. The size of the core-set can be influenced by the 'size' of the claim made—for example, the extent to which it seeks to overturn a small or large part of the conventional theories: see Trevor Pinch, *Confronting Nature: The Sociology of Solar-Neutrino Detection* (Dordrecht: Reidel, 1986)—and the availability of resources which limit the ability of scientists to participate in the debate at all: see Bruno Latour, *Science in Action: How to Follow Scientists and Engineers through Society* (Milton Keynes, Bucks., UK: Open University Press; Cambridge, MA: Harvard University Press, 1987).

19. Thomas F. Gieryn, 'Boundary-Work and the Demarcation of Science from Non-Science: Strains and Interests in Professional Ideologies of Scientists', *American Sociological Review*, Vol. 48 (1983), 781–95; T. F. Gieryn, *Cultural Boundaries of Science: Credibility on the Line* (Chicago, IL & London: The University of Chicago Press, 1999).

20. These remarks were made by Steven Yearley, who kindly allowed us to identify him as one referee of an earlier draft of this paper.

21. And this has been known for a long time by those convinced by the arguments supporting moral relativism; moral relativism does not lead to moral anarchy, but to the sad acceptance that, beyond a certain point, moral judgements cannot be justified, but are nevertheless right—one just has to take responsibility for them.

22. Or it might be that the appropriate circle of judgement for a work of art is still wider than the trained critics, and hence the claim that 'I may not know much about art, but I know what I like', is not entirely frivolous. Indeed, some art is intended to make a fool of circles of specialist critics, or to cause us to reflect on the nature of the establishment. But, setting all that aside, should we feel happy with: 'I may not know much about science, but I know what I like'?

23. This is not to say that once upon a time the public, or at least those who witnessed experiments, were not more important to the process of science. And it is not to say that such rights are not being increasingly demanded. It is this latter process in which we are interested.

24. In an unpublished paper to the conference on 'Democratisation Socialised', held at Cardiff University (25–28 August 2000), Harry Collins argued that a demarcation criterion between science and art could be found in the relationship between the intentions of the

author of a paper/work and the interpretation of the consumer—and that in scientific paper-writing, the author's intention must always be to limit interpretative licence, whereas in some forms of art or poetry, it might well be to provoke an unanticipated response or interpretation. Though our main three-fold classification—no special expertise, interactional expertise, and contributory expertise—was initially chosen because it is already present in the discourse and practice of social scientists, it has begun to feel less arbitrary as the argument has developed. The distinction seems to 'pop up all over the place' once one starts to think about these matters. In this case, pressed upon us by our referee, it seems the obvious way to think about the relationship between artists and critics.

25. Steven Shapin, 'The Politics of Observation: Cerebral Anatomy and Social Interests in the Edinburgh Phrenology Disputes', in Roy Wallis (ed.), *On the Margins of Science: The Social Construction of Rejected Knowledge, Sociological Review Monograph* No. 27 (Keele, Staffs., UK: Keele University Press; London: Routledge & Kegan Paul, 1979), 139–78.

26. By 'Lysenkoism and the like' we mean cases where state power is used to over-rule scientific conclusions that are subject to broad consensus within the international scientific community. We note that all but the most and least radical of scientific commentators decry, for example, the involvement of the tobacco companies in supporting scientific research aimed at certain conclusions.

27. The degree of 'visibility' of the politics is not, by itself, a good criterion of 'intrinsicness' or 'extrinsicness' of the politics, since degree of visibility is contingent on historical events and contexts (we thank Charles Thorpe for this point). The *criterion* of intrinsicness has to be the extent to which scientists, or other commentators, would willingly endorse the input of politics into the science. To play the Western science 'language game' (and this whole paper stands and falls on an agreement to play it) means being unwilling to endorse, publicly, an input of political influence into science. However irreducible the political input, the politics must remain intrinsic if it is Western science that is being done. As we explain in the Appendix, this means that there are two ways to look at modern 'standpoint theories'. One way is to see the input of new classes of expert, such as women, as experts on women, as a way of reducing already existing political biases so as to increase the integrity of the science. The other way is to see them as insisting that science is a product of its political milieu, that there are different sciences based on different political viewpoints, and that the influence of the 'standpoint' should be explicit and extrinsic. As we indicate, to argue in the second way is to abandon the language of Western science, something which we stand against.

28. But the compartmentalization is analytically vital. The difference between SSK's descriptions and its prescriptions seem to be at the root of certain earlier heated debates. The prescriptions, as in the case of the justice system, are a matter of knowing how to act appropriately within a set of institutions or 'language games'. Misunderstanding the difference between the analyst's 'is' and the analyst's 'ought' has led to some ghastly confusions: see, for example, Pam Scott, Evelleen Richards and Brian Martin, 'Captives of Controversy: The Myth of the Neutral Social Researcher in Contemporary Scientific Controversies', *Science, Technology, & Human Values,* Vol. 15, No. 4 (Autumn 1990), 33–57; Martin, Richards & Scott, op. cit. note 16; H. M. Collins, 'In Praise of Futile Gestures: How Scientific is the Sociology of Scientific Knowledge?', *Social Studies of Science,* Vol. 26, No. 2 (May 1996), 229–44. One might say that Scott, Richards and Martin, having noticed that SSK has shown that politics

is intrinsic to science, believe it should be made extrinsic also. We disagree. When one moves upstream into the area of prescription, one must be aware that one no longer has the analytic privileges and advantages accorded to those who remain downstream. Likewise, staying downstream is incompatible with overt prescription, because symmetry is central to downstream analysis.

29. This phrase is due to Collins, Changing Order, op. cit. note 17, 145. The idea has been modified and extended by Donald MacKenzie, who points out that uncertainty and opposition can increase as science enters the policy-making sphere: D. MacKenzie, 'The Certainty Trough', in Robin Williams, Wendy Faulkner and James Fleck (eds), *Exploring Expertise: Issues and Perspectives* (Basingstoke, UK: Macmillan, 1998), 325–29.

30. We talk here of the cognitive debate. As Latour (op. cit. note 18) has argued, there are many factors that make scientific disputes more or less settled in practice.

31. Thomas S. Kuhn, *The Structure of Scientific Revolutions* (Chicago, IL: The University of Chicago Press, 1962; rev. 2nd edn, 1970); Harry Collins and Trevor Pinch, *The Golem: What You Should Know About Science* (Cambridge: Cambridge University Press, 1st edn, 1993; Cambridge & New York: Canto, 2nd edn, with new afterword, 1998). In the afterword to the second edition of *The Golem*, evidence is used to show the relationship between textbook accounts and other accounts of the foundations of relativity.

32. For example, the UK government's response to the possibility of a link between BSE in cattle and CJD in humans was orchestrated around these ideas, and government statements invariably took the line that there was no risk, or that beef was completely safe: see Barbara Adam, *Timescapes of Modernity: The Environment and Invisible Hazards* (London & New York: Routledge, 1998); B. Adam, 'The Media Timescapes of BSE News', in Stuart Allan, Barbara Adam and Cynthia Carter (eds), *Environmental Risks and the Media* (London & New York: Routledge, 2000), 117–29. The same concern for certainty was also reported in Brian Wynne's study of the sheep-farmers: B. Wynne, 'May the Sheep Safely Graze? A Reflexive View of the Expert-Lay Knowledge Divide', in Scott Lash, Bronislaw Szerszynski and Brian Wynne (eds), *Risk, Environment & Modernity: Towards a New Ecology* (London: Sage, 1996), 44–83; and, more recently, it can be seen in the response to concerns about the safety of the measles-mumps-rubella (MMR) vaccine given to young children and in the possible dangers posed to service men and women, and presumably civilians in war zones, by the use of depleted uranium (DU) ammunition.

33. What is meant is that, like Harold Garfinkel's famous breaching experiments, scientific controversies highlighted the rules of scientific behaviour and their ambivalences: see Harold Garfinkel, 'A Conception of, and Experiments With, "Trust" as a Condition of Stable Concerted Actions', in O. J. Harvey (ed.), *Motivation and Social Interaction* (New York: Ronald Press, 1963), 187–238.

34. See also Jean-Marc Lévy-Leblond, 'About Misunderstandings About Misunderstandings', *Public Understanding of Science*, Vol. 1, No. 1 (January 1992), 17–21.

35. See, for example, Gieryn, opera cit. note 19.

36. In other words, as indicated above, expertise is being treated in the way it would be treated under 'knowledge science'.

37. We are led to ask this question after, on the advice of a referee, re-reading Turner, op. cit. note 2.

38. Our claim in respect of astrology is not that it has never been used to contribute to decision-making at a variety of levels, but that very few of its proponents confuse it with science, any more than they would confuse the sayings of an oracle with science.

39. Turner, op. cit. note 2.

40. Part of our job could be described as helping to realize such continuities in expertise as continuities in social and cognitive networks.

41. Wave Two studies show that many of the arguments used by scientists to exclude some whole field or other from scientific consideration are based on risible or disingenuous oversimplifications of the way their own sciences work, but this is not to make the other fields valid: H. M. Collins and Trevor J. Pinch, 'The Construction of the Paranormal: Nothing Unscientific is Happening', in Wallis (ed.), op. cit. note 25, 237–70. The stress on the orthogonal nature of decisions about fields and decisions about expertise within fields, and the subsequent setting out of Figure 8, emerged from the discussion at Cornell University in November 2001, mentioned above.

42. For a discussion of expertise and experience, see Peter Dear, *Discipline and Experience: The Mathematical Way in the Scientific Revolution* (Chicago, IL & London: The University of Chicago Press, 1995).

43. To anticipate a potential question, what we will call 'interactional competence' in an expertise can be a Type I interactional competence, even though it is not itself a full-blown Type I expertise.

44. There is a terminological difficulty here. Turner classifies expertise, rather than competence within an expertise. We want to talk about competence within an expertise. Unfortunately, the possession of certain expertise is also seen as a sign of competence, as when we say that certain humans are 'more competent at sports' than others if they possess more sports expertise. We don't think the terminological untidiness causes any great problems, however, as the meaning should always be clear from the context.

45. For example, into tacit and explicit knowledge, with two different more detailed classifications within these broad categories: see, for example, the work of Collins cited in notes 6 and 15. With another author, he has also divided human abilities into 'polimorphic' and 'mimeomorphic': Collins & Kusch, op. cit. note 6, passim. Probably the most currently well-known classification of expertise is that due to Hubert and Stuart Dreyfus (H. L. Dreyfus and S. E. Dreyfus, *Mind Over Machine: The Power of Human Intuition and Expertise in the Era of the Computer* [New York: Free Press, 1986]), but the Dreyfus model is not appropriate for answering the kinds of question we pose here.

46. Researchers in the sociology of scientific knowledge have long understood how difficult it is to employ research assistants, precisely because the skills needed to do the research are not the generic skills of the broadly trained social scientist, but must include interactional skills in the substantive topic of the field study. Here our starting point in the esoteric sciences is felicitous. Sociologists who do not study the esoteric sciences may not be so familiar with these distinctions, and may find them less immediately persuasive, but these distinctions are useful ones nevertheless. As was pointed out at the discussions at Cornell University in November 2001, this classification is very broad, and it may be that more refined classifications are needed. Nevertheless, this classification is all that is necessary to 'hammer in a piton'.

47. Collins experienced complete failure in his attempts to acquire interactional competence in the field of amorphous semiconductors.

48. Collins acquired enough competence to make significant published contributions to the field of the investigation of paranormal metal bending: B. R. Pamplin and H. M. Collins, 'Spoon Bending: An Experimental Approach', *Nature*, Vol. 257 (4 September 1975), 8. Of course, an identical defence could be made of the nature of science, and is made in the *tu quoque* argument. That is to say, in our work, we act as though there is such a thing as science. But this presents no problem, so long as our relativism is of the methodological kind. Likewise, there is nothing in this argument to prevent analyses based on methodological relativism in respect of expertise. We are just demonstrating another way to go about things.

49. Brian Wynne, 'Sheep Farming after Chernobyl: A Case Study in Communicating Scientific Information', *Environment*, Vol. 31, No. 2 (1989), 10–15, 33–39; Wynne, op. cit. note 32; B. Wynne, 'Misunderstood Misunderstandings: Social Identities and Public Uptake of Science', in Alan Irwin and Brian Wynne (eds), *Misunderstanding Science? The Public Reconstruction of Science and Technology* (Cambridge, New York & Melbourne: Cambridge University Press, 1996), 19–46.

50. Here we do not discuss the power relationships and protection of vested interests. Through our discussions, we merely want to use academic argument to lessen the impact of these interests in future incidents of this sort, by lessening their legitimacy.

51. None of this is to claim that making established Scientists listen will be easy. For the AIDS case, see Steven Epstein, 'The Construction of Lay Expertise: AIDS Activism and the Forging of Credibility in the Reform of Clinical Trials', *Science, Technology, & Human Values*, Vol. 20, No. 4 (Autumn 1995), 408–37. There is a wider question about the extent to which the treatment activists represented the whole gay community, let alone the still more heterogeneous group of people suffering from AIDS.

52. We are ignoring, for the purposes of our argument, the very obvious fact that the managers are also likely to be much better scientists than any visiting sociologist.

53. As well as the technical abilities remarked on in the quotation and the previous note.

54. Though in the case in question, some scientists thought that the referral was from too distant a site. They thought that high-energy physics, from where the managers came, gave them a misleading picture of the skills required to do interferometry. General Groves, who managed the Manhattan Project, was an interesting case who would seem to contradict this argument: see Charles Thorpe and Steven Shapin, 'Who Was J. Robert Oppenheimer? Charisma and Complex Organization', *Social Studies of Science*, Vol. 30, No. 4 (August 2000), 545–90.

55. Peter L. Berger, *Invitation to Sociology* (Garden City, NY: Anchor Books, 1963), passim; Collins & Yearley, op. cit. note 14.

56. The problem of translating between self-contained cultures, 'paradigms' or 'forms-of-life', is an old one: see, for example, H. M. Collins and Trevor J. Pinch, *Frames of Meaning: The Social Construction of Extraordinary Science* (Henley-on-Thames, UK: Routledge & Kegan Paul, 1982). In the history of science, it has been alluded to under the heading of 'trading zones': see Peter Galison, *Image and Logic: A Material Culture of Microphysics* (Chicago, IL: The University of Chicago Press, 1997), passim. Skills of journalists are compared with those of sociologists in Phillip M. Strong, 'The Rivals: An Essay on the Sociological Trades', in Robert Dingwall and Philip Lewis (eds), *The Sociology of the Professions: Lawyers, Doctors and Others* (London: Macmillan, 1983), 59–77.

57. These judgements are not dissimilar to those made by scientists within the scientific community. Thus Lewis Wolpert has said that 'scientists must make an assessment of the

reliability of experiments. One of the reasons for going to meetings is to meet the scientists in one's field so that one can form an opinion of them and judge their work': L. Wolpert, 'Review of *The Golem*', *Public Understanding of Science*, Vol. 3, No. 3 (July 1994), 328–29, at 329. We will go on to discuss the relationship between our concept and similar issues discussed by Brian Wynne in 1992 and 1993: B. Wynne, 'Public Understanding of Science Research: New Horizon or Hall of Mirrors?', *ibid.*, Vol. 1, No. 1 (January 1992), 37–43; B. Wynne, 'Public Uptake of Science: A Case for Institutional Reflexivity', *ibid.*, Vol. 2, No. 4 (October 1993), 321–37. We, however, distinguish between specialist and ubiquitous expertises.

58. Poor social judgements are the problem with those who believe in, say, newspaper astrology *as a scientific theory*. They are making a social mistake: they do not know the locations in our society in which trustworthy expertise in respect of the influence of the stars and planets on our lives is to be found.

59. For a similar argument in respect of the rejection of claims about the existence of gravitational waves, see H. M. Collins, 'Tantalus and the Aliens: Publications, Audiences and the Search for Gravitational Waves', *Social Studies of Science*, Vol. 29, No. 2 (April 1999), 163–97.

60. Increasing the potential for debates about who is in and who is out—a typical boundary problem.

61. To make the point from the opposite side, so-called 'junk scientists', such as many of those who are called as expert witnesses in court rooms, often have paper credentials, but are not counted as experts by their peers.

62. The term 'phase' is used here in the materials-science sense—as in a 'phase diagram' for a material—rather than in the time-sequence sense.

63. Epstein, op. cit. note 51; Steven Epstein, *Impure Science: AIDS, Activism, and the Politics of Knowledge* (Berkeley, Los Angeles & London: University of California Press, 1996).

64. In the case of the sheep farmers, there was probably never a nexus.

65. If Figure 6 is taken to represent the Cumbrian case, there would be no solid-line nexus at all between the core-set and the 'pocket'. The dotted-line nexus would stay where it is, however—the sheep farmers should have been in the core-set from early in the game.

66. For a full account, see H. M. Collins, 'Public Experiments and Displays of Virtuosity: The Core-Set Revisited', *Social Studies of Science*, Vol. 18, No. 4 (November 1988), 725–48.

67. This sentence is not as naive as it appears. Compare what has been said about the Edinburgh phrenology case. We are not trying to suggest any hard and fast distinction between science and politics, nor are we suggesting that these tests and their interpretations could have been carried out completely 'objectively'. What we are suggesting is that the way in which the political sphere encroached on the technical sphere in these cases was clearly illegitimate under almost any analysis of science. There is no difficulty in making prescriptive statements about it.

68. Collins was able to demonstrate the incompetence of audiences of university personnel by showing them a film of the crash, and asking them to criticize it without prompting. They always failed to notice the visible features that had been pointed out by Greenpeace's experts.

69. This is not to say that there are not groups of experience-based experts in different aspects of the safety of the transport of nuclear fuel in the population as a whole. For example, there are pockets of experience-based expertise concerning the degree of radioactiv-

ity on sections of rail (and sidings) used for railway transport. But these people are experts and, by that fact alone, not ordinary.

70. In the case of the train crash, the experts who pointed out the deficiencies of the test came from Greenpeace; in the case of the aircraft crash, the experts (who were represented on a subsequent TV programme) were from ICI—the manufacturers of AMK.

71. As Michael Bloor argues: see M. Bloor, 'The South Wales Miners Federation, Miners' Lung and the Instrumental Use of Expertise, 1900–50', *Social Studies of Science*, Vol. 30, No. 1 (February 2000), 125–40, at 126.

72. There is also the danger that this form of account takes us back to the sociology of error, in which deviant science is explained in a different way to 'proper' science.

73. The correct analysis varies from case to case, but we suspect that the motivation is most often of the first kind, as the stage magicians do not (and are not expected to) adopt the norms of the scientific community, such as honesty and openness. In either case, the welcoming of magicians into the heartland of science makes the point about the permeability of professional boundaries.

74. See, for example, Eric Von Hippel, *The Sources of Innovation* (New York: Oxford University Press, 1988). The role of these lead users is not entirely unproblematic, however, and, as Phil Agre has argued, can lead to the neglect of novice users in the design of technology: consult Agre's website: http://dlis.gseis.ucla.edu/pagre/. The result of this is that inefficient designs, particularly of IT interfaces, become embedded social practices, as manufacturers and lead users overlook the increasingly complex training and restructuring that is needed to make the machines work: see http://commons.somewhere.com/rre/2000/RRE.notes.and.recommenda19. html.

75. Wiebe E. Bijker, Thomas P. Hughes and Trevor Pinch (eds), *The Social Construction of Technological Systems: New Directions in the Sociology and History of Technology* (Cambridge, MA: MIT Press, 1987); W. E. Bijker, *Of Bicycles, Bakelites, and Bulbs: Toward a Theory of Sociotechnical Change* (Cambridge, MA: MIT Press, 1995).

76. Thinking about planning debates brings out two other kinds of ability that belong below the line in the diagram. There is the ability of the middle-class protestors and professional lobbyists, who know how to present an argument and how to penetrate the appropriate networks; and there is the skill of the activists who know how to cause the authorities the maximum inconvenience and expense by climbing trees, burrowing into tunnels in the ground, and so forth. We could thus add these types of ability to the discriminatory and translation skills we identified earlier.

77. See, for example, Michael Lynch and David Bogen, *The Spectacle of History: Speech, Text, and Memory at the Iran-Contra Hearings* (Durham, NC: Duke University Press, 1996); Sheila Jasanoff, *The Fifth Branch: Science Advisers as Policymakers* (London & Cambridge, MA: Harvard University Press, 1990); S. Jasanoff, *Science at the Bar: Law, Science, and Technology in America* (Cambridge, MA & London: Twentieth Century Fund & Harvard University Press, 1995); Brian Wynne, *Rationality and Ritual: The Windscale Inquiry and Nuclear Decisions in Britain*, BSHS Monograph No. 3 (Chalfont St Giles, Bucks., UK: British Society for the History of Science, 1982); Roger Smith and Brian Wynne (eds), *Expert Evidence: Interpreting Science in the Law* (London: Routledge, 1989).

78. As we will explain in the Appendix, Silvio Funtowicz and Jerome Ravetz misleadingly refer to this kind of situation as 'post-normal', whereas it is simply 'pre-normal': see S. O.

Funtowicz and J. R. Ravetz, 'Science in the Post-Normal Age', *Futures*, Vol. 25, No. 7 (September 1993), 739–55.

79. Karl Popper, in his *The Poverty of Historicism* (London: Routledge & Kegan Paul, 1957), used the term 'historicist' to refer to teleological theories which assume a progressive historical trend. We do not discuss progressiveness, only sciences that deal with long-term unique changes.

80. This kind of science has been examined by Barry Barnes in the context of economic decision-making: B. Barnes, *The Nature of Power* (Cambridge/Oxford, UK: Polity Press/ Basil Blackwell, 1988), passim.

81. There is a certain symmetry here: just as the scientific community is the appropriate location for disposing of political influence as it impinges on the construction of knowledge, so the polity is the appropriate locus for decisions about the societal response to uncertain knowledge.

82. For example, do household conservation policies increase or decrease the output of greenhouse gases when one takes into account the environmental cost of collection and processing of recyclable waste? For a discussion of the role of SSK in urban energy policies, see Robert J. Evans, Simon Marvin and Simon Guy, 'Making a Difference: Sociology of Scientific Knowledge and Urban Energy Policies', *Science, Technology, & Human Values*, Vol. 24, No. 1 (Winter 1999), 105–31; for economic policy as social technology, see Robert Evans, *Macroeconomic Forecasting: A Sociological Appraisal* (London: Routledge, 1999).

83. Apologies to Malcolm Ashmore for this un-ironic use of the word 'wrong': M. Ashmore, 'Ending Up On the Wrong Side: Must the Two Forms of Radicalism Always Be at War?', *Social Studies of Science*, Vol. 26, No. 2 (May 1996), 305–22.

84. See Edward W. Lawless, *Technology and Social Shock* (New Brunswick, NJ: Rutgers University Press, 1977), at 418–25.

85. The Greenpeace version of this story is available on their website at: http://www .greenpeace.org/~comms/toxics/dumping/jun20.html.

86. See James C. Petersen and Gerald E. Markle, 'Politics and Science in the Laetrile Controversy', *Social Studies of Science*, Vol. 9, No. 2 (May 1979), 139–66.

87. Bent Flyvbjerg discusses the Aristotelian concept of 'phronesis', which is a form of practical wisdom in a moral setting; prudence and wisdom capture some of its flavour: see B. Flyvbjerg, *Making Social Science Matter: Why Social Inquiry Fails and How It Can Succeed Again* (Cambridge: Cambridge University Press, 2001), passim. Unfortunately, the concept is somewhat slippery and includes components both of the political and of experience. To use the concept with confidence in this discussion, one would first need to redescribe natural science, using the term in the light of what we have learned about science over the last decades. Our paradigm case—the post-Chernobyl Cumbrian sheep farmers as discussed by Wynne—would not seem to benefit from the introduction of the term 'phronesis'. The point is that the sheep farmers had technical knowledge of sheep ecology, not prudent understanding of how to act in a situation requiring ethical judgement, which is an essential element in Flyvbjerg's usage.

88. On artists in the media, see Frank Muir, *A Kentish Lad* (Reading, Berks., UK: Corgi Books, 1997). Muir explains the mass defection of programme-makers from London Weekend Television when the Board of Directors sacked their talented boss. He says (324–25): 'There was no contact at all between the board and the creative side of the company . . .

Lord Campbell told us that in his experience all management was the same. 'You unit heads may think that managing talented producers and performers raises special problems but I have been in sugar all my life and I can assure you that the management of people in television is precisely the same as the management of sugar workers.' On scientists, see Turner, op. cit. note 2, and David H. Guston, 'Evaluating the First US Consensus Conference: The Impact of the Citizens' Panel on Telecommunications and the Future of Democracy', *Science, Technology, & Human Values*, Vol. 24, No. 4 (Autumn 1999), 451–82; and for scientists and government, D. H. Guston, *Between Politics and Science: Assuring the Integrity and Productivity of Research* (Cambridge: Cambridge University Press, 2000).

89. Lawless, op. cit. note 84.

90. The Phillips Report was critical of the way in which scientific advice is solicited, interpreted and used. In particular, caveats inserted in the original advice were not given sufficient weight, contradictory evidence was discounted, and the initial recommendations were not reviewed often enough. The full report is available on the internet at http://www .bse.gov.uk. See also Anne Murcott, 'Not Science but PR: GM Food and the Makings of a Considered Sociology', *Sociological Research Online*, Vol. 4, No. 3 (September 1999); A. Murcott, 'Public Beliefs about GM Foods: More on the Makings of a Considered Sociology', *Medical Anthropology Quarterly*, Vol. 15, No. 1 (March 2001), 1–11.

91. House of Lords Science and Technology Committee, Science and Society (London: HMSO, 2000); European Union White Paper on Governance: Broadening and Enriching Public Debate on European Matters, *Report of the Working Group on Democratising Expertise and Establishing Scientific Reference Systems*, available on the internet at http://www .cordis.lu/rtd2002/science-society/governance.htm; Loka Institute (http://www.loka.org), 'telecommunications and democracy' (April 1997); 'genetically engineered foods' (February 2002). The Loka Institute website provides links to reports on over 40 consensus conferences held in over a dozen countries.

92. House of Lords, op. cit. note 91, paragraph 5.48. Guidance on how government departments should put these principles into practice are given in the Office of Science and Technology (OST) publication, *Guidelines 2000: Scientific Advice and Policy Making*, available on the internet at http://www.dti.gov.uk/ost/aboutost/guidelines.htm, and in the *Code of Conduct for Written Consultations* produced by the Cabinet Office: http://www.cabinet-office.gov.uk/servicefirst/2000/consult/code/ConsultationCode.htm.

93. Wellcome Trust and the Office of Science and Technology, *Science and the Public: A Review of Science Communication and Public Attitudes to Science in Britain* (London: Wellcome Trust & OST, 2000), 8.

94. *NSF Science and Engineering Indicators, 2000*, quote at page 8–13 of on-line PDF version, available at http://www.nsf.go/sbe/srs/seindoo/start.htm.

95. *Eurobarometer 52.1: The Europeans and Biotechnology* (Brussels: EU, 2000), available via the internet from http://europa.e.int/comm/dg1o/ep/eb.html.

96. Perhaps surprisingly, the support for scientists was higher amongst younger people, defined as those aged between 15 and 24; it was 79%, higher than that for the sample as a whole.

97. Recent examples in Britain include: Dr Harold Shipman, a former GP in Manchester who is currently in prison after being found guilty of murdering over a dozen of his patients, and being implicated in the deaths of many more; the scandals at the Bristol Children's

Hospital, where doctors continued to operate, despite much higher death rates and the concerns of their colleagues; and the retention of children's organs by pathology labs without their parents' knowledge.

98. There are clearly some inconsistencies here. For example, professors receive a different rating to scientists, though it is not clear what the distinction is, as it is possible to be both. Similarly, news readers score significantly more highly than journalists, despite the fact that what they read is the product of journalistic endeavour (and many of them are trained and experienced journalists).

99. Michelle Corrado, 'Trust in Scientists', paper presented at the Annual Meeting of the British Association for the Advancement of Science (Glasgow, 3–7 September, 2001); also available at the Mori website: http://www.mori.com.

100. *The Advisory and Regulatory Framework for Biotechnology: Report from the Government's Review* (London: HMSO, 1999), quote at paragraph 36; available via the internet from: http://www.dii.go.uk/ost/rmay99/Bioreport_1.htm.

101. The survey data supporting this observation is summarized in Mori, *The Role of Scientists in Public Debate: Full Report* (London: Wellcome Trust & Mori, 2000), available via the internet from: http://www.wellcom.ac.uk/en/1/mismscnesos.html.

102. Interview by Brian Wynne given to River Path Associates, who were conducting research on science communication for the British Association: The River Path Report, 'Now for the Science Bit—Concentrate!', was published in 1997, and is available via the internet from: http://www.riverpath.com/library/.

103. Anne Kerr, Sarah Cunningham-Burley and Amanda Amos, 'The New Genetics and Health: Mobilizing Lay Expertise', *Public Understanding of Science*, Vol. 7, No. 1 (January 1998), 41–60, at 48.

104. Robin Grove-White, Phillip Macnaghten and Brian Wynne, *Wising Up: The Public and New Technologies* (Lancaster, UK: Centre for the Study of Environmental Change, Lancaster University, 2000).

105. Steven Yearley, 'Computer Models and the Public's Understanding of Science', *Social Studies of Science*, Vol. 29, No. 6 (December 1999), 845–66; COPUS, *To Know Science is to Love It? Observations from Public Understanding of Science Research* (London: COPUS & the Royal Society, no date); Irwin & Wynne (eds), op. cit. note 49; Ian Hargreaves and Galit Ferguson (eds), *Who's Misunderstanding Whom? Bridging the Gap Between the Public, the Media and Science* (Swindon, UK: Economic and Social Research Council [ESRC], 2000).

106. Alan Irwin, Alison Dale and Denis Smith, 'Science and Hell's Kitchen: The Local Understanding of Hazard Issues', in Irwin & Wynne (eds), op. cit. note 49, 47–64.

107. Geoffrey Evans and John Durant, 'The Relationship Between Knowledge and Attitudes in the Public Understanding of Science in Britain', *Public Understanding of Science*, Vol. 4, No. 1 (January 1995), 57–74; Welsh, op. cit. note 7.

108. A full report based on this survey is available from: http://www.vcu.edu/lifesciencessurvey.

109. Wynne, op. cit. note 32; Steven Yearley, 'Making Systematic Sense of Public Discontents with Expert Knowledge: Two Analytical Approaches and a Case Study', *Public Understanding of Science*, Vol. 9, No. 2 (April 2000), 105–22.

110. For a brief statement of the 'deficit model', and a discussion of its weaknesses, see Alan Irwin, 'Science and its Publics: Continuity and Change in the Risk Society', *Social Stud-*

ies of Science, Vol. 24, No. 1 (February 1994), 168–184, at 170–72. See also Simon Locke, 'Golem Science and the Public Understanding of Science: From Deficit to Dilemma', *Public Understanding of Science*, Vol. 8, No. 2 (April 1999), 75–92, esp. the references listed at 90, note 14.

111. As we explain in the main text, Turner (op. cit. note 2) takes the Problem of Legitimacy to be one based in political philosophy.

112. Ulrich Beck, *The Risk Society: Towards a New Modernity* (London: Sage, 1992), 166 & passim.

113. Anthony Giddens, *The Consequences of Modernity* (Cambridge, UK: Polity Press, 1990), esp. 124–34.

114. Welsh's research on the nuclear industry clearly demonstrates that there was organized opposition to nuclear power from the moment it was first proposed: Welsh, op. cit. note 7.

115. Michael Mulkay, *Science and the Sociology of Knowledge* (London: George Allen & Unwin, 1979), 11. Cited works are: Karl Mannheim, *Ideology and Utopia* (New York: Harcourt, Brace & World, 1936); K. Mannheim, *Essays on the Sociology of Knowledge* (London: Routledge & Kegan Paul, 1952).

116. See, for example, Robert K. Merton, *The Sociology of Science* (Chicago, IL & London: The University of Chicago Press, 1973). See also Michael Mulkay, 'Norms and Ideology in Science', *Social Science Information*, Vol. 15 (1976), 637–56. Here Mulkay questions the adequacy of the norms as we move into the period of Wave Two.

117. Lewis Wolpert, *The Unnatural Nature of Science: Why Science Does Not Make (Common) Sense* (London: Faber & Faber, 1992); Paul R. Gross and Norman Levitt, *Higher Superstition: The Academic Left and its Quarrels with Science* (Baltimore, MD & London: Johns Hopkins University Press, 1994); P. R. Gross, N. Levitt and Martin W. Lewis (eds), *The Flight From Science and Reason, Annals of the New York Academy of Sciences*, Vol. 775 (24 June 1996), i–xi, 1–593; Richard Dawkins, *Unweaving the Rainbow: Science, Delusion and the Appetite for Wonder* (London: Penguin, 1999); Noretta Koertge (ed.), *A House Built on Sand: Exposing Postmodernist Myths About Science* (Oxford: Oxford University Press, 2000). A more recent attempt to develop a reasoned dialogue about the nature of science can be found in Jay Labinger and Harry Collins (eds), *The One Culture? A Conversation About Science* (Chicago, IL: The University of Chicago Press, 2001).

118. Jasanoff, opera cit. note 77.

119. Jasanoff (1995), op. cit. note 77, 215.

120. Craig Waddell, 'The Role of Pathos in the Decision-Making Process: A Study in the Rhetoric of Science Policy', *Quarterly Journal of Speech*, Vol. 76 (1991), 381–400; C. Waddell, 'Reasonableness Versus Rationality in the Construction and Justification of Science Policy Decisions: The Case of the Cambridge Experimentation Review Board', *Science, Technology, & Human Values*, Vol. 14, No. 1 (Winter 1989), 7–25. For a more critical account of the CERB's deliberations, see Rae S. Goodell, 'Public Involvement in the DNA Controversy: The Case of Cambridge, Massachusetts', *ibid.*, Vol. 4 (Spring 1979), 36–43.

121. Of course, it is possible to question the sense in which the members of the CERB 'represented' the population of Cambridge. Clearly they were unusual in that they were selected for, and then chose to be involved in, a complex scientific and technical controversy. For an account by a CERB member, see Sheldon Krimsky, *Genetic Alchemy* (Cambridge, MA: MIT Press, 1982); for a comprehensive account of the whole debate, see Susan Wright,

Molecular Politics: Developing American and British Regulatory for Genetic Engineering, 1972–1982 (Chicago, IL & London: The University of Chicago Press, 1994).

122. Waddell argues that whilst the rhetoric of science may be of facts and not emotion, the transcripts of the proceedings clearly reveal that advocates of both sides combined rational arguments (*logos*) with the other two elements of rhetoric (*ethos* or integrity, and *pathos* or emotional argument). In the end it was the combination of all three, with concrete examples of sick children being cured by a committed physician as a result of the research, that seems to have won the day.

123. Alan Irwin, *Citizen Science: A Study of People, Expertise and Sustainable Development* (London & New York: Routledge, 1995), 17–21 & passim.

124. How this participation is to be accomplished is, of course, another matter. The next section discusses the difficult choices farm workers and their representatives faced when they were, eventually, invited to participate in a US conference on 2,4,5,T.

125. Examples include: Evelyn Fox Keller, *Reflections on Gender and Science* (New Haven, CT: Yale University Press, 1985); E. F. Keller and Helen E. Longino (eds), *Feminism and Science* (Oxford: Oxford University Press, 1990); Sandra Harding, *Is Science Multicultural?: Postcolonialisms, Feminisms and Epistemologies* (Bloomington: Indiana University Press, 1998); Donna Haraway, *Modest_Witness@Second_Millennium.FemaleMan@Meets_OncoMouse™: Feminism and Technoscience* (New York: Routledge, 1997).

126. Described in Epstein, op. cit. note 63.

127. Arksey, op. cit. note 5; Bloor, op. cit. note 71.

128. Arie Rip, Thomas J. Misa and Johan Schot (eds), *Managing Technology in Society: The Approach of Constructive Technology Assessment* (London & New York: Pinter, 1995).

129. Michael Gibbons, Camille Limoges, Helga Nowotny, Simon Schwartzman, Peter Scott and Martin Trow, *The New Production of Knowledge: The Dynamics of Science and Research in Contemporary Societies* (London, Thousand Oaks, CA & New Delhi: Sage Publications, 1994).

130. For an exchange of views on this issue, see the papers contained in Pickering (ed.), op. cit. note 14. The burden of our paper, of course, is to resurrect boundaries between expert and non-expert in order to resolve the Problem of Extension.

131. Wynne (1992), op. cit. note 57, 39.

132. Wynne (1993), op. cit. note 57, 328.

133. Collins, op. cit. note 6; Collins & Kusch, op. cit. note 6; H. M. Collins, 'Socialness and the Undersocialized Conception of Society', *Science, Technology, & Human Values*, Vol. 23, No. 4 (Autumn 1998), 494–516.

134. James C. Petersen, 'Citizen Participation in Science Policy', in J. C. Petersen (ed.), *Citizen Participation in Science Policy* (Amherst: University of Massachusetts Press, 1984), 1–17.

135. This is not to deny that the relevant agencies may not have embraced these changes as enthusiastically as they might. In practice, therefore, although procedural changes have enabled public representatives to participate in the development of policies, most of the reforms have focused on making data and information available. As a result, they do not fundamentally challenge the existing definitions of who/what is an expert. In other words, they remain based on, and do little to challenge, a deficit model in which expertise is a resource denied to the socially and economically disadvantaged and abundantly available to the pow-

erful. Participation therefore depends on redistributing expertise, but does not assume that it lies outside the scientific community. (See Petersen, op. cit. note 134, 26.)

136. Petersen, ibid., 6–7.

137. See, for example, Welsh, op. cit. note 7, passim. It is worth noting, however, that the importance attached to this is perhaps peculiar to the USA, and that culture's longstanding distrust in its own institutions. The contrast between US systems and those employed in France is clearly brought out in Theodore Porter, *Trust in Numbers: The Pursuit of Objectivity in Science and Public Life* (Princeton, NJ: Princeton University Press, 1995), esp. Chapters 6 and 7.

138. See Lawless, op. cit. note 84.

139. Information about the MPC's membership, past and present, can be found on the Bank of England website: http://www.bankofengland.co.uk.

140. Certainly, we would expect Mr Greenspan, and other central bankers, to consult widely, but there is no suggestion that anyone other than this group should weigh up the evidence and make the decisions.

141. Rip, Misa & Schot (eds), op. cit. note 128; Arie Rip, 'Controversies as Informal Technology Assessment', *Knowledge: Creation, Diffusion, Utilization*, Vol. 8, No. 2 (December 1986), 349–71.

142. Rip, ibid., 363–4.

143. Ibid.

144. EU Workshop on 'Democratising Expertise' and Establishing a European Scientific Reference System (Brussels, 30 March 2001); report available from: http://europa.eu.int/comm/governance/areas/group2/index_en.htm. This is also one of the recommendations of the Phillips Report into the BSE crisis: see note 90.

145. In addition, the participants at the EU workshop were concerned about the point at which science goes public. EU scientists were uncomfortable with following a US model, in which all meetings and debates are public, and would prefer private discussions to prepare their positions beforehand. Concerns were also expressed about the resources consultation might consume, how the 'cacophony of voices' would be dealt with and how 'consultation fatigue' could be avoided. Dorothy Nelkin's work, however, suggests that these latter problems are not usually as significant as scientists expect: see, for example, D. Nelkin, 'Science and Technology Policy and the Democratic Process', in J. C. Petersen (ed.), op. cit. note 134, 18–39.

146. The parliamentary debate on embryonic stem cell research in Britain may be an example implicitly referred to here, where scientific advisers gave technical information to both sides of the debate.

147. Cf. Evans, op. cit. note 82; see also Guston (1999), op. cit. note 88.

148. Sandra Harding, *The Science Question in Feminism* (Ithaca, NY: Cornell University Press, 1986), 250; cited in Isobelle Stengers, *The Invention of Modern Science* (Minneapolis & London: University of Minnesota Press, 2000), at 20–21.

149. For example, Harding, op. cit. note 125.

150. Harry Collins and Trevor Pinch, *The Golem at Large: What You Should Know About Technology* (Cambridge: Cambridge University Press, 1998), at 7–29.

151. Arksey, op. cit. note 5, 174.

152. Locke, op. cit. note 110; Collins & Pinch, opera cit. notes 31, 56 & 150.

153. Jon Turney, 'Review of *The Golem at Large*', *Public Understanding of Science*, Vol. 8, No. 2 (April 1999), l39–40, at 140.

154. Wynne (1993), op. cit. note 57, 333.

155. Funtowicz & Ravetz, op. cit. note 78, 751.

156. Yearley, op. cit. note 105; Yearley, op. cit. note 109.

157. For example, Yearley notes that the risks due to GM crops could be classed as 'high' or 'very high'. It may be, however, that the NUSAP notation developed by Funtowicz and Ravetz in other publications goes some way to addressing these concerns. See: Silvio O. Funtowicz and Jerome R. Ravetz, *Uncertainty and Quality in Science for Policy* (Dordrecht: Kluwer, 1990).

SUPPOSE THAT TWO groups of expert mathematicians disagree about a complex mathematical question—say, whether Princeton mathematician Andrew Wiles really did solve "Fermat's Last Theorem," which no mathematician had been able to prove since Louis Fermat first propounded it about 360 years ago. These experts have had an opportunity to hear one another's reasons for their competing conclusions about Wiles's proof, and neither group is convinced by the other. How might we decide which of the two groups is making the correct mathematical judgment? Here's a suggestion: Convene a group of twelve or so nonmathematicians, give them an opportunity to hear from representatives of each of the competing groups of mathematicians, and have the nonmathematicians decide whether Wiles's proof really succeeded. If the truth of the matter was among one's chief concerns, would this decision procedure seem sound? There is serious reason to doubt it. The most obvious problem with such a procedure is that it seems to turn the decision about this disputed, highly complex question in the science of mathematics over to those who are least competent to answer it.

Many legal systems, including the state and federal systems of the United States, use decision procedures that are disturbingly close to the one just imagined, procedures in which nonexpert judges and juries are called upon and authorized to evaluate expert scientific testimony. This Essay's goal is to offer a sustained critical analysis of the legal rules and doctrines that create and administer this procedure. Expert scientific information is relevant to, even decisively important in, a rapidly growing percentage of decisions throughout civil and criminal law. Most judges and juries, however, are not sufficiently familiar with relevant scientific fields to be able independently and reliably to bring scientific information to bear on their decisions. Instead, they must solicit and defer to the judgments of expert scientific witnesses.

Moreover, almost inevitably in litigated cases in which expert scientific evidence is offered, nonexpert judges and juries are presented, not with one authoritative "voice" of scientific truth, but instead with competing scientific expert witnesses who testify to contrary or even contradictory scientific propositions. Lacking the information necessary to make cogent independent judgments about which of the competing scientific experts to believe, nonexpert legal decision makers choose among the experts by relying on such indicia of expertise as credentials, reputation, and demeanor. Thus, even the act of soliciting and deferring to expert scientific judgment requires nonexperts to use a reasoning process—the process of selecting the experts, deciding which expert to believe when the experts compete, and, finally, deciding how to use the believed expert's information in resolving the central dispute being litigated.

In 1998, I published "Scientific Expert Testimony and Intellectual Due Process" in the *Yale Law Journal*,[1] in which I argued that America's current legal system requires judges and jurors to defer *arbitrarily* to expert witnesses. In the seven years that have passed, no sustained reform of the treatment of expert witnesses has taken place. Accordingly, this Essay, which is derived from that earlier article, presents a current, critical challenge to the legitimacy of current legal practice. Constrained as it is by space, this Essay cannot, and does not, replicate the entire argument of the earlier article. Instead, as I elaborate below, this Essay focuses on what expert witnesses are meant to provide, and why this role is impossible given decision makers' lack of sophistication.

Drawing on work in jurisprudence, epistemology, philosophy of science, and theories of practical reasoning, as well as on doctrines and leading cases on scientific expert evidence, this Essay carefully models the reasoning process by which nonexpert legal reasoners defer to scientific experts in the course of applying a law to individual litigants. Drawing on this model, I argue for three central conclusions. Taken together, these conclusions have far-reaching consequences for virtually all legal systems in which nonexpert legal decision makers confront expert scientific testimony.

First, the Essay argues that in order to avoid making an epistemically arbitrary choice about which of the competing scientific experts ought to be believed, a person must understand (in a special sense discussed in the text) the cognitive aims and methods of science. But nonexpert judges and juries lack just that kind of understanding, which is why they rely instead on other indicia of expertise, such as credentials, reputation, and demeanor. Second, nonexpert judges' and juries' lack of understanding of the cognitive aims and methods of science and their reliance on such indicia of expertise as credentials, reputation, and demeanor to choose between competing scientific experts thus yield only epistemically arbitrary judgments. Third, the conclusions that nonexpert judges and juries ultimately reach by relying significantly on expert scientific testimony are often also epistemically arbitrary and are therefore not justified from a legal point of view.

In sum, I argue that values to which legal systems are and ought to be committed actually condemn one of the most firmly entrenched evidentiary methods currently in place. As scientific theories continue to become more specialized, complex, and relevant to a widening range of cases, this incoherence between normative aspiration and actual doctrinal and institutional procedure will increasingly threaten the legitimacy of nonexperts' legal decisions. I conclude the Essay by arguing that the current regime fails to afford "intellectual due process" to litigants whose cases involve expert testimony.

I. Expertise:
Concepts and Basic Problems for the Nonexpert

A. Knowledge, Warranted Beliefs, and Degrees of Epistemic Competence

In my 1998 article, I defined expert (and theoretical authority) in terms of *epistemic competence*, which in turn I described using the term 'understanding.' Some explanation of the concepts of *epistemic* and *understanding* are now due. The distinctive mark of the *epistemic* is the concern with warranted belief. That is, the epistemic point of view is the view of a reasoner whose overriding cognitive goal is to acquire *warranted beliefs*. 'Warranted belief' here is a placeholder. Different epistemologies fill in this place in different ways. Some declare that truth is the sole overriding cognitive goal for a properly epistemic point of view,[2] while others allow a broader range of cognitive aims to play a role, even when they are not valued as instrumental means to achieving the end of discovering truth.[3] In identifying the epistemic point of view with the cognitive goal of achieving warranted beliefs, I offer no strong commitment to any particular epistemic axiology, although it does seem that exclusively "vericentric" epistemologies (those that treat discovery of truth as the overriding epistemic goal) run a significant risk of misdescribing the epistemic structure of many epistemic practices, including the practice of science.

The concept of *understanding* should be an important part of any full epistemological theory. In contemporary epistemology, it has received relatively little express treatment compared to headliner concepts like *knowledge* and *justification*, and the treatments it has received have tended to focus specifically on linguistic understanding. Miles Burnyeat has begun a promising, more general line of inquiry into understanding, one very much consonant with the approach I am taking in this Article.[4] It comes in his discussion of the Platonic and Aristotelian conception of the relation between knowledge on the one hand, and synoptic, explanatory understanding on the other:

> The important difference between knowledge and understanding is that knowledge can be piecemeal, can grasp isolated truths one by one, whereas understanding always involves seeing connections and relations between the items known. "The only part of modern physics I understand is the formula '$E = mc^2$'" is nonsense. "The only part of modern physics I know is the formula '$E = mc^2$'" is merely sad.[5]

Burnyeat speaks of the philosopher who, like Plato, "wants to assimilate knowledge to rational understanding."[6] Though assimilating knowledge to understanding or, perhaps even better yet, replacing the concept of knowledge as the focal point of epistemology with the concept of understanding, has its attractions,[7] I have no ambition in the present work to take on so large a task. My theoretical needs are far more modest, attempting to explicate only a notion of epistemic competence that captures and helps to explain the nature of the cognitive capacity involved in possessing scientific expertise. For that purpose, the notion of understanding as the possession of a widening, explanatory,[8] synoptic grasp serves quite well.[9]

My 1998 article contained a discussion of point of view and axiology, in which I extended Larry Laudan's model of rationality into a general account of rational enterprises and their distinctive aims, methods, and factual judgments. Laudan calls his model of scientific justification "reticular" to mark the way in which the aims, methods, and beliefs of science are mutually supporting and explaining, and how *each* type (aim, method, factual judgment) can occasion a change in each of the other types. A chosen cognitive aim, for example, can change or dictate choice of method—as when a psychologist chooses the double-blind experiment method—because she believes it will serve the cognitive aim of truth in psychological experiment. Note that this example also illustrates the way in which a *factual judgment* helps guide the choice of method—for it is a factual judgment in the efficacy of the double-blind experiment (and the substantially lesser efficacy of single-blind experiment) that leads her to choose *that* as a method of producing truths. Similarly, a factual judgment about the realizability of an aim can compel a change in the aim—for it makes little sense to pursue an aim that one decides, as a matter of *factual* judgment, is unachievable. Each "node" of the net can have a justificatory and explanatory impact on each of the others, This model of justification, suitably extended from science to intellectual discipline, is a holistic account that emphasizes the centrality of reflective adjustment in epistemic justification, giving it much the same prominence as in the account of exemplary argument I have offered elsewhere.[10] And it is a model that gives content to the kind of latticework of explanatory relations present in the *understanding* of an expert discipline. To have epistemic competence in such a discipline is, I suggest, to be *capable of grasping and manipulating this kind of reticular structure of aims, methods, and factual judgments in an expert discipline.* It is precisely the lack of this kind of understanding in nonexpert legal reasoners that casts doubt on their capacity to rely legitimately on expert scientific testimony in reaching practical decisions.

One more point is important for my later assessments of cases like *Daubert,* and for the overall conclusion of this Article. Epistemic competence in an expert discipline comes in degrees; it is not an all-or-nothing "switch." This is perhaps not surprising. Is it not a familiar fact that some mathematicians, logicians, physicists, economists, geneticists, and so forth are more skilled at grasping and manipulating the aims, methods, and factual judgments of their respective expert disciplines than are other experts in the same disciplines? Surely Isaac Newton was a more epistemically competent physicist than Isaac Asimov. By the same token, we should recognize that there is no bright line separating expertise from nonexpertise—just as there is no bright time line or light line separating night from day, even though there is clearly a difference between night and day. Not all experts are equally epistemically competent in their disciplines, nor are all nonexperts equally incompetent with regard to a given expert discipline.

B. Practical Epistemic Deference and Theoretical Judgment

As I have noted above, I use "practical epistemic deference" as an abbreviation for deference *by* a nonexpert practical reasoner, like a scientifically untrained judge or jury, *to* a scientific theoretical *expert.* This is the reasoning process that is my central concern. Of course, not every instance of epistemic deference by a practical reasoner to a theoretical expert is deference by a *nonexpert* practical reasoner; an epistemically competent practical reasoner could defer to an epistemic equal or near equal. I restrict "practical epistemic" deference to nonexpert practical reasoners only for the sake of abbreviation.

A "theoretical judgment" is a judgment about what one ought to believe from an epistemic point of view. Although theoretical judgments are a central part of scientific inquiry, not only scientists make them. Religious beliefs and pseudoscientific judgments, such as those of astrology or necromancy, as well as judgments in the literary and plastic arts, are also "theoretical" judgments in this sense.[11]

C. The Nonexpert's Selection and Competition Problems

The nonexpert faces at least four distinct problems, which I will call "selection problems." To explain them, I begin with a simple (and vague, but nonetheless sufficient) definition. Call H_c the overall hypothesis of a case presented to a practical reasoning authority (like a judge or jury). H_c might be, for example, the prosecution's claim that Jones committed the murder; or the plaintiff's claim that Smith breached the contract. (H_c will usually be an "ultimate" issue in the case, and a mixed question of law and fact.) Call each H_i an individual hypothesis whose conjunction with all the other H_i implies H_c. (H_c need not also imply the H_i.) An evidentiary proposition (which itself can be logically simple or complex), call it e, is *objective evidence* for some H_i just in case H_i is better warranted given the truth of e than it is given the falsity of e. An evidentiary proposition e is *rationally pertinent to* some H_i just when it constitutes objective evidence of that H_i. (That is, 'objective evidence for' and 'rationally pertinent to' are synonyms.) An evidentiary proposition e is rationally pertinent to the overall hypothesis of a case, H_c, just in case e is rationally pertinent to some H_i (which, as defined, is in the set of propositions that imply H_c). This definition is a slight generalization and amalgamation of the concepts of relevance and materiality found in the common law of evidence as well as a basic definition of objective evidence found in the epistemological literature.[12] A virtue of the definition's vagueness is that it remains neutral among various standards of epistemic appraisal and attendant levels of confidence.

When a nonexpert judge or judge and jury (who may divide decision making labor in the familiar ways, relying on judgments of admissibility, relevance and materiality, sufficiency, and weight) must decide whether to consult a putative scientific expert in the course of deliberating about H_c, that nonexpert faces four "selection problems": (1) determining which of the intellectual enterprises that might yield expert testimony is a *science*; (2) determining who is a *scientist* capable of using her science in a manner that satisfies the standard of epistemic appraisal and the attendant level of confidence that the practical reasoner has established; (3) determining which of the intellectual enterprises that might yield expert testimony is a science that is *rationally pertinent* to the case (that is, to H_c); (4) in cases in which there is significant doubt occasioned by task (3), determining who is capable of answering (3) in a way that can identify an expert scientific discipline capable of satisfying the chosen standard of epistemic appraisal and the attendant level of confidence.

"Competition" in expert testimony occurs when two experts testify to evidentiary propositions that are either contrary or contradictory.[13] The fact that much scientific expert testimony is competitive in this way presents a particular problem and puzzle for systems in which nonexperts make decisions that rely on expert testimony. When experts disagree about the truth of some evidentiary proposition e, the nonexpert must decide whom to believe on the scientific issue. But, *ex hypothesi*, the nonexpert does not have sufficient competence in the expert discipline to be able to make the choice on substantive grounds, so how can the nonexpert make that choice? If we assume honesty on the part of each expert, this can seem especially puzzling in that it may look like we are expecting *greater* ability to discern the scientific truth from the nonexpert than we are from the expert. This is so because the rules of evidence that govern admissibility of expert testimony (i.e., what testimony the nonexpert jury or judge is allowed to hear) do not presuppose that all the experts are testifying to the truth. This we know in part from the basic fact that judges "qualify" experts who the judges know will testify to contrary or contradictory propositions—even when those judges are also fully aware that two evidentiary propositions that are mutually contrary or contradictory cannot both be true. But if a judge does not expect that every scientific expert is testifying to the truth (and thus, on the assumption that the expert is testifying honestly, the judge is not expecting every expert to *know* the truth either), then is the judge expecting the nonexpert judge or jury actually to do *better* than the expert at discerning the truth? It might seem so, because obviously each expert has an opportunity to hear what the opposing expert will say, and thus has as much opportunity to revise his own contrary beliefs as the jury has to form its beliefs.

There are intra- and extra-disciplinary versions of the problem of competition. The intra-disciplinary version arises when two experts within the same field, e.g., epidemiology, testify in contrary or contradictory ways about some issue, The extra-disciplinary version arises when experts from different fields (e.g., philosophy and psychiatry) testify in contrary or contradictory ways about some issue. (I assume that fields, roughly academic fields, have real epistemic significance, and are not merely matters of university administration.) There are also what I shall call problems of *actual competition* and *implied competition.* Actual competition occurs when the testimony of two or more competing experts is actually admitted at trial. Implied competition occurs when there is an expert opinion on a particular subject or topic that exists "extra-camerally," and that, if it had been admitted, would have created a problem of actual competition. One can also imagine a kind of *diachronic competition* in expert testimony, in which claims that experts support at one time become challenged by experts at a later time. The basic problem about competition is how a nonexpert faced with competing expert testimony about some evidentiary proposition can decide which of the competing experts (intra-disciplinary or extra-disciplinary, actual or implied) to believe without being able to assess the substantive merits of the competing arguments they offer to support their evidentiary judgments.

II. Knowledge Versus Justified Belief: What Do Nonexperts Want from Experts?

A. What the Law Desires: Acquiring Justified Beliefs from Experts

One can assess a given state of mind as simple *belief,* or as *knowledge,* or as a *belief that is well grounded* or *rational* or *justified*—and so on. These categories might be arrayed on a spectrum (for example, with simple belief at one end, knowledge at the other, and some of the additional terms of epistemic evaluation somewhere between), or might be explicated as more or less separate categories, each with its own criteria. It is the category of *knowledge* that has dominated the attention of modern epistemologists. In recent decades, this attention has often taken the form of considering the modern "classical" criteria of knowledge as *justified true belief*, and looking for additional or different criteria of the concept that are capable of handling Edmund Gettier problems and the like.[14] More recently, epistemologists have offered theories containing robust explications of other terms of epistemic evaluation (though some of these have in turn been offered principally as accounts of some of the classic tripartite criteria of knowledge), such as *justification, coherence, reliability,* and *evidence.*

I am concerned in this Article with the epistemic competence of nonexpert judges and juries when, in the course of making a legal judgment, they assess putatively relevant and material scientific information. But in order to assess their epistemic competence, *I* myself must select the proper term of epistemic evaluation in terms of which to assess that competence. The term I have settled on *is justified belief*, for three basic reasons. First, it appears to be the central concept of epistemic evaluation that concerns jurists who themselves attend closely and critically to the process by which scientific expert evidence enters legal decision making. Second, *knowledge*, however that concept is cashed out, seems too demanding for a system that consciously solicits *competing* expert scientific testimony (as just noted, judges routinely admit testimony by experts that is mutually contradictory, with an awareness that two contradictory factual claims cannot both constitute knowledge). Third, there turns out to be an interesting and fertile connection between epistemic justification and the justification required for legal legitimacy.

I shall amplify these three reasons a bit. Over the past several decades, jurists attentive to the doctrines and institutions of evidence have become increasingly concerned about scientific expert testimony. Something, they think, is lost along the epistemic chain from scientific research to factfinder belief. What is it that they think gets lost? It might seem that their concern is with *scientific knowledge* possessed by scientists serving as expert witnesses but lost in the process of transmission through testimony to nonexpert judges and juries. As a result of this loss, they fear, nonexperts are relying on unjustified beliefs about scientific information when they render final decisions. Despite much talk of "knowledge" and "scientific knowledge" in some of the leading (American) court decisions and litigators' and scholars' arguments, I maintain that it is not actually *knowledge* with which these jurists are concerned, but rather *justified belief*. (I might put the point this way: Although many jurists seem to have *de dicto* concerns about "knowledge" and "scientific knowledge" and often conduct their analyses in just those terms, from a philosophical point of view their *de re* concern is really only with justified belief.) *Daubert* is a superb example.

Recall that in *Daubert* the Supreme Court considered the proper method by which federal judges are to evaluate proffers of scientific evidence in deciding whether to admit that evidence for consideration by scientifically nonexpert factfinders. The Court framed the issue of the admissibility of scientific evidence by focusing on the term *knowledge*, since that term is central to Federal Rule of Evidence 702, the rule at issue in *Daubert*. Interpreting the "legislatively-enacted Federal Rules of Evidence as [it] would any statute," the Court proceeded to ascertain what the term *knowledge* meant in this rule partly by using the "plain meaning" method of statutory interpretation (a method that, as in *Daubert* itself, often amounts to little more than looking a term up in a dictionary).[15] The Court's analysis is worth quoting at length:

> The subject of an expert's testimony must be "scientific knowledge." The adjective "scientific" implies a grounding in the methods and procedures of science. Similarly, the word "knowledge" connotes more than subjective belief or unsupported speculation. The term "applies to any body of known facts or to any body of ideas inferred from such facts or accepted as truths on good grounds." Webster's Third New International Dictionary 1252 (1986). Of course, it would be unreasonable to conclude that the subject of scientific testimony must be "known" to a certainty; arguably, there are no certainties in science. See, e.g., Brief for Nicolaas Bloembergen et al., as Amici Curiae 9 ("Indeed, scientists do not assert that they know what is immutably 'true'—they are committed to searching for new, temporary, theories to explain, as best they can, phenomena"); Brief for American Association for the Advancement of Science et al., as Amici Curiae 7–8 ("Science is not an encyclopedic body of knowledge about the universe. Instead, it represents a process for proposing and refining theoretical explanations about the world that are subject to further testing and refinement . . ."). But, in order to qualify as "scientific knowledge," an inference or assertion must be derived by the scientific method. Proposed testimony must be supported by appropriate validation—i.e., "good grounds," based on what is known. In short, the requirement that an expert's testimony pertain to "scientific knowledge" establishes a standard of evidentiary reliability.[16]

Is the Court really concerned with scientific *knowledge*? The Court does take quite seriously the *phrase* 'scientific knowledge,' but it is clear on reflection that this cannot quite be the same concept that has concerned traditional epistemology—at least not those epistemological theories for which *truth* is a necessary condition of knowledge. Although the Court and the traditional epistemologist agree that "the word 'knowledge' connotes more than subjective belief or unsupported speculation,"[17] it is obvious that, under the Court's interpretation, Rule 702 does *not* presuppose that every expert is testifying to the truth. Otherwise, the Court would not allow scientific experts to testify to contrary or contradictory propositions.[18] Instead, the Court offers a more expansive explication of 'knowledge,' one that embraces "any body of known facts or . . . *any body of ideas inferred from such facts or accepted as truths on good grounds. . . .* Proposed testimony must be supported by appropriate validation—i.e., 'good grounds,' based on what is known."[19] (The Court also concludes that *certainty* is not a necessary condition of "scientific knowledge.")[20] In a nutshell, the *concept* of epistemic assessment with which the Court is concerned when interpreting the *term* 'knowledge' and the *phrase* 'scientific knowledge' is that of a *judgment that is supported by good reasons.*

We may confirm this interpretive judgment (our interpretive judgment, that is, regarding the Court's interpretation of the term 'knowledge' in the Federal Rules of Evidence) by attending to the brief written for the respondents in *Daubert,* which clearly had a substantial shaping influence on the Court's analysis. Like the Court's opinion, that brief also emphasized the distinction between *knowledge* in the stronger (philosophical) sense and *judgment supported by good reasons.* Indeed, it amplifies this distinction and, even more to the present point, acknowledges that competing testimony can all be admitted under the same requirement of "scientific knowledge" in Rule 702:

> [E]ven when offering an "opinion," a scientific expert must be testifying "[]to" "scientific knowledge."
>
> These words naturally mean that an expert must be testifying to more than merely his or her own view on a scientific issue. Rule 702 does not allow an expert to offer "beliefs" or "hypotheses" or "theories" or "claims" or "assertions" or "evidence" or "testimony"; instead, it limits expert testimony to "knowledge." Even by itself, *"knowledge" ordinarily requires, at a minimum, appropriate validation for the proposition*—i.e., "good grounds" for the belief based on what is known. See Webster's Third New International Dictionary 1252 (1986). And when used as part of the phrase "scientific, technical, or other specialized knowledge," *the term naturally refers to grounds that are deemed good by the relevant scientific, technical, or other specialized field*—that is, to claims that are validated, or derived, according to the accepted standards in the relevant field. See J. Kourany, Scientific Knowledge: Basic Issues in the Philosophy of Science 112 (1987) (hypotheses "must prove their mettle" to become "part of scientific knowledge"). Rule 702's language, naturally read, thus requires validation to the extent possible—i.e., a foundation, or good reason for acceptance as valid— based on established standards in the expert's field.[21]

While this brief's analysis admirably identifies the term of epistemic assessment that is relevant to rules and institutions of evidence, it also highlights a deep puzzle about nonexperts' epistemic deference to experts. In effect, the brief argues that when the putatively scientific evidence proffered by a party is so weak as not to be "validated, or derived, according to the accepted standards in the relevant field," then the judge should not even allow the evidence to be presented to the nonexpert jury. But when evidence supporting contrary or contradictory propositions *is* supported by *"grounds that are deemed good by the relevant scientific, technical, or other specialized field,"* then the nonexpert judge or jury is to make the decision as to which of those competing and well-supported claims is to be accepted for purposes of the legal decision at hand. That is, on this brief's view and apparently on the view of the *Daubert* Court itself, when qualified epistemically competent experts disagree, the decision as to who is *correct* is to be given by the judge to the *least* epistemically competent institutional actor, the nonexpert judge or jury. Again, we are driven to ask, what is being expected or demanded of the nonexpert legal reasoner in assessing scientific testimony? The *Daubert* opinion and at least some sources on which it relies seem to have it thus: When the evidence is so weak that no reputable scientist in the field would endorse it, prevent the nonexpert from hearing it (and from hearing that no reputable expert would endorse it); but when the best scientific theories and methods underdetermine the result, let the nonexpert decide who is correct. How can an epistemically responsible decision emerge from that rule?[22]

We shall return to this puzzle later. For now, I restate the basic descriptive observation I have made in this section. What centrally concerns lawyers, scholars, and judges with regard to the cogency of scientific expert testimony is not whether the expert has—or can transmit to the nonexpert—*knowledge* in the strong philosophical sense, but rather whether the expert has and is in a position to be able to transmit to the nonexpert *a belief that is supported by good reasons.* I think it not inaccurate to go further and say that what concerns these jurists is not the epistemic concept of *knowledge*, but rather that of *justified belief.* (Obviously a good deal more evidence from the writings of lawyers, scholars, and judges would be needed to support this claim more conclusively, but I hazard the judgment that such evidence is readily available.).

I shall go beyond my descriptive claim about what type of epistemic assessment it is that concerns evidence jurists. Regardless of whether jurists concerned with the cogency of scientific expert testimony *are* specifically focusing on belief supported by good reasons—that is, on *justified belief*—rather than on knowledge in the philosophical sense, I now suggest that that is the concept with which they *should* be concerned. Why the weaker concept only? To anticipate later discussion, while it seems that science has much of importance to tell the law about matters that are rationally pertinent to a great many legal decisions, it is also clear that scientific truth is elusive. That science is not an epistemic monolith or a univocal oracle is trite learning. In a great many cases to which some particular angle of the scientific point of view is rationally pertinent, and at every "level" of the scientific point of view (axioiogical aims,

methods, particular judgments), there is room for skilled, learned, and reasonable scientists to disagree. Were a legal system to set its rules of procedure
and evidence—the rules guiding "legal epistemology"—so as to insist on only
knowledge (with truth as a necessary condition), the law would vastly deprive
itself of counsel it *needs* to make legal decisions sufficiently *epistemically*
legitimate to be *legally* legitimate, The law is wise not to have its epistemic
reach so far exceed its grasp. Justified belief is all it does and all it should seek
to have transmitted from the scientist-witness to the nonexpert judge or jury.

But can even that more modest goal be achieved?

B. Testimony as a Source of Justified Belief

I turn now to a more focused investigation of whether the nonexpert can make
the epistemically cogent judgments that legitimate legal decision making requires—holding off, for the moment, a discussion of the source and nature of
the criteria of legitimacy. In accord with the conclusions reached in Section
I.C, the precise overall philosophical question involved here is this: Can a nonexpert practical legal reasoner acquire justified beliefs about scientific propositions and their rational pertinence even in the face of selection and competition problems? I am attempting to answer that question by explaining and
modeling the reasoning process that underlies epistemic deference to experts
about expert subject matters. What kinds of theoretical insights are available
to help with this inquiry? Insights may be discerned, I suggest, in two types of
philosophical analysis. One is generic epistemological analysis of *testimony* as
a source of knowledge or justified belief. The other is more specific analysis by
philosophers of the epistemic dimensions of *expert testimony*. I consider some
of each of these contributions. One terminological device is worth using in this
brief survey. Because these discussions do not consistently distinguish between
knowledge and *justified belief*, and because for my current purposes the distinction is not important, I shall use 'KJB' to refer indiscriminately to knowledge and justified belief.

1. Testimony

I begin with some philosophers' generic analyses of the epistemology of testimony. Surprisingly little explicit and sustained treatment of testimony is found
in the epistemological literature, given how fecund a source of KJB it seems to
be. In recent years, however, philosophical treatments of testimony have multiplied, so that the subject will soon likely take its rightful place as a major domain of epistemology. Much of the contemporary debate focuses on whether
testimony can be an independent source of KJB, on par with memory, percep-

tion, and inference, or whether instead testimony derives whatever epistemic integrity it has from being reducible to other familiar sources of KJB.[23]

Hume offered a brief but influential account in his essay *Of Miracles*.[24] Hume's account is reductionist. He treats the integrity of testimonial KJB as dependent on empirical confirmation; indeed, his treatment of testimony is a fairly straightforward application of the thoroughgoing empiricism he developed in the *Treatise*[25] and other works.[26] He begins by observing that "there is no species of reasoning more common, more useful, and even necessary to human life than that which is derived from the testimony of men and the reports of eyewitnesses and spectators."[27] "[O]ur assurance in any argument of this kind," he maintains,

> is derived from no other principle than our observation of the veracity of human testimony and of the usual conformity of facts to the report of witnesses. It being a general maxim that no objects have any discoverable connection together; and that all the inferences which we can draw from one to another are founded merely on our experience of their constant and regular conjunction, it is evident that we ought not to make an exception to this maxim in favor of human testimony whose connection with any event seems itself as little necessary as any other. . . . The reason why we place any credit in witnesses and historians is not derived from any *connection* which we perceive *a priori* between testimony and reality, but because we are accustomed to find a conformity between them.[28]

Such an account has a ready explanation of both the reach and the limits of credulity that one ought, from an epistemic point of view, to extend to beliefs acquired from testimony: "[W]hen the fact attested is such a one as has seldom fallen under our observation, here is a contest of two opposite experiences, of which the one destroys the other as far as its force goes, and the superior can only operate on the mind by the force which remains."[29]

Several contemporary philosophers have been attracted to accounts of the epistemic integrity of testimony that have both elements of the Humean account: empiricism and reductionism. W. V. Quine and J. S. Ullian, for example, endorse an account not unlike Hume's, save for the particular naturalistic and holistic flavor familiar from Quine's general epistemology:

> [W]hen we hear an observation sentence that reports something beyond our own experience, we gain evidence that the speaker has the stimulation appropriate for its utterance, even though that stimulation does not reach us. Such, in principle, is the mechanism of testimony as an extension of our senses. It was the first and greatest human device for stepping up the

observational intake. Telescopes, microscopes, radar, and radio astronomy are later devices to the same end.[30]

Quine and Ullian also provide a naturalistic discussion of the legitimacy of a hearer's presumption that a "testifying" speaker is telling the truth—telling the truth both when uttering observation sentences, whose truth the hearer can check relatively easily[31] and when uttering nonobservation sentences, which a hearer often cannot easily check, and thus for which "the danger of mistaken testimony soars."[32] Quine and Ullian suggest that, without the presumption by a hearer of testimony that testifying speakers are telling the truth and without an extension of that presumption even to nonobservation sentences, testimony could not be a significant source of KJB. But they also readily acknowledge that the epistemic integrity of such presumptions is open to serious doubt, since knavery and fallibility attend both observational and nonobservational testimony. Even so, they argue that hearers may be justified in having some presumptive confidence that witnesses are telling the truth, a justification Quine and Ullian locate in natural features of language and language learning.[33]

Just as there is a natural tendency toward veracity in speakers, argue Quine and Ullian, there is for related reasons a natural (apparently charity-based, though they do not say so explicitly) tendency toward credulity in hearers. On the other hand, credulity is not, and should not rationally be, unlimited. Rationality reigns in credulity. On Quine and Ullian's view (as on Hume's), the empirical grounding of the integrity of testimonially acquired beliefs mandates tight constraints on rational trust in beliefs acquired from testimony:

> Veracity is generally admirable, if not always prudent; but credulity, in more than modest measure, is neither admirable nor prudent. . . . The courtroom is worthy of the attention of anyone who is inclined toward taking too much of what he is told at face value. It teaches a stern lesson. People disguise the truth in certain situations, whether out of deviousness, self deception, ignorance, or fear. They also, of course, misremember, misjudge, and misreason.[34]

Neither Quine and Ullian nor Hume explicitly addresses the question of *expert* testimony. The difference between general testimony and scientific testimony can be quite epistemically significant, especially for a Humean account according to which the integrity of testimony depends on the hearer's ability independently to confirm or disconfirm what he has been told (at least, in a sufficiently large sample of cases). Simple, nontechnical (including observational) reports are easiest for a nonexpert to confirm. A nonexpert cannot independently and directly check complex theoretical propositions that do not have simple observational consequences or whose observational consequences themselves require complex training even to be recognized as such. Whatever checking the nonexpert can manage must rely on indirect devices like demeanor, credentials, and reputation.[35]

Elizabeth Fricker offers a sustained epistemology of testimony that emphasizes the importance of a hearer's assessment of a witness's sincerity and competence.[36] Although she does not expressly discuss expert testimony, her emphasis on epistemic competence is illuminating and helpful here. The main relevant points of her account are these. First, it is important to distinguish between *global* and *local* reductionism in accounts of testimony. Global reductionism has two forms: (1) the thesis that it is possible to reduce all testimonial KJB to more familiar, fundamental, and less problematic epistemic sources and principles, such as perception, memory, and inference; and (2) the distinct thesis that the epistemologist must make this reduction in order to account for the epistemic integrity of testimonially acquired KJB. Hume seems to endorse both of these global reductionist theses;[37] I surmise, but am less confident here, that Quine and Ullian do too.[38] Local reductionism also has two forms: (1) that it is possible to reduce some, but not necessarily all, testimonial KJB to more familiar, fundamental, and less problematic epistemic sources and principles; and (2) that the epistemologist must reduce *some*, but not necessarily all, testimonial KJB in order to account for its epistemic integrity.

Fricker maintains, rightly, that a great deal of what we are inclined intuitively to acknowledge as KJB comes from testimony, and that many of those beliefs *cannot* be reduced to other sources of KJB. There is powerful support for this view—much of it marshaled by Tony Coady in his extended critique and rejection of the reductionist position.[39] Against Hume, for instance, Coady argues that one cannot empirically consult one's personal experience to check to see whether testimony is reliable because there is too much testimony and too small a personal observation base. Moreover, much of what one would do to verify or falsify testimonial evidence is itself suffused with testimonial information; perhaps even the conceptual framework in which one describes the world and its objects and institutions has been acquired by testimony. But, as Fricker stresses, antireductionist arguments like Coady's fail to distinguish global and local reductionism, and that weakens their analysis of the independent epistemic merits of testimony. In arguing against Coady's version of antireductionism, Fricker observes that Coady denies only the possibility of *global* reductionism. Flicker allows that we seem to have no choice but to accept much testimonial information on "simple trust" in our developmental stages, in which we acquire language and concepts from parents, teachers, and peers.[40] But we *do* have a choice once we have come of epistemic age. At that later developmental stage, she claims, it is both feasible and indeed rationally obligatory for mature epistemic actors to adopt a stringently critical stance toward testimony. Posed in this critical stance, the hearer of testimony is rationally obliged to focus on the testifier's *sincerity* and *competence*.

In an effort to model the reasoning process of a hearer confronted with testimony (an effort in much the same spirit as my own in this Article, but with a different focus), Fricker claims that "common sense semantics" endorses an inference rule that allows an interpreter to move from the fact that a testifying witness has spoken a proposition to an endorsement of the proposition itself.[41] That inference rule, which she considers to be an "an analytic truth about the speech act of assertion," is as follows: (S asserted that P at t and S was sincere, and was competent at t with respect to $P \rightarrow$ P) where 'S is competent with respect to P at t' $=_{df}$ 'at t, S believes $P \rightarrow P$'.[42]

In Fricker's view, a hearer is rationally obliged to *presume* both sincerity and competence, but those presumptions are rather easy to overcome. Sincerity "can be assumed unless there are signs of its lack," but "the hearer must always be scrutinizing the speaker for telltale signs of its absence."[43] Similarly, Fricker allows that a presumption of competence "may be assumed as the default setting."[44] As she acknowledges, Davidsonian considerations suggest that some "charitable" presumption by the hearer of the speaker's competence (and sincerity) may be part of the nature of language; and Coady, Fricker's main target, leans very heavily on such considerations in his account. Nevertheless, Flicker's empiricist approach leads her (as it led Hume and Quine and Ullian) to conclude that a properly *rational* presumption of competence is sharply limited:

> [D]espite the conceptual constraints on interpretation it remains an *empirical question* whether particular speakers, on particular occasions, are either competent or sincere; one which a self-consciously rational belief-former will wish to have positive evidence about, before he believes what he is told. Correspondingly, as to what structure of justification must support a testimony-belief, it is right to insist that a hearer must be in a position to know that the speaker is sincere and competent, and that his being so requires his possession of some particular evidence pertaining to the case at hand.[45]

Fricker reveals just how weak she believes this presumption of competence is, and perforce, how weak are the "conceptual" requirements that a hearer be charitable toward testimony and testifiers whose competence may be presumed:

> [W]ith respect to a subclass of telling only, viz. those with subject matters for which commonsense psychological knowledge licenses one to expect the speaker to be competent about them: such as her name, where she lives,

what she had for breakfast, what is in clear view in front of her, and so forth. Again, the speaker [sic][46] must be sensitive to indicators of its lack. . . . A hearer who engages in [this interpretative task] does not believe what she is told uncritically, and she has *empirical* grounds for her trust in her informant.[47]

To the claim that a stronger presumption of competence is rationally warranted—say, a presumption of competence in the absence of special evidence to the contrary—Fricker responds that "[t]he proportion of utterances which are made by speakers who are either insincere or incompetent is far too high for this to be an attractive policy."[48]

Fricker argues for *local* reductionism. We cannot globally reduce testimonially acquired KJB because we cannot escape the necessity of relying on simple trust in our formative years (though even some of the beliefs acquired at this stage are subject to later coherentist confirmation). But we can, in our mature years, *locally* reduce testimonially acquired propositions to other forms of KJB—perception, inference, and memory—by performing the critical presumptive interpretive exercise outlined above.

Fricker's account has much to recommend it. Its insights are especially valuable for their focus on a hearer's need to bring a robustly critical stance toward both implicit and explicit claims of speaker competence. As I have argued, such a critical stance is built into the law's epistemology in the form of rules of evidence and procedure that proceduralize the caution of the factfinder toward expert (and other) factual claims. Fricker's fairly persuasive account suggests that this kind of proceduralized caution attends *every* rational encounter with testimony and not just testimonial encounters in the highly formalized institutional setting of a courtroom. If she is right, legal rationality in this domain is continuous with rationality more generally. Fricker's account is also attractive for its cogent orchestration of three strong intuitions. First, one cannot, *pace* Hume, confirm by means of one's own nontestimonial experience *all* the testimony one has heard. Second, even many of the testimonially acquired beliefs that cannot be confirmed by other means, nevertheless, do seem to produce KJB. Third, there is a limit to rational credulity, where uncritical trust spills into warrantless gullibility.

Fricker's account does leave important questions unanswered that must be addressed before her account can provide an adequate account of the rationality of trust in testimony. She leaves us to wonder, for example, how much of what we *think* is testimonially generated KJB is *actually* KJB, because her presumption about speaker competence applies only to a relatively trivial "subclass of tellings" (e.g., the speaker's name, where she lives, what she had for breakfast, etc.). Surely the real power of the intuition that KJB arises from testimony comes from its ability to explain the vast majority of our beliefs that are far more complex and that we seem not to be in a good position to confirm.[49] Fricker's account leaves largely unexplained how and whether we can acquire KJB under these conditions.

Perhaps it is alertness to this difficulty that leads Fricker to make the sweeping claim that "the key to the epistemology of testimony is: disaggregate."[50] She elaborates, "Disaggregate both regarding the question of whether and when we may rightly trust without evidence, and regarding the empirical confirmation of speakers' trustworthiness."[51] This disaggregation principle does not offer any final answers to the foregoing question about the status of a great many apparently unverifiable, testimonially generated beliefs. (Can these be real sources of KJB? Is there some way to verify them short of becoming experts in the domain of these propositions?) But taken as a regulatory rule for the project of investigating the epistemic integrity of testimony, this maxim seems unassailable. Some such principle is guiding my inquiry here into deference by nonexpert practical reasoners to scientific experts. The foregoing discussion should suggest just this: Whatever *general* account might be given of the epistemic virtue of testimony as a source of KJB, it cannot be assumed that *all* testimonial knowledge has the same virtues. There are special features of scientific knowledge that make it particularly difficult to see how they can be sources of KJB for nonexpert practical reasoners who are constrained by certain norms of epistemic accuracy and practical legitimacy.

2. Expert Testimony

I turn now from *general* philosophical theories of *testimony* to accounts that focus on *expert* testimony more specifically. Some (for example, those offered by Hilary Putnam and John Hardwig)[52] focus on deference to experts in everyday life; these I refer to as "collectivist" accounts of legitimate epistemic deference to experts. Others (for example, the accounts offered by Kenny and Coady)[53] focus more specifically, and more pointedly for my purposes, on expert testimony in courts. I refer to these as "extra-cameral" accounts of legitimate epistemic deference to experts. Both types of account defend the strongly held intuition that nonexperts can acquire KJB from experts—even in the face of the selection and competition tasks discussed above.[54] I shall consider how adequate these defenses are.

"Collectivist" accounts of expert testimony suggest that to understand how and why epistemic deference to experts can yield KJB,[55] the theory of KJB itself must depart significantly from the "individualist" epistemological model that has long dominated philosophy. That standard model is individualist in that it treats knowledge solely as the property or capacity of individual knowing minds, and what is good, reliable, warranted, or true in the way of belief is analyzed from the individual knower's point of view. Although there are important exceptions, the dominance of this model cuts across foundationalist, coherentist, internalist, and externalist theories of knowledge.

By contrast, the collectivist model maintains that creating and transmitting KJB can be, and is, a collective enterprise, rather than solely an individual one. Collectivist theorists maintain that KJB is far too complex for an individualist account to explain fully. Hardwig, a leading recent expositor of this view, puts the intuition in a compelling way:

> I find myself believing all sorts of things for which I do not possess evidence: that smoking cigarettes causes lung cancer, . . . that mass media threaten democracy, . . . that my irregular heart beat is premature ventricular contraction, that students' grades are not correlated with success in the nonacademic world. . . . The list of things I believe, though I have no evidence for the truth of them is, if not infinite, virtually endless. And I am finite. Though I can readily imagine what I would have to do to obtain the evidence that would support any one of my beliefs, I cannot imagine being able to do this for *all* of my beliefs, I believe too much; there is too much relevant evidence (much of it available only after extensive, specialized training); intellect is too small and life is too short.[56]

There is indeed reason to believe that an individualist model cannot explain how we, as nonexperts, acquire KJB about a great many of the things we firmly believe we do hold *as* KJB. There is even reason to question whether the individualist model can adequately explain the KJB putatively possessed by scientific experts. Consider, for example, that empirical scientists routinely and increasingly depend on computers in constructing, verifying, and falsifying their theories. But they are usually not able themselves to verify or even to comprehend—without a good deal of additional training that they probably do not have the time to acquire and do not feel the need to obtain—the theoretical work that guides computer scientists when they build the computers and write the programs on which the empirical scientists rely. Thus, these scientists may be said to rely on—i.e., epistemically defer to—the work of computer scientists without understanding the details of the computer scientists' work. Even so, one would have to traverse a fair way down the path of skepticism to conclude that these deferring scientists do not have KJB about, or epistemic competence within, their own fields just because of their limited understanding of the computer science that underlies the technology on which they rely.[57]

What might a nonindividualist account of KJB look like? Hilary Putnam sketches such a theory in some brief but rich remarks about what he calls the "division of linguistic labor."[58] Putnam addresses the question of how experts and nonexperts in a given society manage to use "expert" terms—terms that come within the special province of expert theoretical disciplines, such as the empirical sciences—with the same *meaning* even though the nonexperts are not themselves competent to know the meaning of those terms. How, for example, can nonscientist members of modern English-speaking cultures routinely use the terms 'water,' 'gold,' and 'beech tree' without knowing the chemistry, physics, or biology required to understand what such terms really refer to (so as to be able, say, to distinguish "fool's gold" from gold)?

Putnam proposes a "sociolinguistic hypothesis" to explain this phenomenon. There is, he claims, a "division of linguistic labor" in which most members of the society acquire such words as 'gold' and 'water' as part of their general vocabulary, while only a subclass of those members acquire the expert methods of recognizing whether a given item is within the scope of those words or not. That is, even though not himself an expert, an epistemic pedestrian can use "expert" terms with the same meaning as an expert:

> He can rely on a special subclass of speakers. The features that are generally thought to be present in connection with a general name—necessary and sufficient conditions for membership in the extension, ways of recognizing if something is in the extension ('criteria'), etc.—are all present in the linguistic community *considered as a collective body.* . . .
>
> . . . .
>
> Every linguistic community exemplifies the sort of division of linguistic labor just described: that is, possesses at least some terms whose associated 'criteria' are known only to a subset of the speakers who acquire the terms, and whose use by other speakers depends upon a structured cooperation between them and the speakers in the relevant subsets.[59]

Putnam's brief discussion offers at least the start of a promising account of how a community as a whole can be said to have expert knowledge even when not every member has the requisite expertise. In a community that includes both experts and lay people, the expert might be said to exercise epistemic authority over the meaning of terms whose meaning is discerned only by the use of specialized expert methods.[60]

Hardwig takes the collectivist theory much further than does Putnam, making some remarkable concessions to his individualist challengers along the way. Hardwig entertains, and comes quite close to endorsing, the idea that a nonexpert, *B*, can acquire *knowledge* of some proposition *p* (and not the more inclusive KJB on which I have been focusing) from an expert, *A*, even under the following conditions: (1) *B* has not performed the inquiry capable of providing evidence for the truth of proposition *p*; (2) *B* is not competent, and could never become competent, to do so; (3) *B* is not competent to assess the merits of the reasons that expert *A* offers for his opinion; and (4) *B* cannot understand what *p* means.

In Hardwig's view, it makes sense even under these conditions to say both that B's belief is rationally justified *and* that *B* knows that *p*. His principal argument for the first of these remarkable conclusions (regarding rationally justified belief) is just that

> we *must* say that *B*'s belief is rationally justified—even if he does not know or understand what *A*'s reasons [for believing that *p*] are—if we do not wish to be forced to conclude that a very large percentage of beliefs in any complex culture are simply and unavoidably irrational or nonrational.[61]

He makes the same basic argument regarding knowledge, pointing to common cases of scientific practice in which "each researcher is forced to acknowledge the extent to which his own work rests on work which he has not and could not . . . verify":[62] "Unless we maintain that most of our scientific research and scholarship could *never*, because of the cooperative methodology of the enterprise, result in knowledge, I submit that we must conclude that the *p* is known in cases like this," in which "each researcher is forced to acknowledge the extent to which his own work rests on . . . work which he has not and could not . . . verify."[63]

There is much intuitive appeal in the collectivist view. It does seem that nonexperts can, and do, acquire KJB concerning a great many propositions about complex scientifically articulated subject matters whose warrant can be ascertained only by the use of specialized tools that they do not possess and cannot realistically acquire.[64] But collectivist accounts like those sketched by Hardwig and Putnam are just that—sketches—and they too (Hardwig's most clearly) offer up epistemological pills that are hard to swallow. Basically, Hardwig offers a collectivist argument that has the structure of a *modus tollens*:

> (1) *If* justified epistemic deference to experts is not possible, *then* a very large percentage of beliefs in any complex culture is irrational or nonrational;
> (2) It is not the case that a very large percentage of beliefs in any complex culture is irrational or nonrational;
> *Therefore*,
> (3) Justified epistemic deference to experts is possible.

Is this argument compelling? Clearly Hardwig believes (as does Putnam) that justified epistemic deference is possible. But Hardwig's *modus tollens* leans very heavily on a parade of epistemic horribles: "If justified deference is not possible, then see how irrational our culture would be!" Although such a parade delights the skeptic and strikes fear in the heart of every epistemically responsible citizen, *modus tollens* cannot suffice where explanation is lacking. Aside from standing on some strong intuitions, for all Hardwig and Putnam have explained, we have no less justification for running the inference this way:

> (1) *If* justified epistemic deference to experts is not possible, *then* a very large percentage of beliefs in any complex culture is irrational or nonrational;

(2) Justified epistemic deference to experts is not possible;

Therefore,

(3) A very large percentage of beliefs in any complex culture is irrational or nonrational.

One philosopher's *modus tollens*, as the way would have it, is another's *modus ponens*. Collectivist arguments like Hardwig's and Putnam's point to a conclusion that seems compelling: In a complex society, the epistemic whole is greater than the sum of its parts, and the collective *as such* can know, or at least believe justifiably, far more than any member of the collective could possibly know individually. They point to something we think is true, something that many would like to believe is true, but without offering any explanation of how it could be true. Yet it is precisely the account of *how* collective knowing is possible that we seek to provide in our philosophical explanation.[65] Without more, these collectivist theories do not provide the necessary analysis.

To be sure, Hardwig's intuition that testimony on such matters *does* yield KJB, even under his striking conditions of hearer incompetence, is compelling, at least prima facie. But is Hardwig entitled, as a matter of epistemological debate, to his *modus tollens* presumption, so that the burden of proof is on the person who would deny that scientific testimony to a starkly incompetent hearer (as Hardwig concedes many of us to be vis-à-vis the testimony we receive) can yield KJB in that hearer? In my view, Hardwig is not entitled to that presumption, Although the intuitions on which Hardwig leans so heavily are somewhat compelling, we also have compelling *experiential* reason to believe that a good deal of testimony, including testimony by experts, is false. Surely Fricker and Quine and Ullian are on target in this respect. We have too much evidence of incompetence, dissembling, and epistemically distorting bias for the burden of argumentative proof to be set as Hardwig seems to claim. We need not even enter the courtroom to flood ourselves with memorial evidence of the "stern lesson" that "[p]eople disguise the truth in certain situations, whether out of deviousness, self-deception, ignorance, or fear. They also, of course, misremember, misjudge, and misreason."[66]

In any event, lest my specific focus here on expert *scientific* testimony become lost in more general epistemological considerations, I leave my reaction to collectivist accounts with this contention: Hardwig's *modus tollens* can be of no help to a practical legal reasoner who is faced with *actually competing* scientific experts.[67] Short of making radical revisions to the logical principles they are willing to accept, nonexperts cannot believe all scientific experts when those experts testify in contradictory or contrary ways.

Other epistemological accounts of expert scientific testimony, specifically those that focus on testimony in legal settings, either do not endorse, or at least do not focus on, collectivist epistemic assumptions. Instead, their central concern is the manner and setting in which expert testimony is presented to non-experts in a contentious litigative context.[68] Proponents of what I call "extra-cameral" accounts believe that although there are obvious problems with the orderly, epistemically justified reliance on expert testimony in adversarial legal systems, these are not problems that *inhere* in the attempt to transfer expert information from experts to nonexperts. Instead, these are problems that occur only when expert testimony is presented in an adversarial context. Accordingly, these accounts propose to remove expert testimony from the courtroom (the *camera*) and to have nonexperts defer to such testimony as neutral testimony by relatively disinterested experts.

Anthony Kenny provides a philosophical theory of this type, one that is specifically directed to the question of the reliability of expert evidence in courts. Kenny begins his analysis by specifying four criteria that a putatively expert scientific discipline must satisfy in order actually to be an expert *scientific* discipline. First, it must be consistent.[69] Second, it must be methodical.[70] Third, it must be cumulative.[71] Fourth, it must be predictive and therefore falsifiable, in a special sense.[72]

Having articulated the conditions that a discipline must satisfy to qualify as "scientific," Kenny considers the problem of competing expertise in the courtroom. With regard to novel (putatively) scientific expert methods, he observes, parties usually battle out in court the question whether the discipline to which the method belongs is really a science. He argues that because judges and juries lack the institutional competence to judge whether a new field really is a science, "the courtroom is not the best place, and the adversary procedure is not the right method, to decide what is and what is not a science."[73] Like many legal commentators,[74] Kenny maintains that, within reasonable limits of time and money, expert testimony ought to be an effort to discern the truth, but that the adversary system overwhelms the effort.[75] This leads to his principal recommendation:

> To remedy the abuses in the giving of expert evidence we should:
>   . . . .
>   . . . remove from the courts the decision as to whether a nascent discipline is or is not a science capable of providing expert evidence. A register should be set up of such disciplines, and those claiming to have developed a new science should seek admission to the register. . . . The essential thing is that the matter should be decided not by a judge or banister in haste, but by experts in adjacent disciplines at leisure.[76]

Kenny's register would function as a kind of super-credential, the very existence of which would lend *credence* to the expert judgments issuing from those experts who were duly registered. But his and other extra-cameral proposals[77] can neither surmount nor resolve problems of selection and competition. They perhaps relocate those problems to a different link along the epistemic chain from expert to nonexpert, but they do not resolve the problems or explain how the nonexpert practical reasoner can handle selection and competition in a nonarbitrary manner. That is, an extra-cameral proposal like Kenny's would succeed only in relocating the competition among experts from within the courtroom, where the question is either whose testimony to *admit* and subsequently whose to *credit* among those whose testimony has been admitted, to some antechamber in which the question will be *who among the competing experts is to be appointed to the external commission?* What criteria will be used to select them, and how could this credential advance the process of reliable deference by a nonexpert to an expert? Without more, we have no reason to believe that courts (or, for that matter, legislatures) are more competent to make that judgment reliably than we have reason to believe that judges and juries can reliably defer to experts inside the courtroom.

C. POINT: THE FOUR POSSIBLE ROUTES OF WARRANTED EPISTEMIC DEFERENCE BY NONEXPERTS TO EXPERTS

I have just canvassed a few epistemological accounts of testimony in general and a few accounts of epistemic deference to scientific experts in particular, but found little help in explaining how the nonexpert practical reasoner can handle the problems of selection and competition described above. I now change tack. I shall consider directly the four principal reasoning mechanisms that nonexperts, unpossessed of epistemic competence, seem forced to deploy when they defer epistemically to experts. One is the nonexpert's *epistemically substantive judgment* about the scientific evidence in question. A second is what we may call *general canons of rational evidentiary support*. A third is the expert's *demeanor*, either as he appears before the nonexpert in person or as indicated by such quasi-literary marks as the tone and authoritative style of written submissions to the court.[78] The fourth is the expert's *credentials*. I shall consider these four methods in the order just listed.

1. First Route: Substantive Second-Guessing in Practical Epistemic Deference

Substantive second-guessing of the expert's judgment seems an unlikely route to rationally cogent epistemic deference. After all, the nonexpert turns to the expert precisely because the former does not have the substantive training, and consequent capacity for expert judgment, that the latter has. Indeed, the more a nonexpert relies on his own substantive assessment of scientific evidence, the less one can be said to *defer epistemically*. Though this method of nonexpert quasi-deference is thus odd to consider at all, I do so for two reasons. One is that, on the model of epistemic deference developed earlier in this Article, nothing in principle rules out epistemic deference from epistemic equals or near-equals, so that nothing in principle prevents a suitably epistemically qualified practical reasoner from second-guessing the expert on the merits of his testimony. Such second-guessing does narrow the scope of the deference *as* deference, but on my account, deference is a matter of degree and not an all-or-nothing affair. Even so, considering this to be a mechanism of *practical epistemic deference* is still odd, because my stipulated concern is with the practical reasoner who is *not* epistemically competent in the scientific subject matter about which the expert testifies, even though the practical reasoner has decided that the subject matter of that testimony is rationally pertinent to a case before him and that information from the expert in that discipline is therefore worth hearing.

But there is a second reason for considering the "second-guessing" as an option for even the nonexpert practical reasoner in evaluating expert testimony: At least some prominent legal systems, including the American federal system, seem to *require* it or at least come asymptotically close to doing so. That is, they seem to require that the nonexpert judge *select* experts, "defer" to experts, or *choose* among competing experts, on the basis of an *epistemically substantive judgment* about the merits of an expert's proffered testimony.

The Supreme Court's *Daubert* opinion is a high-profile culprit here. Simply put, it instructs judges to make their own *independent* judgment about the scientific reliability of proffered expert scientific testimony. Recall that Judge Kozinski was the appellate judge from whose court the *Daubert* case went to the Supreme Court, and to whose court the case was remanded. His words are worth repeating:

> [T]hough we are largely untrained in science and certainly no match for any of the witnesses whose testimony we are reviewing, it is our responsibility [according to the Supreme Court's *Daubert* decision] to determine whether those experts' proposed testimony amounts to "scientific knowledge," constitutes "good science," and was "derived by the scientific method."[79]

To be sure, the Supreme Court's *Daubert* opinion qualifies the task of the judge with the declaration that

> [t]he inquiry envisioned by Rule 702 is, we emphasize, a flexible one. Its overarching subject is the scientific validity—and thus the evidentiary relevance and reliability—of the principles that underlie a proposed submission. The focus, of course, *must be solely on principles and methodology, not on the conclusions that they generate.*[80]

The distinction between methodology and results of methodology is an important one. It is, for example, central to the model of scientific rationality presented by Laudan, on which I have relied in explicating the notion of axiological aims and point of view. Nevertheless, there are at least two reasons that an inquiry into methodology is no less epistemically demanding than an inquiry into the application thereof. First, in a significant percentage of cases, the boundary between methodology and application is indistinct. Second, even when the methodology is sharply distinguishable from its application, it is no less likely to be the *methodology* that presents the epistemically troublesome barrier to the nonexpert's comprehension than it is the theory that underwrites and motivates the methodology.

I take it, then, that a solution to the problems of selection and competition that counsels "make up your own mind, nonexpert" cannot be adequate. Indeed, because this is so obviously an unsatisfactory solution, it seems likely that many judges would be led to convert what is on the surface a substantive inquiry by nonexpert judges—as directed by *Daubert*—into a form of deference based on demeanor and credentials. This is predictable from the armchair, and there is evidence from the practice of courts that this has indeed been happening in at least some federal courts under the *Daubert* regime. Some courts have essentially converted the *Daubert* test into the old *Frye* test, which, in turn, rests on assessing the credibility of persons who have the "credential" of being members of the "scientific community."[81]

2. Second Route: Using General Canons of Rational Evidentiary Support

Sometimes an expert's testimony is afflicted by a kind of rational incoherence that a nonexpert can discern even without training in the expert's field. One of the clearest examples is self-contradiction. Consider *People v. Palmer* (*Palmer II*),[82] a criminal prosecution in which the state's expert medical witness was called to testify about whether the defendant, on trial for having stabbed his victim to death, was sane at the time of the stabbing. The medical expert testified that the bizarre behavior the defendant showed around the time of the stabbing was attributable to the defendant's faking his psychosis, that the defendant *was* "grossly psychotic" on the day of the stabbing, that the defendant "was able to distinguish right from wrong and could conform his conduct to the requirements of the law," and that the defendant "was not mentally ill."[83] Reviewing this testimony, an Illinois appellate court held that the doctor's self-contradictory statements did not provide *any* evidence of the defendant's sanity.[84]

Self-contradiction in testimony, as in other assertions, is a hallmark of rational incoherence. It is not the only kind, however. Consider, for example, the events and legal proceedings leading to the Kentucky Supreme Court's decision in *Potter v. Eli Lilly & Co.*[85] *Potter* arose from a former employee's shooting and killing or wounding of several of his coworkers. The employee had been taking the antidepressant drug Prozac, and the victims or their estates brought a products liability action against Eli Lilly, the drug's manufacturer. The precise issue at the trial was whether Prozac was unreasonably dangerous and defective and whether it *caused* the employee to kill or injure the plaintiffs.[86] At trial there was a considerable amount of expert testimony about whether Prozac caused the defendant to commit the shootings. After hearing all the testimony, the jury decided for the defendant.[87]

A person who is not an expert in medicine or psychopharmacology, and who is called upon to assess expert testimony in a case like *Potter*, could, and indeed should, heed the distinction between a drug's *causing* certain behavior and that drug's *not preventing* that behavior, and impose a suitably heavy burden of proof on any expert's argument that does not respect that distinction or present an explanation in light of it.[88] Similarly, one need not be an expert in psychiatry to discern the rational problem in the *Palmer II* expert's self-contradictory testimony, and the nonexpert court in that case properly discounted it severely. The avoidance of self-contradiction and the mindfulness of the distinction between causation and nonprevention are two examples of what we may call *general canons of rational evidentiary support*. One who is not expert in a given discipline can sufficiently understand and use canons like this to assess expert testimony in an expert discipline.[89] It is beyond this Article's scope to enter into detailed discussion of such canons, though such a discussion would, I think, be of considerable interest and value. Here, it is enough to identify the existence of such canons, with the two quick examples presented above, and consider the extent to which such canons might enable a nonexpert to evaluate expert testimony in a rationally cogent manner.

To answer that question, let us identify a spectrum along which there are varying degrees of *obscurity* in failures of rational evidentiary support. The obscurity of a message is in the ears of its hearer, and the obscurity with which I am concerned here is the obscurity *to* a nonexpert *of* the failure of rational coherence in an expert's testimony. At one end of the obscurity spectrum are those failures that are *least* obscure and are therefore *easiest* for a nonexpert to discern in an expert's testimony. At this end, or close to it, would be the expert's testimony in *Palmer II*—assuming that the nonexperts in that case readily understand (as they seem to) that a person cannot be both "grossly psychotic" and "sane" at the same time. At the other end of the spectrum are those failures that are *most* obscure, and are therefore *hardest* for a nonexpert to spot in an expert's testimony. Much closer to this end would be, for example, the testimony of Dr. Bruce Weir, a statistician and population geneticist who served as a prosecution expert witness in the O. J. Simpson case. In his original testimony, Weir gave the jury an analysis that failed to account for certain DNA characteristics in crucial blood samples, characteristics that were *possibly* but not *definitely* present in some of those samples.[90] The result of this error, as Weir later conceded under cross-examination, was that his analysis of the probabilities that various combinations of randomly selected individuals would have certain DNA types, was inaccurate and biased (in a probabilistic sense) against the defendant.[91] It seems reasonable to regard this failing of probabilistic statistical analysis as a failing in the rational coherence of this expert's testimony, but it is a failing that would require a good deal of sophistication in the expert's field to discern.

With the spectrum identified, we can assess the question: In what percentage of cases is it likely that nonexperts will be able to evaluate rationally expert testimony by deploying general canons of rational evidentiary support? For at least three related reasons, I speculate (and do not claim greater certainty) that it is only a relatively small percentage. First, it seems likely that failures of rational coherence in an expert's testimony will most often be closer to the *obscure* end of the spectrum than to the nonobscure end. *Palmer II* seems the unusual case; Weir's testimony in the Simpson case, the more usual. Second, and relatedly, many expert witnesses are repeat players in the "game" of giving expert testimony. It would not take long for word to get out to the trial bar about experts whose testimony has been so unartful as to appear to a nonexpert, nonobscurely, insufficient to satisfy general canons of rational evidentiary support. Indeed, one would hope that someone "qualified as an expert" by a court would know enough about the substance of her field not to render such testimony. Third, when the failure of the expert testimony does occur toward the more obscure end of the spectrum, it is left to a sufficiently tutored opposing counsel, perhaps aided by his own opposing expert, to point out the failing to the jury; this is precisely what happened with Weir's testimony. But the more obscure the expert's failure of rational coherence, the more comprehension of the expert discipline one must have in order to see that it is a failure. The nonexpert will thus be at a serious epistemic disadvantage in discerning such failures. This problem will be exacerbated in the many cases in which, unlike Weir, the expert *denies* that his testimony suffered from any failure of rational coherence. In those cases, the nonexpert will be little or no better off deploying general canons of rational evidentiary support than he would be in trying to make a substantive assessment of the expert's testimony. For, in such cases, it will be difficult for the nonexpert to judge *on the merits* whether there really is a failure of rational coherence, because it will require considerable information about the expert's discipline even to know whether there was such a failure, and this is precisely the kind of information the nonexpert is, by hypothesis, unlikely to have.[92]

3. Third Route: Evaluating Demeanor in Practical Epistemic Deference

As I discussed in Section II.B, several philosophers have remarked on the importance of a speaker's demeanor as an epistemic tool by which a hearer assesses and evaluates that speaker's testimony, Fricker makes the rebuttable presumption of sincerity on the part of a witness a centerpiece of her account.[93] Were Quine and Ullian to offer a more thoroughgoing account of testimony consistent with Quine's general epistemology (an ambition they seem to have in the joint work),[94] they would surely argue that it is both possible and necessary for a language-learner to distinguish sincere assertions of belief by an expert from insincere ones. Hume, too, was sensitive to the dynamics of demeanor in his treatment of testimonially acquired KJB.[95]

There is indeed a proper epistemic role for judgments of demeanor in the assessment of testimony, both expert and nonexpert. One might even go so far as to conclude that a hearer's capacity to assess accurately the demeanor of a witness is a necessary condition of the acquisition of KJB from that witness. The hearer should be able to discern whether the speaker is dissembling, and, when she is dissembling, the hearer will usually decline to endorse the testimony reported. Even when the hearer discerns that the speaker is dissembling, however, he need not reject her testimony, for it is possible that the prevaricating witness may bungle the effort to lie in a way that still allows the testimony to provide useful evidence.

But assessment of demeanor is unlikely to be accurate enough *in general* to provide a basis for an explanation of how a nonexpert can acquire KJB from an expert. Demeanor is an especially untrustworthy guide where there is what we might call a lucrative "market" for demeanor itself—demeanor has "traded" at high prices since the days of the sophists and finds exceptionally robust business in adversarial legal systems. When judges and juries use demeanor as a test for the credibility of expert evidence, they face this severe difficulty: Epistemic warrant and persuasiveness diverge, *especially* when the "persuadee" has too limited an epistemic capacity to be able to assess competently the epistemic warrant of testimony *independently* of the criteria that make an expert seem persuasive.

Demeanor is perhaps the chief of these criteria, as Aristotle explained.[96] He maintained that there are three principal means of persuasion—appeal to reason (*logos*), appeal to emotion (*pathos*), and appeal to the character of the speaker (*ethos*)—and that, of these three, *ethos was the most effective*, more effective than appeal to reason (not surprising, perhaps) or appeal to emotion (surprising, but the point does survive careful reflection). The demeanor of an expert witness is precisely what Aristotle referred to as the *ethos* of that "speaker." But as theorists of KJB, we have no reason to believe that an expert witness's persuasive demeanor has any particular connection to the epistemic warrant for what the witness asserts. This is the basic epistemic obstacle to the use of demeanor as a route to legitimate practical epistemic deference in legal systems (like the American) in which there is a market for demeanor. Judges, lawyers, and commentators are thoroughly aware that lawyers choose expert witnesses at least as much because they will appear to a jury to be competent as because (in the lawyer's judgment)[97] the experts actually are competent. A brief look at some of the actors in this market is instructive.[98]

A 1967 survey of judges, lawyers, and doctors in the Los Angeles area found that "[o]ver three-quarters of the attorneys responding indicated that some factor other than medical expertise—usually an impressive 'courtroom manner'— often determines the choice of an expert witness."[99] An article in a litigator's trade magazine describes the selection of experts as follows:

> Usually, I like my expert to be around 50 years old, have some grey in his hair, wear a tweedy jacket and smoke a pipe. . . .
>
>   . . . .
>
> You must recognize the jurors have prejudices, and you must try to anticipate those prejudices. . . .
>
> *Some people may be geniuses, but because they lack training in speech and theater, they have great difficulty conveying their message to a jury.*[100]

Another trade magazine instructs that an expert witness must "exude confidence, create empathy, and seem and be completely sincere and convincing."[101] A law review article similarly notes that "the selection process involves more than securing an expert who will render a favorable opinion. The credibility and persuasiveness of an expert are equally important concerns."[102] Another author adds that "[t]he best expert is one . . . who has . . . qualities that give a certain 'glow' to an otherwise acceptable position. In fact, the best expert testimony in the world may be utterly useless unless it is presented by someone whose other attributes can add a ring of truth to it."[103] Lawyers are of course being prudent when they choose experts who will favor their side *and* who will do so with convincing demeanor. But because, *ex hypothesi*, nonexpert factfinders do not have epistemic competence, they are not in an epistemic position to be convinced by substantive arguments—that is, they cannot be convinced by what Aristotle called the reason (*logos*) behind an expert judgment because they cannot understand those substantive arguments.[104] In an adversarial setting, the criterion of being convincing to the nonexpert subordinates the criterion of being competent to produce accurate expert scientific judgments.[105] My point here is not that the market for demeanor is biased in selection toward false or ignorant experts. Rather, I am suggesting that given what the market is selecting for (something that will convince jurors who are not competent to judge substantive scientific argument), there is no reason to believe that epistemic warrant has any particular connection with what chosen experts will say.

4. Fourth Route: Evaluating Credentials

Credentials (along with reputation, which I treat as a species of credentials) is probably the most important device relied upon by philosophers and jurists to explain how practical epistemic deference can yield KJB. Collectivist and extra-cameral accounts of epistemic deference share the assumption that nonexperts can use credentials to acquire KJB from experts. This same basic assumption also motivated the *Frye* rule, a rule that many evidence scholars still support. When coupled with a fairly accurate capacity to use an expert's demeanor as a guide to his sincerity, the nonexpert's judgment of credentials seems the most promising means by which to explain how a nonexpert can acquire KJB from an expert scientist.

I will now investigate this mechanism to see whether it is likely to produce KJB in a nonexpert. I assume for the purpose of this investigation that the epistemic device of credentials is coupled to that of demeanor, and that the nonexpert who uses demeanor "evidence" is a fairly accurate judge of when the expert is being sincere. The simple reason for these related assumptions is that if indiscernible prevarication were a habit among experts, credentialism alone would be incapable of generating KJB. By assuming that the credentialed expert is sincere and is accurately perceived as sincere by the nonexpert, I am able to consider credentialism as a potential route to KJB in the setting in which credentials are likely to be most helpful.

The use of credentials (assuming sincerity) as a route to practical epistemic deference has both strengths and weaknesses. I begin by mentioning some of the weaknesses and then setting these against the strengths. There are at least three closely related reasons that use of credentials seems unlikely to provide the nonexpert practical reasoner with KJB: regress, question begging, and underdetermination. These reasons are so closely related that traversing the path along any one of them soon leads to one or both of the others.

A. THE REGRESS PROBLEM

Regress can involve one or more of the *selection* problems, or one or more of the *competition* problems identified in Section I.C. Recall that the selection problems are: (1) determining which of the intellectual enterprises that might yield expert testimony is a *science*; (2) determining who is a *scientist* capable of using her science in a manner that satisfies the standard of epistemic appraisal and the attendant level of confidence that the practical reasoner has established; (3) determining which of the intellectual enterprises that might yield expert testimony is a science that is *rationally pertinent* to the case; and (4) in cases in which there is significant doubt occasioned by task (3), determining who is capable of answering (3) in a way that can identify an expert scientific discipline capable of satisfying the chosen standard of epistemic appraisal and the attendant level of confidence. For the purposes of this discussion, it is not important to determine exactly which of these selection problems the nonexpert faces, for the regress problem can arise regarding each one (and several, though perhaps not all, combinations of them).

The basic problem of *competition* is how a nonexpert can rationally decide which of the competing experts (whose competition is intra-disciplinary or extra-disciplinary, actual or implied) to believe when the nonexpert is not competent to assess the substantive merits of the experts' competing arguments. When experts testify to contrary or contradictory propositions, the nonexpert must decide whom to believe on the scientific issue. But, *ex hypothesi*, the nonexpert does not have sufficient competence in the expert discipline to be able to make the choice on substantive grounds, so on what rational basis can the nonexpert make that choice?

A solution to these related problems commonly offered by both jurists and philosophers is to maintain that nonexperts can and do acquire KJB from experts by relying on *credentials*. Kenny's solution to the problem of expert testimony, for example, is a kind of super-credentialism.[106] Although Putnam and Hardwig are less explicit about it, their nonexpert epistemic communalists would presumably consult credentials to identify the "scientist" whose referential terms are to be deferred to (Putnam) or who is the expert to be accorded deference in the collective (Hardwig).[107] This same basic approach underlies the *Frye* rule that *Daubert* displaced, and even after *Daubert* many federal and state courts have gravitated back to it.[108] Typical of post-*Daubert* endorsement of *Frye*'s "credentialist" solution to the problem of expert evidence is this assertion by an evidence scholar:

> *Science is the only source of its own reliability.* Anything less than complete deference to the weight of credible scientific opinion concerning the reliability of scientific evidence means going outside science—to the judge or jury to resolve a scientific dispute. The resulting judgment cannot be scientific and therefore we cannot honestly speak of the evidence as having "scientific" reliability. . . .
>
> . . . .
>
> . . . [T]he "real" issue is whether good scientists consider the evidence reliable at this time.[109]

These commentators argue that the more modest task that the credentialist solution sets for judges[110] is much better suited to their limited capacities to understand complex scientific evidence.[111] Whereas *Daubert* requires courts to judge whether some given evidence *is* scientifically reliable, what I call the "*Frye* solution" is to have judges ask whether scientists *think* that the evidence is reliable.

But *which* credentials indicate membership in the scientific community? An Ed.D.? A Th.D.? A Ph.D. from a correspondence school? A degree in "creation science"? Analogously, would a Ph.D. in philosophy from such an unlikely place as the University of Pittsburgh[112] be worthy of respect were one looking for philosophical expert testimony? Clearly, in an age in which formal credentials have significant market value in many different kinds of markets, the non-expert needs guidance through the thicket of would-be experts, wannabe experts, and magic elixir mongers. Here, a new solution presents itself (actually, once again, it is the old solution relocated): Have the nonexpert judge or jury consult some kind of "meta-expert" (someone who is an expert *about* expertise in a given area) for a list or specification of the credentials that a nonexpert could reliably use to pick out competent experts in the scientific field. But then how can the nonexpert rationally identify the proper "meta-experts"? On what basis is the nonexpert to identify those meta-experts who have KJB *about* the proper credentials? Must not the nonexpert rely on credentials (including reputation) to identify the appropriate meta-experts as well?

Thus, it seems that the "*Frye* solution"—ask those with the credentials of science whether an expert has the credentials of science—threatens to slide into an epistemically unworkable regress. Using credentials only pushes the inquiry back a step without resolving the basic problem. If nonexpert judges and juries are not competent to judge the content of expert information, how are they going to be competent to judge credentials of those who would give expert information? If the answer is to ask credentialed "meta-experts" what the proper credentials are, the regress has begun. Yet the "*Frye* solution," so common in one form or another in analyses by jurists and philosophers, cannot let the regress slide on infinitely *if* these analysts truly intend to explain how the nonexpert's use of credentials can serve as a *possible* or *actual* means of acquiring KJB from an expert. Nonexperts certainly cannot indefinitely continue to ask expert after expert about proper credentials. Somehow, the regress must be stopped.

B. THE QUESTION-BEGGING PROBLEM

Can the nonexpert stop the regress without vicious question begging? In many—I do not say all—cases, credentials will be of little help, for they will either reproduce the problem (leading to regress) or cause the nonexpert to settle on one competing expert without having a good reason (thereby begging the question). We have seen how the credentializing solution (like the *Frye* solution) can lead to regress. Such a solution can also lead to question begging, as the following example indicates.

McLean v. Arkansas Board of Education[113] was one of the early federal creation science cases dealing with the constitutional merits of a statute that mandated equal classroom time for evolution and "scientific creationism." Relying on expert testimony, the district judge concluded that "scientific creationism" could not pass muster under the Establishment Clause because *it was not science* but a religious doctrine that public schools could not promulgate. But assuming that the judge was not an expert on science (or, for that matter, philosophy or religion), whom should he have asked about the identity of the appropriate scientific expert on whether creation science is real science—a creation scientist or a Darwinian? Either answer begs the question.[114] Moreover, to make matters at the very least somewhat more difficult for the nonexpert, "creation scientists" are quite careful to proclaim their own credentials *as scientists*. Indeed, the very use of the terms 'creation science' and 'Christian science' are efforts at establishing credentials. Philip Kitcher criticizes creation scientists for just this kind of "credential mongering":

> Creationist claims about credentials look better when presented in soft focus. . . . On closer inspection, the "21 scientists who believe in Creation" [listed in a creationist publication] hardly constitute a distinguished panel of experts on the origins of life: three hold doctorates in education; two are theologians; five are engineers; there is one physicist, one chemist, a hydrologist . . . one entomologist, one psycholinguist, and someone who holds a doctorate in Food Science Technology; finally, there are two biochemists . . . an ecologist, a physiologist, and a geophysicist. While the last five may have some expertise in related areas, the credentials of the others are utterly irrelevant to many of the questions Creationists address. The "authority" of these men should not convince us that there is a scientifically reputable alternative to a major *biological* theory. The word of just any "scientist" is not enough. I am prepared to bet that Creationists, like the rest of us, take care to consult the *appropriate* experts. I doubt that they take their sick children to the vet.[115]

One ought to concede to Kitcher that it would be far too quick a skepticism to conclude from the possible fact of disagreement that no one can ever know the truth of the matter or have decisively good reasons that overcome other reasons, and I do *not* draw that conclusion here. That is, nothing in my argument denies that there are good, or even decisive, arguments on such questions as whether creation science is "real" science.

But my current concern is not whether there are compelling arguments— *expert* arguments such as might be made by a philosopher of science—for the view that creationism is only ersatz science. Rather, my concern is with a nonexpert's epistemic capacity to use credentials as a rationally warranted means of resolving that debate. Kitcher himself helps to make my point on this separate question. After offering the remarks quoted above, Kitcher argues that "the crucial issue is not whether some people who possess doctoral degrees *say* that there is a case for Creationism, but whether they are *right* in saying so. To settle that issue, it is wiser to look at the evidence itself."[116] Very well, but now we have come full circle, for looking at the *nontestimonial* evidence to decide who, among competing "credentialed" experts is right, is *precisely* what the nonexpert is incapable of doing. From the point of view of the nonexpert, relying on Kitcher or Stephen Jay Gould as experts is no less question begging than relying on "Dr. Henry Morris" who proclaims himself to be "recognized as one of America's greatest authorities on scientific creationism," who is "[a]rmed with three earned degrees (including a Ph.D.) in the sciences," and who "served as department head or professor at four famous institutions, Louisiana University, the University of Minnesota, Rice University, and Virginia Polytechnic Institute."[117] My only contention is that nonexperts are not in a position to judge *who* is in a position to judge such matters without begging the question. Whether such question begging is vicious, harmless, or even virtuous, awaits discussion of the role of nonarbitrariness as a practical aim governing the legal reasoner's epistemic deference to experts. I should also hasten to add that, though my example may be an unusual one, it is actually not unrepresentative of the kind of epistemic predicament in which the nonexpert practical reasoner typically finds himself.[118]

C. THE UNDERDETERMINATION PROBLEM

It is really underdetermination that gives rise to both the problem of regress and the problem of question begging rehearsed above. Underdetermination creates a difficulty for credentialist approaches to explaining the legitimacy of epistemic deference to experts: When the credentials of the experts are, to the eyes of the nonexpert, evenly matched for *all the nonexpert justifiably believes*—that is, when they underdetermine the credibility of the competing witnesses—it is very difficult to see how credentials could provide an epistemically legitimate method the nonexpert can use to resolve selection and competition problems.

D. COUNTERPOINT: THE ANTI-SKEPTICAL RESPONSE AND THE DIALECTICAL IMPASSE

The foregoing skeptical considerations,[119] however, may seem too quick, too cheap, too thin. It is unduly skeptical to deny that nonexperts can *ever* use credentials as a tool to acquire KJB on the basis of epistemic deference. Surely, even nonexperts can judge credentials well enough to handle what might be called the *crank factor*,[120] also familiarly recognized as the Flat Earth Society phenomenon. Presumably the admission of crank science is, after all, what *Daubert* seeks to reduce by having the trial court judge play a "gatekeeping" role, using his assessment of scientific reliability as a criterion of admissibility. The skeptic, however, will insist in response: But how often can a nonexpert *justifiably* dismiss an expert's credentials as cranky? Even if the nonexpert's use of credentials could help weed out cranks by weeding out egregiously uncredentialed putative experts, in a great many cases the credentials of experts will remain—to the eyes of the nonexpert—evenly matched.

Let us suppose with the anti-skeptic that there is an epistemically valuable role for credentials in a nonexpert's assessment of expert testimony. What might that role be? Coady provides a modest and reasonable answer:

> We have certifying bodies and institutions and their various certificates and, typically, the courts require that the [expert] witness be shown to have some relevant certification from such bodies. Doubt can arise, of course, about the credentials of supposedly expert institutions . . . but usually the courts do not doubt such credentials. Were they to require for every such certifying body some proof of its credentials, it is hard to see what could be forthcoming, other than more of the same.[121]

Universities, colleges, and scholarly associations are the leading examples of "certifying institutions" that play the credentialing role. Certainly in our everyday lives and work we make many judgments on the basis of credentials from these institutions. And certainly many feel that this manner of credentialism works fairly well in those spheres. Our teeth get their cavities replaced; our computers get their hard drives fixed (and built in the first place) and run programs that perform wonders—wonders verifiable to all but Cartesian skeptics; our microwave ovens verifiably cook food in very few seconds; our cars and planes verifiably carry us at high speeds or altitudes; our home entertainment systems verifiably deliver larger and larger sights and sounds from smaller and smaller systems of audio and visual information. We are aware, indeed, too often reminded, that behind each technological miracle stands a network of credentialed inventors, technicians, and theorists. All of this is powerful evidence that something is not rotten in the world of credentialed epistemic authorities.

Still, questions remain. Coady frankly acknowledges one problem with such certifying institutions—their "inbuilt tendencies towards intellectual conservatism and towards the monopolization of social control over knowledge conceived of as a kind of commodity."[122] That is certainly an important problem, assuming that commodified monopolization of this sort has an epistemically distorting effect, as it surely can. (It is in effect a problem of competition, either intra- or extra-disciplinary, actual or implied.) How is the nonexpert to become aware of, or resolve, this particular problem of competition? Coady offers little in the way of a persuasive answer. He suggests that "the courts should give initial credence to the verdicts of such bodies as universities on the issue of bogus science, though they should be *prepared to hear argument about the matter if it can be produced.*"[123] But what kind of argument should they be prepared to hear, and how would they assess it if they heard it? What is Coady supposing the nonexpert *can* do in making such an appraisal—that she can do a *better* job of adjudicating between competing experts (including would-be monopolists and those seeking to raise the barriers to epistemic entry) than the experts themselves can do? As long as judges and juries lack epistemic competence, what reason do we have to believe that they will be able to make a justified judgment about the arguments they are "prepared to hear" that contravene the judgments of the certified experts?

Before moving on, let me sum up the problems that remain for credentialism as a proposal for solving the related problems of selection, competition, and underdetermination. Coady's solution is to have nonexperts give prima facie weight to credentials—principally, memberships in reputable universities and learned societies. This solution could work only if the weight of credentials is clearly on one side; but often the nonexpert justifiably believes that credentials are not so clearly weighted. The underdetermination difficulty with credentials is that when the jury hears contradictory testimony from two experts who seem to have even credentials, these credentials cannot yield even prima facie weight. Coady argues that we can endow credentials with prima facie weight by stamping them with the imprimatur of a leading university or learned society. Three problems remain for this suggestion. First, it brings back the related regress and question-begging problems: Being a famous university in general is not the same as being a leading university in a particular field (the University of California at San Francisco may not be a "leading university," but, we are *told*, it is a leading center for medical research). We can construct an analogy of proportion—UCSF : medical research :: University of Pittsburgh : philosophical research. Indeed, many such examples can be found. In order for the nonexpert to find out which institutions were "leading" in the proper way, she would have to find an expert already endorsed by one, thus either begging the question or regressing. Another problem lingers even after the credentialist solution. It is not clear how that purported solution can handle cases of *intra-disciplinary* competition of credentialed experts. Such competition occurs when two sets of witnesses who purport to be expert in the same area (e.g., statistics, economics, genetics, epidemiology) *and* whose credentials are evenly matched testify to contradictory or inconsistent propositions. Nor is it clear how credentialism can handle cases of *extra-disciplinary* competition of credentialed experts. This type of competition occurs when two sets of witnesses who do *not* purport to be expert in the same area, but whose credentials are evenly matched, offer testimony that is in some way mutually undermining. Kenny provides a charming example, arguing that the concept of the "irresistible impulse," brought into the criminal law by expert psychiatric testimony, is philosophically incoherent and thus unworthy of any serious credence in a courtroom:

> The only remedy for this state of affairs will presumably be for the prosecution to call a philosopher to testify that there cannot be any such thing as an irresistible impulse, and therefore the accused cannot have acted on one, any more than he can have murdered a married bachelor or stolen a square circle. The desperate nature of this proposal will, I hope, bring home vividly the indefensibility of the present state of the law.[124]

And if Kenny, as credentialed as he is in the academic world, can lob such extra-disciplinary firebombs at psychiatry, imagine what the inarguably well-credentialed Paul Feyerabends and Richard Rortys of the world could do to challenge the KJB-producing capacities of physics, biology, and mathematics, were their expert testimony admitted in a trial. Extra-disciplinary challengers like these well-credentialed scholars might not be correct on the merits, but one must wonder how a nonexpert, using the credentials of academic prestige as her guide, could make a justified judgment to reject their arguments. (This kind of competitive testimony would, for the most part, have to be implied, since such testimony is rarely admitted. But is it *justifiably* excluded by a nonexpert judge?)

We reach a dialectical impasse, an aporia—our escape blocked by opposing educated intuitions. On one side is the intuition that credentials *must* be sufficient to enable KJB to arise from practical epistemic deference. On the other side is the (to my mind) equally strong philosophical "intuition" that without epistemic competence, credentials quickly crumble under scrutiny into a pile of regressive, question-begging, underdeterminative heaps of unjustified belief. The way through this impasse, I suggest, is by studying with meticulous care, and *rationally reconstructing*, the step-by-step reasoning process a nonexpert judge or juror must use in order to assess expert scientific testimony in the course of reaching a conclusion in a legal dispute. In my 1998 paper, I showed how the inference process known as *abduction* plays a critical role in that reasoning process. And it is some crucial features of abductive inference that will lead us out of the impasse I have identified. I then presented a model of the reasoning process the nonexpert uses to assess expert testimony, explaining the role of abduction along the way.

Although it is possible to construct a theoretical model under which decision makers can work through the dialectical impasse discussed above,[125] decision makers will, in practice, be crippled by their inability rationally to decide between experts on the basis of their testimony or their credentials. And this arbitrariness will only grow as complex scientific theories continue to grow ever more common and prominent in litigation.

Arbitrary decisions offend due process, whether they are based on racial animus, impossible legal standards, or, as here, on ignorance of the necessary background on which to make informed judgment. As I have argued at length elsewhere,[126] the way in which decision makers evaluate expert testimony offends the principle of "intellectual due process," which, if not enshrined in the constitution, nevertheless animates our legal regime and reflects our basic notions of fairness.

The only solution (actually, it is a family of solutions) I see requires that one and the same legal decision maker wear two hats, the hat of epistemic competence and the hat of practical legitimacy. That is, whether it is a scientifically trained judge or juror or agency administrator, the same person who has legal authority must also have epistemic competence in relevant scientific disciplines. In an age in which culture will increasingly take advantage of the massive intellectual power of science, this is not too high a price for the legal system to pay to satisfy its own just intellectual aspirations.

NOTES

1. 170 YALE L. J. 1535 (1998).

2. Bonjour, for example, appears to emphasize truth in this way: "the distinguishing characteristic of epistemic justification is . . . its essential or internal relation to the cognitive goal of truth. It follows that one's cognitive endeavors are epistemically justified only if and to the extent that they are aimed at this goal, which means very roughly that one accepts all and only those beliefs which one has good reason to think are true." Laurence Bonjour, *The Structure of Empirical Knowledge* (1985), 8.

3. *See, e.g.,* CATHERINE Z. ELGIN, CONSIDERED JUDGMENT (1996).

4. *See* M. F. Burnyeat, *Wittgenstein and Augustine De Magistro,* 61 PROC. ARISTOTELIAN SOC'Y SUPPLEMENT 1, 20 (1987).

5. Id.

6. Id.

7. Catherine Elgin is making a serious inroad on this project. See ELGIN, *supra* note 3.

8. Explanation is important to the conception of understanding I endorse here. Again, Burnyeat puts its well: "Plato, like Aristotle, makes it a condition on knowing or understanding that *p* that one grasps the explanation of *p*. This of course involves seeing the connection between *p* and a whole lot of other propositions, but it is not mere connectedness so much as explanatory connectedness that counts, and it is by way of this thought that Plato and Aristotle reach the conclusion that knowledge in the full sense, i.e., understanding requires the synoptic grasp of a whole-field." Burnyeat, *supra* note 4, at 21.

9. The view of understanding as possession of a synoptic grasp of explanatory relations is not uncommon among philosophers who explicitly discuss the concept. For example, Neil Cooper argues that "understanding is concerned with relations and connections" and

that "[i]t is possible to have knowledge of a bitty or superficial kind, while we only have understanding when we relate or connect bits of knowledge with other bits in a more or less coherent whole." Neil Cooper, *Understanding*, 68 PROC. ARISTOTELIAN SOC'Y SUPPLEMENT 1, 3–4 (1994). To like effect is Catherine Elgin's more ambitious project of articulating an understanding-oriented, rather than a knowledge-oriented, epistemology: "'Understanding' is a better term for the epistemic achievement that concerns us here. Not being restricted to facts, understanding is far more comprehensive than knowledge ever hoped to be. We understand rules and reasons, actions and passions, objections and obstacles, techniques and tools, forms and functions and fictions, as well as facts. We also understand pictures, words, equations, and patterns. Ordinarily these are not isolated accomplishments; they coalesce into an understanding of a subject, discipline, or field of study . . . . Understanding a particular fact or finding, concept or value, technique or law is largely a matter of knowing where it fits and how it functions in a matrix of commitments." ELGIN, *supra* note 3, at 123.

10. *See* Scott Brewer, *Exemplary Reasoning: Semantics, Pragmatics, and the Rational Force of Legal Argument by Analogy*, 109 HARV. L. REV. 923/96 (1996).

11. *See* NELSON GOODMAN, WAYS OF WORLDMAKING 1–7 (1978).

12. *See, e.g.,* Peter Achinstein, *Concepts of Evidence*, 87 MIND 22 (1978). As Achinstein discusses, this conception of evidence is "objective" in that e's being evidence for H_i depends neither on anyone believing e or H_i nor on anything about their relation. *See id.* at 23.

13. Logically and epistemologically weaker forms of competition, such as "incoherence," are possible, but the more easily articulable and understood forms of contradiction and competition will suffice for my analysis.

14. *See* Edmund Gettier, *Is Justified True Belief Knowledge?*, 23 ANALYSIS 121 (1963).

15. Daubert v. Merrell Dow Pharms., Inc. 509 U.S. 579, 587, 590 (1993).

16. *Id.* at 589–90.

17. *Id.* at 590.

18. *Cf. supra* Section I.C (discussing "competition" among experts).

19. *Daubert*, 509 U.S. at 590 (internal quotation marks omitted). Strictly speaking, one cannot, on pain of circularity, explicate the concept of knowledge by referring to "any body of known facts," unless one goes on to give a noncircular explication of what a known fact is. *Id.* In context, the Court may escape this problem if what it is really saying is that "when the Rules of Evidence speak of 'knowledge,' they are actually referring to beliefs supported by good reasons." Even so, the Court speaks a bit too broadly in the quote in the text above. Surely it is not the case that *any* "body of ideas inferred from 'known' facts" satisfies the criteria of even the Court's weaker conception of "knowledge." *Id.* For there are, of course, a great many invalid and otherwise unacceptable types of inference possible from acknowledged known truths. What the Court really has in mind are *good* inferences (those that are "accepted as truths on good grounds," "supported by appropriate validation—i.e., 'good grounds,' based on what is known," and the like). *Id.*

20. *See id.* at 590.

21. Brief for Respondent al 14–15, *Daubert* (No. 92–102) (emphasis added) (second alteration in original). The brief adds the following footnote: "It is, of course, perfectly conceivable for each of several competing scientific (or other expert) claims to be validated to the extent possible at any given time. The very reason that scientists can disagree, and that sci-

entific knowledge advances as it does, is that what is known at a particular moment does not uniquely predetermine all answers to new questions. Accepted standards and available evidence thus may not rule out either of two competing, but well-reasoned, conclusions. By the same token, however, they do rule out some answers. What validation means, therefore— all it can mean, given the ever-evolving body of knowledge—is *good reason for acceptance as true*, based on what is known at the time." *Id.* at 15 n.8. To similar effect are several other passages in the brief. The brief asserts, for example, that Rule 702 requires that "the specific testimony of each expert *have an adequate foundation,* judged by the accepted standards of the expert's field," *id.* at 12 (emphasis added); that "[i]t is the judge's fundamental duty under the Rules to screen evidence for admissibility to ensure that the body of evidence *provides a rationally reliable basis for judgment,*" *id.* (emphasis added); that "by restricting expert testimony to 'scientific, technical, or other specialized knowledge [that] will assist the trier of fact,' Rule 702 demands that an expert's testimony be *well-grounded in the standards generally followed in his or her field for validating—establishing the truth of—assertions of the type offered,*" *id.* (emphasis added) (quoting FED. R. EVID. 702); and that "[t]he critical term 'knowledge' demands more than individual belief or speculation, but instead refers to *inferences or assertions that are grounded in the standards reliably used to support such claims,*" *id.* at 14 (emphasis added).

22. I pose this question, with a different example, at the beginning of the Article.

23. A useful discussion of reductionist and antireductionist views of testimony is found in Elizabeth Fricker, *Telling and Trusting: Reductionism and Anti-Reductionism in the Epistemology of Testimony,* 104 MIND 393 (1995) (reviewing C. A. J. COADY, TESTIMONY: A PHILOSOPHICAL STUDY [1992]).

24. DAVID HUME, *Of Miracles, in* AN INQUIRY CONCERNING HUMAN UNDERSTANDING 46 (Charles W. Hendel ed., The Bobbs-Merrill Co. 1955) (1748).

25. DAVID HUME, *An Abstract of a Treatise of Human Nature, in* AN INQUIRY CONCERNING HUMAN UNDERSTANDING, *supra* note 24, at 186.

26. *See, e.g.,* HUME, *supra* note 24.

27. *Id.* at 119.

28. *Id.* at 119–21.

29. *Id.* at 121.

30. W. V. QUINE & J. S. ULLIAN, THE WEB OF BELIEF 33–34 (1970).

31. *See id.* at 34 ("Observation sentences, taken narrowly, are comparatively foolproof. That is what makes them the tribunal of science.").

32. *Id.*

33. *See id.* at 35. They explain: "Truthfulness is essential, in large part, to the survival of language itself. . . . Our learning of the primitive vocabulary of observation sentences consists, after all, in our learning to associate it with the appropriate sensory stimulations. Small wonder then if those same stimulations dispose us in the future to affirm the properly associated observation sentences. Lying is an effortful deviation from the conditioned response."

34. *Id.* at 36–37.

35. I explore these devices shortly. *See infra* Section II.C.

36. *See* Elizabeth Fricker, *The Epistemology of Testimony,* 61 PROC. ARISTOTELIAN SOC'Y SUPPLEMENT 57, 73 (1987).

37. *See* HUME, *supra* note 24, at 119–24.

38. *See* QUINE & ULLIAN, *supra* note 30, at 33–34.

39. COADY, *supra* note 23, at 79–100.

40. Fricker argues: "Each one of us, in becoming the adult master of our commonsense scheme of things, has been through a historical process of development during which her attitude towards her teachers and other informants was one of simple trust. . . . Bearing in mind the role of teaching by others whom we trust unquestioningly in our learning of language (which is not separate from learning about the world), this seems inevitable." Fricker, *supra* note 23, at 401. Fricker also suggests that the putative KJB acquired at this stage can be confirmed in coherentist fashion as the stock of our KJB from all sources, including perception, memory, and inference, grows. She continues: "[A] belief first acquired through testimony very often gains support later on both through corroboration by other testimony, and through its coherence with what we learn from perception, and the empirical theory we base on this." *Id.* at 410.

41. Fricker, *supra* note 36, at 72–73.

42. *See id.* at 73.

43. Fricker, *supra* note 23, at 405.

44. *Id.*

45. Fricker, *supra* note 36, at 77 (emphasis added).

46. I assume this is a misprint for "hearer."

47. Fricker, *supra* note 23, at 405 (emphasis added). How good the empirical grounds are depends, of course, on what kinds of empirical reasons there are for the presumption in favor of sincerity and competence. Fricker does not offer much discussion of those grounds in the pieces I have been discussing, but has more extended discussion elsewhere.

48. Fricker, *supra* note 36, at 75.

49. John Hardwig vigorously advances this point about the complexity of the testimony that we intuitively believe produces KJB in hearers. *See infra* notes 56–68 and accompanying text (discussing Hardwig's account).

50. Fricker, *supra* note 23, at 407.

51. *Id.*

52. *See* HILARY PUTNAM, *The Meaning of "Meaning,"* in 2 MIND, LANGUAGE AND REALITY 215 (1975); John Hardwig, *Epistemic Dependence*, 82 J. PHIL. 335, 335 (1985).

53. *See* COADY, *supra* note 23; Anthony Kenny, *The Expert in Court*, in THE IVORY TOWER: ESSAYS IN PHILOSOPHY AND PUBLIC POLICY (1985).

54. *See supra* Section I.C.

55. I continue to use "KJB" for "knowledge or justified belief" for the same basic reason offered at the outset of this section.

56. Hardwig, *supra* note 52 at 335. Since Hardwig does not deny that testimony is a form of evidence, and he clearly has testimonial evidence for the listed beliefs, we must take his assertion in the first sentence to mean that he possesses no *nontestimonial* evidence for them.

57. Hardwig offers a nice example of (what he calls) epistemic dependence among physicists *See id.* at 346–47.

58. Putnam, *supra* note 52, at 227.

59. *Id.* at 228.

60. Putnam uses the term "division of linguistic labor" because of the particular version of realist semantics he was urging at the time. One may just as well call it a division of *epistemic* labor, as he himself implicitly acknowledges. *See id.* ("This division of linguistic labor rests upon and presupposes the division of nonlinguistic labor, of course. . . . [W]ith the increase of division of labor in the society and the rise of science, more and more words begin to exhibit this kind of division of labor.")

61. Hardwig, *supra* note 52, at 339.

62. *Id.* at 348.

63. *Id.*

64. *See supra* notes 55–57 and accompanying text. This is the intuition to which I think Fricker is insufficiently attentive.

65. Nozick offers an account of one mode of philosophical explanation as the explanation of the possibility of some state (knowledge, justice, truth) in the face of other apparently true propositions that deny its possibility. *See* ROBERT NOZICK, PHILOSOPHICAL EXPLANATIONS 8–24 (1981).

66. QUINE & ULLIAN, *supra* note 30, at 37.

67. *See supra* at 117.

68. Many of these accounts presuppose some kind of epistemic collectivism, or at least cohere better with it than with individualism, for the reasons Hardwig highlights. *See* Hardwig, *supra* note 52.

69. *See* Kenny, *supra* note 53, at 49 (stating that "different experts must not regularly give conflicting answers to questions which are central to their discipline" though there may be differences about borderline cases).

70. *See id.* at 50 ("[T]here will be agreement about the appropriate procedures for gathering information within the discipline. A procedure carried out by one expert to reach a particular conclusion is one which must be capable of duplication by any other expert.")

71. *See id.* ("[T]hough any expert must be able to repeat the results of others he does not have to: he can build up on foundations that others have built.")

72. *See id.* ("It need not necessarily predict the future (paleontology does not). But it must predict the not yet known from the already known. . . .") Note that both Kenny and the *Daubert* opinion articulate four criteria of science, and they overlap fairly closely on three of them: testing, peer review, and general acceptance. Kenny has no explicit version of the rate-of-error criterion the Supreme Court endorses. *Compare* Daubert v. Merrell Dow Pharms., Inc., 509 U.S. 579, 594 (1993), *with* Kenny, *supra* note 53, at 49–50. Note also that Kenny offers his criteria as "four criteria which are necessary conditions for a discipline to be scientific," Kenny, *supra* note 53 at 49, while the court assiduously avoids treating any of its factors as either necessary or sufficient conditions, *see Daubert*, 509 U.S. at 593.

73. Kenny, *supra* note 53 at 51–52.

74. Among evidence scholars, John Langbein articulates this position well: "At trial, the battle of experts tends to baffle the trier, especially in jury courts. If the experts do not cancel each other out, the advantage is likely to be with the expert whose forensic skills are the more enticing. The system invites abusive cross-examination. Since each expert is party-selected and party-paid, he is vulnerable to attack on credibility regardless of the merits of his testimony. A defense lawyer recently bragged about his technique of cross-examining plaintiffs' experts in tort cases. Notice that nothing in his strategy varies with the

truthfulness of the expert testimony he tries to discredit." John H. Langbein, *The German Advantage in Civil Procedure*, 52 U. CHI. L. REV. 823, 836 (1985). Langbein then quotes an article on trial strategy which reads: "A mode of attack ripe with potential is to pursue a line of questions which by their form and the jury's studied observation of the witness in response, will tend to cast the expert as a 'professional witness.' By proceeding in this way, the cross-examiner will reap the benefit of a community attitude, certain to be present among several of the jurors, that bias can be purchased, almost like a commodity." *Id.* (quoting Joseph Ryan. Jr., *Making the Plaintiff's Expert Yours*, FOR. DEF. Nov. 1982, at 12, 13). Langbein concludes: "Thus, the systematic incentive in our procedure to distort expertise leads to a systematic distrust and devaluation of expertise. Short of forbidding the use of experts altogether, we probably could not have designed a procedure better suited to minimize the influence of expertise." *Id.*

75. Kenny notes: "[T]he adversary system does not fit well with the use of experts to assist the court. It leads to dangers that the experts will be more concerned to assist one or other party to win their case than to assist the court to arrive at the truth." Kenny *supra* note 53, at 61.

76. *Id.* at 61–62. While Kenny's proposal is limited to judging the question whether a new discipline is a science, the basic problem he identifies affects competing testimony *within* acknowledged scientific fields no less than competing testimony about what is a science. (The former was arguably the situation, for example, in *Daubert.*) But that omission can be remedied, and other writers in the "extra-cameral" camp do so. Kenneth Culp Davis, for example, has proposed that Congress create "a research organization outside the [Supreme] Court to make studies at the Court's request," so that the Court could assign questions of legislative fact "to a qualified staff for a study or investigation." Kenneth Culp Davis, *Judicial, Legislative, and Administrative Lawmaking: A Proposed Research Service for the Supreme Court*, 71 MINN. L. REV. 1, 9, 15 (1986).

77. These include neutral commissions, such as Kenny proposes, special research services, and the like. *See* Kenny, *supra* note 53, at 61–62.

78. Not infrequently, expert "testimony" is offered in written form (for example, through affidavits). This is especially likely to be true when the nonexpert judge, performing the *Daubert* "gatekeeping" function, is assessing the evidence in order to make the threshold decision about admissibility.

79. Daubert v. Merrell Dow Pharms., Inc., 43 F.3d 1311, 1316 (9th Cir. 1995).

80. Daubert v. Merrell Dow Pharms., Inc., 509 U.S. 579, 594–95 (1993) (emphasis added) (footnote omitted).

81. *See, e.g.*, Zuchowicz v. United States, 870 F. Supp. 15 (D. Conn. 1994); *see also Developments in the Law—Confronting the New Challenges of Scientific Evidence*, 108 HARV. L. REV. 1481, 1514 (1995) (discussing *Zuchowicz* and its application of *Daubert*). Although this may be within the spirit of *Daubert*, it is inconsistent with its letter, since *Daubert* expressly refused to make any one criterion of scientific reliability necessary and expressly allowed that the *Frye* criterion was a permissible one.

82. 543 N.E.2d 1106 (Ill. App. Ct. 1989).

83. *Id.* at 1107.

84. *See id.* at 1109. In an earlier ruling reached largely because of the weakness of the state expert's testimony, the Illinois appellate court that decided *Palmer II* had held that the defendant was not guilty by reason of insanity, thus overturning a jury verdict of guilty but in-

sane. *See* People v. Palmer (Palmer I) 487 N.E.2d 1154 (Ill. App. Ct. 1985). At issue before the appellate court in *Palmer II* was whether the Double Jeopardy Clause of the U.S. Constitution permitted Illinois to retry the defendant after the court's ruling in *Palmer I*. The court held that the state expert's self-contradictory testimony was so weak that the earlier ruling went to the *sufficiency*, and not merely to the weight of that testimony, and thus that the defendant could not be tried consistent with double jeopardy principles. *See Palmer II*, 543 N.E.2d at 1109.

85. 926 S.W.2d 449 (Ky. 1996).

86. *See id.* at 451.

87. The trial and appellate courts ultimately disposed of the case on the basis of a settlement the parties reached before the jury rendered its verdict. *See id.*

88. Jury deliberations are not reported, and there is no indication of the extent to which the jury relied on this distinction in deciding in favor of the manufacturer Eli Lilly.

89. Socrates—at least Plato's Socrates—deployed general canons of rational evidential support to question, challenge, and embarrass many of the "experts" he encountered in Athens. *See* PLATO, *The Apology, in* THE LAST DAYS OF SOCRATES 96 (H. Tredennick trans., 1969).

90. For discussion, see D. H. Kaye, *The DNA Chronicles: Bad Numbers, Good Lawyering, and a Better Procedure, available in* 1995 WL 564589.

91. *See id.*

92. In most of my analysis, I assume for the sake of argument that experts are sincere and testify in good faith, but we should not wholly overlook those who are not. Consider the following speech, by an engineer who often served as an expert witness to other prospective expert witnesses, showing them the ropes: "The way I counteracted the thing, I used another technique. I used the technique as [sic] science as a foreign language. I made a statement to the attorney that absolutely nobody could understand. Now, what it amounts to, it's going to terminate the cross-examination, and it's going to terminate it in a hurry. I want the jury to understand what I say when I feel there are certain conditions. Under direct examination, the jury understands everything that I say. Under cross-examination, there are some things I will allow the jury to understand and there are some things which I will not allow the jury to understand. If you don't want the jury to understand something, then what you do is you answer the question precisely, you see. If somebody is working with a form of inertia, why I use a form of inertia. I say, "Do you mean the second bolt above the first bolt," you know. Just get into something which is a very precise way of saying something. The interval of minus infinity to plus infinity of X times X, X^2, and you know the—no one is going to be able to do much with that kind of thing. And he says, "Can you simplify it?" You say, "See, there's too much simplification already. This is the only way that I can state it to you so there will be no misunderstanding." Sanchez v. Black Bros. Co., 423 N.E.2d 1309, 1320 (Ill. App. Ct. 1981) (emphasis omitted) (determining that a trial court's refusal, during cross-examination of a manufacturer's expert witness, to permit questioning about this speech was reversible error).

93. *See* Fricker, *supra* note 23.

94. *See* QUINE & ULLIAN, *supra* note 30, at 33–41.

95. *See* HUME, *supra* note 24, at 120 ("We entertain a suspicion concerning any matter of fact when the witnesses . . . deliver their testimony with hesitation or, on the contrary, with too violent asseverations.").

96. *See* ARISTOTLE, RHETORIC 75 (A. Jebb trans., Cambridge Univ. Press 1925).

97. Notice that when the judgment of the lawyer is also nonexpert, as it often is, the lawyer may not even be in a good position to decide on substantive grounds which experts are in fact competent. That is, the lawyer's relative lack of epistemic competence compounds the problem of possible divergence between warranted assertion and persuasiveness.

98. I draw on material collected in Samuel R. Gross, *Expert Evidence,* 1991 WIS. L. REV. 1113, 1126–36. Similar materials are gathered in Langbein, *supra* note 74, at 835–41.

99. Note, *The Doctor in Court: Impartial Medical Testimony,* 40 S. CAL. L. REV. 728 728-29 (1967). It is not clear whether the authors of the survey were attentive to lawyers' relative inability to judge the epistemic merits of experts' testimony.

100. Hyman Hillenbrand, *The Effective Use of Expert Witnesses,* BRIEF, Fall 1987, at 48, 49 (emphasis added).

101. Stephen E. Nagin, *Economic Experts in Antitrust Cases,* LITIGATION, Winter 1982, at 36, 37.

102. Thomas V. Harris, *A Practitioner's Guide to the Management and Use of Expert Witnesses in Washington Civil Litigation,* 3 U. PUGET SOUND L. REV. 159, 161 (1979).

103. Albert Momjian, *Preserving Your Witness's Stellar Testimony: How To Qualify Your Expert to the Court,* FAM. ADVOC., Summer 1983, at 8, 8.

104. *Cf. supra* Section I.A (discussing understanding and epistemic competence).

105. Samuel Gross makes a related but slightly different point: "The confident expert witness is less likely to have been chosen because she is right, than to have been chosen because she is confident whether or not she is right." Gross, *supra* note 99, at 1134.

106. *See* Kenny, *supra* note 53, at 61–62.

107. *See supra* notes 56–69 and accompanying text.

108. *Daubert* made the *Frye* test one of its four factors of scientific reliability. Some federal courts have resuscitated the *Frye* test while ostensibly applying the *Daubert* rule, and many state courts have expressly rejected the *Daubert* rule and expressed continuing allegiance to the *Frye* test when applying state rules of evidence. *See Developments in the Law— Confronting the New Challenges of Scientific Evidence, supra* note 81, at 1514 n.40 (citing cases).

109. Paul S. Milich, *Controversial Science in the Courtroom:* Daubert *and the Law's Hubris,* 43 EMORY L. J. 913, 923–24 (1998) (emphasis added).

110. I say "judges" because at the stage of litigation to which *Frye* is relevant, the question is one of *admissibility.* Obviously, at a later litigative stage, factfinding juries can also rely on credentials in assessing the weight to be given to evidence that a judge has seen fit to admit.

111. *See, e.g.,* Milich, *supra* note 109, at 918–20.

112. Most jurors presumably would give less weight to a degree from Pittsburgh than from, say, Harvard, since they would be ignorant of Pittsburgh's lofty reputation in philosophy.

113. 529 F. Supp. 1255 (E.D. Ark. 1982) *aff'd,* 723 F.2d 45 (8th Cir. 1983).

114. Of course, there are versions of this dilemma that pit traditional science (the science of "established" universities and learned societies) against "nontraditional" science more directly. There are cases in the American courts, for example, dealing with the desire of Christian Scientist parents not to allow their gravely ill children to receive conventional medical care. *See, e.g.,* Newmark v. Williams, 588 A.2d 1108 (Del. 1990). These cases turn in part on whether the child is actually receiving adequate medical care. Whom should the judge ask

about whether Christian Science healing methods have due regard for the traditional medical facts, or indeed whether the traditional medical facts are the only relevant facts to be known—the Christian Scientist or a member of the traditional medical establishment?

115. PHILIP KITCHER, ABUSING SCIENCE: THE CASE AGAINST CREATIONISM 179 (1982).

116. *Id.*

117. HENRY M. MORRIS, THE REMARKABLE BIRTH OF PLANET EARTH at back cover (1972), *quoted in* KITCHER, *supra* note 115, at 178–79.

118. One example (and there are many like it) is found in the following case, which addressed the same issue as in *Daubert*, namely whether Bendectin could cause birth defects: "Have the plaintiffs established by a preponderance of the evidence that ingestion of Bendectin at therapeutic doses during the period of fetal organogenesis is a proximate cause [i.e. does it in a natural and continuous sequence produce injuries that would not have otherwise occurred] of human birth defects? . . . The jury unanimously answered no. Judge Rubin denied a post-trial motion for j.n.o.v. by the plaintiffs because '[b]oth sides presented testimony of eminently qualified and highly credible experts who differed in regard to the safety of Bendectin.' The great weight of scientific opinion, as is evidenced by the FDA committee results, sides with the view that Bendectin use does not increase the risk of having a child with birth defects. Sailing against the prevailing scientific breeze is the DeLucas' expert Dr. Alan Done, formerly a Professor of Pharmacology and Pediatrics at Wayne State University School of Medicine, who continues to hold fast to his position that Bendectin is a teratogen. In spite of his impressive curriculum vitae, Dr. Done's opinion on this subject has been rejected as inadmissible by several courts." DeLuca v. Merrell Dow Pharms., Inc., 911 F.2d 941, 945–46 (3d Cir. 1990) (citations omitted). In a footnote, the court added: "Dr. Done served as a Special Assistant to the Director for Pediatric Pharmacology of the FDA's Bureau of Drugs from 1971 to 1975. In this role, Done aided in the provision 'of FDA input on research involving children and fetuses, and development of guidelines for pre-clinical safety evaluations of drugs for use in children and in pregnancy. . . .' He also participated in publishing a paper called 'General Guidelines for the Evaluation of Drugs to be Approved for Use during Pregnancy and for the Treatment of Infants and Children,' in conjunction with the American Academy of Pediatrics in 1974. Dr Done's opinion that Bendectin is a teratogen largely rests on inferences he draws from epidemiological data, most of which he contends are the same that was utilized by the experts, including the FDA committee, to whom Merrell Dow cites to bolster its contention that Bendectin does not cause birth defects. The principal difference is that Dr. Done analyzes that data using an approach, advocated by Professor Kenneth Rothman of the University of Massachusetts Medical School, that places diminished weight on so-called 'significance testing.'" *Id.* at 946 n.7 (citations omitted); *cf.* K. J. Rothman, A *Show of Confidence*, 300 NEW ENG. J. MED. 1362 (1978) (arguing that significance testing is misleading).

119. Concerns about regress, question begging, and underdetermination are present in ancient skeptical (specifically, Pyrrhorian) challenges to the claim that we can know anything. *See* JULIUS ANNAS & JONATHAN BARNES, THE MODES OF SKEPTICISM 19–30 (1985) (discussing 'ten modes" of Aenisidemus).

120. Judge Posner described the crank factor with typically unvarnished asseveration: "[The witness' testimony was] either [that] of a crank or, what is more likely, of a man who

is making a career out of testifying for plaintiffs in automobile accident cases in which a door may have opened; at the time of trial he was involved in 10 such cases. His testimony illustrates the age-old problem of expert witnesses who are 'often the mere paid advocates or partisans of those who employ and pay them, as much so as the attorneys who conduct the suit. There is hardly anything, not palpably absurd on its face, that cannot now be proved by some so-called "experts."'" Chaulk v. Volkswagen of Am. Inc., 808 F.2d 639, 644 (7th Cir. 1986) (Posner, J., dissenting) (quoting Keegan v. Minneapolis & St. L.R.R., 78 N.W. 965, 966 (Minn. 1899)). One can be forgiven for wondering to what extent Posner's view is driven not so much by a concern for the expert witness's epistemic competence as by a strong disagreement with the expert over the normative attractiveness of the expert's views: "It is not the law, in Wisconsin or anywhere else, that the standard of care is set by the designers of $60,000 automobiles, so that the omission of any safety device found in such automobiles is negligent. . . . The buyer of a Mercedes 560 may be willing to pay extra for minuscule, perhaps wholly theoretical, improvements in safety, but such a buyer's willingness to buy the ultimate refinement in safety technology does not define the standard of care for the whole industry." *Id.* at 644–45. This is perhaps another instance in which the borderline demarcating the expert's zone of competence is blurry and in need of policing. It is also relevant to the question of underdetermination to note that the *Chaulk* majority characterized the expert as "an engineer with expertise in automobile safety." *Id.* at 642.

121. COADY, *supra* note 23, at 282.

122. *Id.* at 286.

123. *Id.* at 287 (emphasis added).

124. Kenny, *supra* note 53, at 56.

125. *See* Scott Brewer, *Scientific Expert Testimony and Intellectual Dual Process*, 107 YALE L. J. 1535, 1634–72.

126. *See id.*, at 1672–79.

4. WHAT IS THE PROBLEM WITH EXPERTS?

STEPHEN TURNER

DISCUSSIONS OF EXPERTISE and expert power typically have 'political implications', but the underlying political thinking that gives them these implications is rarely spelled out.[1] In what follows I will break down the problem of expertise to its elements in political theory. The first problem arises from social theory, and concerns democracy and equality. In the writings of persons concerned with the political threat to democracy posed by the existence of expert knowledge, expertise is treated as a kind of possession which privileges its possessors with powers that the people cannot successfully control, and cannot acquire or share in.

Understood in this way, expertise is a problem because it is a kind of violation of the conditions of rough equality presupposed by democratic accountability. Some activities, such as genetic engineering, are apparently out of the reach of democratic control (even when these activities, because of their dangerous character, ought to be subject to public scrutiny and regulation) precisely because of imbalances in knowledge, simply because 'the public', as a public, cannot understand the issues. This is not to say that democratic actions cannot be taken against these activities, but they are necessarily actions beyond the genuine competence of those who are acting. So we are faced with the dilemma of capitulation to 'rule by experts' or democratic rule which is 'populist'—that is to say, that valorizes the wisdom of the people even when 'the people' are ignorant and operate on the basis of fear and rumour.

The second problem arises from normative political theory. Regarding differences in knowledge as a problem of equality leads in some troubling directions. If we think of knowledge as a quantity, or a good to which some have access and others do not, the solution, admittedly one with practical limitations, is egalitarianization through difference-obliterating education or difference-obliterating access to expertise, for example through state subsidy of experts and the dissemination of their knowledge and advice.[2] But if the differences are better understood as differences in viewpoint rather than differences in quantities of knowledge, then we have another problem. Paul Feyerabend (1978: 73–76) insisted that a programme of extensive public 'science education' is merely a form of state propaganda for a faction, the faction of 'experts'. Thus it is a violation of the basic neutrality of the state, of the impartiality the liberal state must exhibit in the face of rival opinions in order to ensure the possibility of genuine, fair and open discussion. This second issue may seem to be a marginal issue, of interest only to fanatics, but increasingly it has become a practical problem.

The abstract form of the problem is this: if the liberal state is supposed to be neutral with respect to opinions—that is to say that it neither promotes nor gives special regard to any particular beliefs, world views, sectarian positions, and so on—what about expert opinions? Do they have some sort of special properties that sectarian opinions lack? If not, why should the state give them special consideration, for example through the subsidization of science, or by treating expert opinions about environmental damage differently than the opinions of landowners or polluters? If they do have special status, what is it? The special status granted to religious opinion leads to its *exclusion* from the domain of state action. Religion is either not an acceptable subject of state action, or is granted a protected status of limited autonomy, as in the case of the established church in England, in exchange for the church's renunciation of politics. The status of religion has often been proposed as a model for the state's relation to science (Polanyi, 1946: 59; Price, 1965). But it is a peculiar analogy, because the state not only protects and subsidizes science, it attends to the opinions of science, which is to say it grants science a kind of authority, and reaffirms this authority by requiring that regulations be based on the findings of science or on scientific consensus, and by promoting the findings of science as fact.

With respect to religion, then, the state attempts some form of neutrality, if only by separating the two and delegating to churches authority over special topics or special rights. With science, and more generally with 'expert' opinion, it is the opinions themselves that are treated as being 'neutral'. This special status becomes problematic when admittedly 'sectarian' beliefs come into conflict with expert opinion and the non-sectarian neutral character of expert opinion is called into question. Problems of this sort have occurred, for example, in connection with the teaching of creationism throughout the 20th century in the United States. But the issues here are more easily ridiculed than solved. For problems like 'is "creation science" really "science"?' there are no very convincing answers in principle, and no 'principles' on which to rely that cannot themselves be attacked as ideological. Nor is the problem limited to sectarian beliefs. Research on the genetic background of criminals has been denounced as 'racist' and government agencies have been intimidated into withdrawing support. Studies of race and intelligence, similarly, have been attacked as inherently racist, which is to say 'non-neutral'. A letter writer to *Newsweek* writes that 'theories of intelligence, the test to measure it and the societal structures in which its predictions come true are all developed and controlled by well-off white males for their own benefit' (Jaffe, 1994: 26). The idea that science itself, with its mania for quantification, prediction and control, is merely an intellectual manifestation of racism and sexism—that is to say, is non-neutral—is not only widespread, it is often treated in feminist theory as a given. There is a more general problem for liberalism that arises from this: if the liberal state is supposed to be ideologically neutral, how is it to decide what is and is not ideology as distinct from knowledge?

THE TWO ISSUES TOGETHER

If the two issues, equality and neutrality, are each taken on their own terms, these two problems can be discussed in a mundane political way: the solution to the problem of experts uncontrolled by democracy is to devise controls such as the citizens councils on technology that have been started in Denmark; the solution to a public incapable of keeping up with the demands of the modern world is to educate it better, a traditional aim of scientists, economists and others, for whom 'public understanding' is central. The problem with liberal democracy created by expert knowledge doesn't need a fancy solution: we can continue to do what we do now, which is to say 'muddle through'. For example, we may just declare science to be non-sectarian and deal with oddities like creation science by judicial fiat, or decline to fund science, or to permit or reject technology that has controversial implications or arouses the antagonism of 'public interest' groups on the basis of public opinion and political expedience. Taken together, however, the two problems raise a more difficult question: if experts are the source of the public's knowledge, and this knowledge is not essentially superior to unaided public opinion, not genuinely expert, the 'public' itself is presently not merely less competent than the experts but is more or less under the cultural or intellectual control of the experts.

This idea, inspired by Michel Foucault, is perhaps the dominant leitmotif in present 'cultural studies', and informs its historiography: the ordinary consumer of culture is taken to be the product of mysterious forces that constrain them into thinking in racist, sexist and 'classist' ways. These constraints are, so to speak, imbibed along with cultural products. In Donna Haraway's famous (but disputed) example, a person exposed to an 'expert' representation of human evolution at a natural history museum in which the more advanced people have features that resemble modern Europeans, for example, becomes a racist and a sexist (Haraway, 1984–85; Schudson, 1997). It is now widely taken for granted that this kind of effect of expertise is the true realm of 'politics', and that politics as traditionally understood is subordinate to—because it is conducted within the frame defined by—cultural givens, which themselves originate in part in the opinions of 'experts' in the past, such as the presentations of museum dioramas. It is this general form of argument that I wish to examine here from the perspective of liberal political theory, for it is to liberal political theory, not to say to liberal politics, that it poses a challenge.

A standard view of liberalism, perhaps most pungently expressed by Carl Schmitt, sees it as the product of the lessons of the wars of religion of early modern Europe. Liberal politics developed, where it did develop, as the consequence of the adoption of a certain kind of convention: matters of religion were agreed to be outside of the domain of the political. The domain of the political was reduced, *de facto*, to the domain of opinions about which people could agree to disagree, to tolerate, and further agree to accept the results of parliamentary debate and voting in the face of disagreement. It was also implicitly understood that some matters, such as matters of fact, were not subject to 'debate', but were the common possession of, and could be appealed to by, all sides in the course of public discussion. Schmitt made the point that parliamentary democracy depended on the possibility of 'persuading one's opponents through argument of the truth or justice of something, or allowing oneself to be persuaded of something as true or just' (Schmitt, [1926] 1985: 5). Without some such appeal—if opinions were not amenable to change through discussion, and persuasion was simply a form of the negotiation of compromises between pre-established interests—parliamentary institutions would be meaningless shells. What Schmitt saw in parliamentary politics of the Weimar era was that this assumption of parliamentarism no longer held. Rational persuasion with respect to what is true or just had ceased. The parties of Weimar politics, however, were more than mere interest parties. They were 'totalizing' parties, that construed the world ideologically, ordered the life experiences and social life of their members, and rejected the world-views of other parties and all the arguments that depended on these other world-views.

Schmitt believed that the former historical domain of parliamentary discussion, in which genuine argument was possible, had simply vanished. The world of totalitarianism, of the rule of totalizing parties, had begun. He didn't say that the liberal idea of parliamentary government, government by discussion, was wrong in principle. But there is a currently influential argument from principle that has the same conclusion. Stanley Fish has recently claimed that liberalism is 'informed by a faith (a word deliberately chosen) in reason as a faculty that operates independently of any particular world view' (Fish, 1994: 134). Fish denies that this can be anything more than a faith, and concludes that this means that liberalism doesn't exist. This is an argument, in effect, for undoing the central achievement of the modern state and unlearning the lessons of the wars of religion. But it is a curiously compelling argument nevertheless, especially if it is conjoined with the idea that the major products of the modern liberal state have been racial and gender inequity and injustice.

Expert knowledge is susceptible to a variant of this argument: expert knowledge masquerades as neutral fact, accessible to all sides of a debate; but it is merely another ideology. Jürgen Habermas makes this charge implicitly when he speaks of 'expert cultures'. Many other critics of past experts, influenced by

Foucault, have substantiated this claim in great detail. Their point, typically, is that 'expert' claims or presentations of reality by experts have produced discursive structures—'ideologies'—that were unwittingly accepted by ordinary people and politicians as fact, but were actually expressions of patriarchy, racism, and the like. Present-day 'expertise' raises the same problems: the difference is that we lack the historical distance to see the deeper meaning of the claims of experts.

If it is true that expert knowledge is 'ideology' taken as fact, the idea of liberal parliamentary discussion is, intellectually at least, a sham. The factual claims on the basis of which parliamentary discussion is possible are exposed as ideological. The true ideological basis of liberalism is thus hidden: it is really what is *agreed to be fact*, and what is agreed to be fact is, some of the time, the product not of open debate but of the authority of experts. The actual discussions in parliament and before the electorate are conducted within the narrow limits imposed by what is agreed to be fact, and therefore, indirectly, by expert opinion. To accept the authority of science or experts is thus to accept authoritative ideological pronouncements. So liberal regimes are no less ideological than other regimes; rather, the basis of liberal regimes in ideological authority is concealed under a layer of doctrinal self-deception.

These two problems—the problem of the character of expert knowledge, which undermines liberalism, and the problem of the inaccessibility of expert knowledge to democratic control—thus combine in a striking way. We are left with a picture of modern democratic regimes as shams, with a public whose culture and life-world are controlled or 'steered' by experts whose doings are beyond public comprehension (and therefore beyond intelligent public discussion), but whose 'expert' knowledge is nothing but ideology, ideology made more powerful by virtue of the fact that its character is concealed. This concealment is the central legacy of liberalism. The public, indeed, is its pitiful and ineffective victim.

Jürgen Habermas gives a version of the social-theoretical argument that suggests why the usual solutions to these two problems, of neutrality and democracy, fail. The argument depends on a characterization of the viewpoint of ordinary people, which he calls 'the internal perspective of the life-world'. This perspective, he claims, is governed by three fictions: that actors are autonomous beings; that culture is independent of external restraints; and that communication is 'transparent', by which he means that everyone can, in principle, understand everyone else (Habermas, [1985] 1987: 149–50). But in fact, he argues, the life-world is the product, at least in part, of external controls, which he calls 'steering mechanisms', operated by experts whose thinking is not comprehensible within the traditions that are part of, and help to constitute, the life-world. There is an unbridgeable cultural gap, in short, between the world of illusions under which the ordinary member of the public operates and the worlds of 'expert cultures' (Habermas, [1985] 1987: 397). The fictions of the life-world themselves prevent its denizens from grasping the manner in which it is controlled.

Robert Merton expresses a related point in a different way, with the notion that professionals and scientists possess 'cognitive authority'. Presumably Habermas has many of the same people in mind when he refers to 'expert cultures', but there is an important difference in the way the two conceive of the problem. Habermas's experts don't exert their influence by persuasion, but rather by manipulating conditions of social existence (especially to manufacture a kind of unthinking satisfaction), which Habermas calls 'colonizing the life-world'. Merton's model, the relation of authority, is more familiar. Merton adds that this authority is experienced as a kind of alien power against which people sometimes rebel—the coexistence of acceptance and rebellion he calls 'ambivalence' (Merton, 1976: 26).

Authority is, in its most common form, a political concept, and it points to the problem posed by expertise for the theory of democracy. Experts are not democratically accountable, but they nevertheless exercise authority-like powers over questions of true belief. Habermas's picture is somewhat different: the experts he seems to be concerned with are not professionals whom we deal with often in a face-to-face way, but policymakers hidden behind a wall of bureaucracy. Whether this difference is significant is a question that I will leave open for the moment. But it points to some difficulties with the concept of expertise itself that need to be more fully explored.

Expertise thus is a more complicated affair than my original formulations supposed. Cognitive authority, whatever it is, seems to be open to resistance and submission, but not to the usual compromises of democratic politics. It is not an object that can be distributed, nor can it be simply granted—so that everyone can be treated as an 'expert' for the sake of service on a committee evaluating risks, for example. To be sure, legal fictions of equality can be extended to such political creations as citizens oversight committees, and legal fictions have non-fictive consequences that may be very powerful. But cognitive authority is likely to elude such bodies, for reasons pinpointed by Habermas: the limitations of the perspective of the life-world preclude communicative equality between expert and non-expert.

Construed in either way, expertise is trouble for liberalism. If experts possess Mertonian cognitive authority, they pose a problem for neutrality: can the state preserve its independent authority, its power to act as neutral judge, for example, in the face of the authoritative claims of experts? Or must it treat them as inherently non-sectarian or neutral? Experts in bureaucracies pose a somewhat different but no less troubling problem. If we think of the distinctively German contribution to liberalism as the idea that official discretionary powers ought to be limited as much as possible—the ideal of a state of laws, not of men—it is evident that 'experts' have an apparently irreducible rôle precisely in those places in the state apparatus that discretionary power exists. Indeed, expertise and discretionary power are notions that are made for one another.

COGNITIVE AUTHORITY AND ITS LEGITIMACY

'Authority' is a peculiar concept to use in conjunction with 'knowledge': in political theory one usually thinks of authority in *contrast* to truth, as for example Schmitt does when he paraphrases Hobbes as saying that authority, not truth, makes law. By authority Schmitt means effective power to make and enforce decisions. Cognitive authority is, in these terms, an oxymoron. If you have knowledge, one need not have authority. But it is a nice analogue to 'moral authority'. And there is of course an earlier, and perhaps more fundamental, notion of *auctoritas* as authorship. The underlying thought is that the 'authority' has at first hand something that others—subjects or listeners—get at second hand. And this is part of the notion of expertise as well: the basis on which experts believe in the facts or validity of knowledge claims of other experts of the same type they believe in is different from the basis on which non-experts believe in the experts. The facts of nuclear physics, for example, are 'facts', in any real sense (facts that one can use effectively, for example), only to those who are technically trained in such a way as to recognize the facts as facts, and do something with them. The non-expert is not trained in such a way as to make much sense of them: accepting the predigested views of physicists as authoritative is pretty much all that even the most sophisticated untrained reader can do. The point may be made very simply in the terms given by Schmitt: it is the character of expertise that only other experts may be persuaded by argument of the truth of the claims of the expert; the rest of us must accept them as true on different grounds than other experts do.

The literature on the phenomenon of the cognitive authority of science (see Gieryn, 1994; Gieryn & Figert, 1986) focuses on the mechanisms of social control that scientists employ to preserve and protect their cognitive authority. The cognitive authority of scientists in relation to the public is, so to speak, corporate. Scientists possess their authority when they speak as representatives of science. And the public judgements of science are of science as a corporate phenomenon, of scientists speaking as scientists. So these social control mechanisms are crucial to the cognitive authority of science, as they are to the professions, Merton's original subject. But what these literatures have generally ignored is the question that arises in connection with the political concept of authority, the problems of the origin of authority. How did cognitive authorities establish their authority in the first place? And how do they sustain it?

If we consider the paradigm case of physicists as cognitive authorities, the answers to this question are relatively straightforward, and it is not surprising that the issue is seldom discussed as problematic. We all know (or have testimony that comes from users or recipients) about the efficacy of the products of physics, such as nuclear weapons, that we do accept, and we are told that these results derive from the principles of physics, that is to say the 'knowledge' that physicists certify one another as possessing. Consequently we do have grounds for accepting the claim of physicists to possess knowledge of these matters, and in this sense our 'faith' in physics is not dependent on faith alone—though, it is important to add, these are not the grounds that physicists themselves use in assessing one another's cognitive claims, or are only a small part of the grounds.

If we take the model that is suggested by this discussion of cognitive authority in science, we have something like this. Expertise is a kind of possession, certified or uncertified, of knowledge that is testified to be efficacious and in which this testimony is widely accepted by the relevant audience. But if this were all there was to expertise, it is difficult to see why anyone would regard claims to expertise or the exercise of expert authority to be a threat to democracy. Authority conceived of as resting in some sense on widely accepted (at least within the relevant audience) testimony to the efficacy of the knowledge that experts correctly claim to possess is itself a kind of democratic authority, for this acceptance is a kind of democratic legitimation.

One might go on to suggest that these authority claims are themselves subject to the same kind of defects as those of democratic political authority generally. Public opinion may be wrong, and mistakenly accept authority that ought not to be accepted. One might cite the relation between theological authorities and audiences of believers as examples of the way in which spurious (or at least mysterious) assertions of special knowledge can come to be regarded as authoritative, and in which highly problematic esoteric knowledge is granted the same sort, or similar kind, of deference as is scientific knowledge. But in the case of theological knowledge we do see something that was perhaps not so clear in the case of the cognitive authority of science, namely that the audiences for authority claims may indeed be very specific, and may not correspond with the public as a whole. And claims made to, and accepted by, delimited audiences may themselves be in error, and subsequently be rejected by these same audiences or by their successors.

Thinking about the audiences of the expert—the audiences for whom the expert is legitimate and whose acceptance legitimates her claims to expertise—illuminates a puzzle in the discourse of the problem of expertise and democracy. Merton and Habermas, it appeared, were not talking about the same kinds of experts. For Merton, the paradigm case was the physician, whose expert advice, say, to cut down on high-fat foods, we receive with ambivalence

(Merton, 1976: 24–25). For Habermas, the 'experts' who steer society from a point beyond the cultural horizon of the life-world, and do so in terms of their own expert cultures (Habermas, [1985] 1987: 397), are themselves a kind of corporate body or audience which has no public legitimacy as expert, and indeed is largely hidden from the public. This model does not fit physics. If the account of the cognitive authority of science that I have given is more or less true, the authority of physics is itself more or less democratically acknowledged, and is thus legitimate in a way that the authority of hidden experts is not. Physics, in short, not only claimed authority, and embodied it in the corporate form of the community of physicists, but this corporate authority has achieved a particular kind of legitimation, legitimation not only beyond the sect of physicists, but acceptance that is more or less universal.

If we begin with this as a kind of ideal-type or paradigm case of a particular kind of legitimate cognitive authority, and call them 'Type I Experts', we can come up with a list of other types in which the character of legitimacy is different. In this list we can include experts of the type discussed by Habermas, who seem not to possess the democratic legitimation of physicists. The easy distinctions may be made in terms of audience and legitimators. As I have suggested, the theologian is an expert with cognitive authority. Like the physicist, the authority of the theologian is legitimated by acceptance by an audience— the audience is simply a restricted one, but a predetermined one. The cognitive authority of the theologian extends only to the specific audience of the sect. We may call these 'restricted audience' experts 'Type II Experts'.

If the first two types of expert are experts for pre-established audiences such as the community of physicists, or a predefined community of sectarian believers, a third, the 'Type III Expert', is the expert who creates her own following. This type shades off into the category of persons who are paid for the successful performance of services. The massage therapist is paid for knowledge, or for its exercise, but payment depends on the judgements of the beneficiaries of that knowledge to the effect that the therapy worked. The testimony of the beneficiaries allows a claim of expertise to be established for a wider audience. But some people do not benefit by massage therapy, and do not find the promises of massage therapy to be fulfilled. So massage therapists have what is, so to speak, a created audience, a set of followers for whom they are expert because they have proven themselves to this audience by their actions. 'Experts' who are considered experts because they have published best-selling books that do something for their audiences, such as Dr Ruth (Westheimer), are experts in this sense as well: they have followings that they themselves created, but which are not general. They have *auctoritas* in the original sense.

Experts of three kinds—those whose cognitive authority is generally accepted, those whose cognitive authority is accepted by a sect, and those whose cognitive authority is accepted by some group of followers—each have a place in the scheme of liberal democracy. The expertise of the physicist is taken to be itself neutral; the state is neutral toward the other two, but in different ways. One can enter politics—Dr Ruth might run for the Senate, for example, or promote some political cause, such as sex education in elementary schools—but it is agreed that the other is to be excluded. The religious sectarian is excluded by way of the concept of the neutrality of the liberal state: the domain of politics is delimited, by agreement, to preclude the state, as the First Amendment puts it, from establishing a religion. But literally 'establishing' religion and at the same time restricting it (on, for example, the model of the established churches of European states), can serve the same purpose of both separating religion from politics and assuring that the boundaries of the domain of the political are decided politically rather than by religious experts. Religious regimes, such as Iran, operate on the premise that the domain of religious authority is decided by religious experts, thus delegating authority to Type II Experts. But this is a conscious and 'public' choice.

There may, of course, be conflicts of a more or less transitory kind between the authority of physicists and political authority—King Canute may attempt to command the tides in the Wash against the advice of his physicists. Conflicts of a less transitory kind—between expert economists and economically inexpert political ideologues, such as those that concerned Joseph Schumpeter ([1942] 1950: 251–68)—might constitute a threat. This is because the economists' expertise is systematically relevant to policy, and that of the physicists is only transitorily so. But if the economists deserve legitimacy as experts with the general public it presumably is because, like Canute, the politicians who ignored their advice would fail to achieve their goals.

The incomplete legitimacy of economists points to a set of interesting issues. Claims to cognitive authority are not always accepted. Economists do agree among themselves, to a great extent, on what constitutes basic competence and competent analysis. There is a community of opinion, and some people who aren't members of the community—that is, the 'public' of economics—accept the community's claims to expertise. But the discipline's claims to corporate authority—claims that would enable any economist to speak 'for' economics on elementary issues, such as the benefits *ceteris paribus* of free trade, in the way that even a high school teacher can speak for physics—may be fragile. The sight of ads signed by several hundred economists is a kind of living demonstration of the distance between the claim to speak representatively found in physics and the claim in economics. In economics, agreement on the basics—long since assured within the community of professionals—still has to be demonstrated by the ancient collective ritual of signing a petition—itself among the most ancient of political documents. Even these near unanimous claims are not always accepted as true: sectarians, textile interests, or a sceptical public may contest them.[3] Moreover, around every core of 'expert' knowledge is a penumbra, a domain in which core competence is helpful but not definitive, in which competent experts may disagree, and disagree because the questions in this domain cannot be decided in terms of the core issues that define competence. Establishing cognitive authority to a general audience is not easy: major achievements, like nuclear weaponry, antibiotics, new chemicals and new technology are the coin of the realm. Policy directives rarely have the clarity of these achievements, and policy failures are rarely as clear as Canute's tidal initiative.

Two Novel Types of Expert

Now consider the following type of 'expert': those who are subsidized to speak as experts and claim expertise in the hope that the views they advance will convince a wider public and thus impel them into some sort of political action or choice. This is a type—the fourth on our list—that appears at the end of the 19th century in the United States, and developed hand-in-hand with the development of philanthropic and charitable foundations. The fifth type, to be discussed below, is a variant of the fourth, or rather a historical development of the fourth. Where the effort to create and subsidize recognized 'experts' failed—typically because their expertise was not as widely accepted or effective as the funders hoped—an effort was sometimes made to professionalize target occupations and to define professionalism in terms of acceptance of the cognitive authority of a particular group of experts. Both types exist nowadays, and in some fields there really is no clear distinction between the two. The difference is in the kind of audience, and in many cases, such as psychotherapists, perhaps, the 'professional' audience is not so different from the public with respect to their actual sources of information and wisdom.

The history of social work provides a good example of a failed attempt to establish a claim of expertise that is further distinguished by the self-awareness of the process of claiming expert status. When the Russell Sage fortune was put to charitable purposes by Sage's widow, her advisors, themselves wealthy community activists, created an organization that attempted to persuade the public to adopt various reforms. The reforms ranged from the creation of playgrounds to the creation of policies for tenement housing and regional plans. Some of the participants were veterans of 'commissions', such as the New York Tenement commission, others were products of the Charity Organization Societies, still others came from social movements, such as the playground movement. What the Foundation did was to subsidize departments with people employed as experts in these various domains. Some of them, such as Mary Richmond, had a great deal of experience, had written books, and were well known. Others were not well known, but learned on the job, and played the rôle of advisor to volunteer groups of various kinds—such as women attempting to promote the construction of playgrounds in their community, who needed advice on what to ask for.

The Russell Sage Foundation had a particular model of how to exert influence, a model that other foundations were to follow. They objected to 'retail' philanthropy, and wished to influence others to commit their own resources to the cause. Playgrounds, for example, were not to be directly financed by the Foundation, as Carnegie had financed libraries. The Foundation offered expertise that local groups could use so as to assure that resources could be mobilized for the cause and used properly. At most, demonstration projects would be directly financed. The means of exerting influence was thus through the creation of public demands; this required means of reaching the public and, at the same time, persuading the public of the validity of the demands.

The Foundation thought it had hit on the ideal device for doing this: the 'Social Survey'. Indeed, surveys were a powerful device, and literally hundreds of surveys—of sanitation, education, housing, race relations, child welfare, crime, juvenile crime, and so on—were done in the period between the turn of the 20th century and the 1930s depression. The particular kind of survey the Foundation was most enamoured with was the comprehensive community survey. They had one great success with community surveying—the Pittsburgh survey—and a few minor successes. What the Pittsburgh survey did was to examine all of the aspects of community life that were of special concern to the 19th-century reform movements, and 'publicize' them. To influence the building of better sewers and a better water system, for example, they included in their public exhibit to the community (one of their primary means of publicizing the results of the survey) a frieze around the top of the hall which illustrated pictorially the number of deaths from typhus in Pittsburgh annually. Some of this effort worked; change did occur.

In the full flush of this success, the leading intellectual figure behind what he called 'The Survey Idea', Paul Kellogg, wrote extensively on the meaning of such surveys, and on the difficulty of persuading others of the expertise of 'social workers', as they styled themselves. The foremost need was to persuade people to pay for expert knowledge. As Kellogg ([1912] 1985: 13) complained:

> ... while many of the more obvious social conditions can be brought to light by laymen, the reach of social surveying depends on those qualities that we associate with the expert in every profession; knowledge of the why of sanitary technique, for example, and of the how by which other cities have wrought out this reform and that. And townsmen who would think nothing of paying the county engineer a sizable fee to run a line for a fence boundary must be educated up to the point where they will see the economy of investing in trained service in social and civic upbuilding.

Kellogg himself said that the task of persuasion would have been easier if there was an event like the Titanic disaster, which dramatized the need for lifeboats and confirmed the warnings of naval engineers. The survey and its publicity, however, were designed to serve the same purpose (Kellogg, [1912] 1985: 17):

> To visualize needs which are not so spectacular but are no less real . . . to bring them into human terms, and to put the operations of the government, of social institutions, and of industrial establishments to the test of individual lives, to bring the knowledge and inventions of scientists and experts home to the common imagination, and to gain for their proposals the dynamic backing of a convinced democracy.

In the end, few communities were educated up to that point—at least to the acceptance of the generic kind of reform expertise that Kellogg and his peers claimed. But to an astonishing extent, the strategy worked, especially in such areas as playgrounds and juvenile justice. Major reforms were enacted on the basis of supposed expert knowledge that was based on little more than the highly developed opinions of the organized reformers themselves.

There is a sense in which this kind of expertise has proven to be a permanent feature of American politics, and now world politics, though in a somewhat different form. Organizations like the Sierra Club can support 'experts' on policy matters, whose expertise is at best part of the penumbral regions of scientific expertise. These 'experts' are not unlike those subsidized by the Russell Sage Foundation in its early years. Their rôle is both to persuade the public of their expertise and, of course, about matters of policy.

What distinguishes these two types of experts is the triad of support, audience and legitimation their rôle involves. Experts of the fourth kind, whose audience is the public, do not support themselves by persuading the public directly of the worthwhile character of their services or advice, as Dr Ruth does, but by persuading potential subsidizers of the importance of getting their message out to the public and accepted as legitimately expert. So, like the economists who seek to be accepted by the public as experts, they too seek public recognition of their expertise. But the expertise they claim is inherently policy-oriented, rather than incidentally so. Kellogg, who played a leading rôle as a publicist for the survey movement, constantly likened the 'social worker' to the engineer. But he rejected the idea that there was any need for a base for engineering knowledge in some sort of social science—the things the 'social worker' engineer knew already about the right way to do things and the right standards to impose were amply sufficient to make public policy. The purpose of the survey was not to advance knowledge but to demonstrate to the public how far below the standards their community was, and thus to spur it into action.

There is a kind of threat to discussion posed by these 'experts' that results from the fact that they are subsidized, and are thus the preferred expert of some funder. The sources of the funding are typically concealed, as is the motivation behind the funding and the process by which the funding occurs. Concealment can serve to lend the claims of an expert a kind of spurious disinterestedness. But these are threats that liberal democracy is used to examining and indeed, in the case of the Russell Sage Foundation, the issue of interests was raised at the start, by no less a figure than Franklin H. Giddings, the leading figure of Columbia University Sociology, and famously raised by Congressmen at the time of the creation of the Rockefeller bequests. The problem, in short, became part of the public discussion, and foundations found ways to deal with the suspicions their activities aroused, not least by genuinely delegating a great deal of the control over the money to boards of notables.

The fifth type of expert is distinguished by a crucial difference in this triad: the fact that the primary audience is not the public, but individuals with discretionary power, usually in bureaucracies. The legitimacy of the cognitive authority exercised by these individuals is not a matter, ordinarily at least, of direct public discussion, because they deal with issues, such as administration, that are not discussed in newspapers until after they become institutional fact, and indeed are rarely understood by reporters, and may be subject to administrative secrecy of some kind. A paradigm case of this fifth kind of expertise is public administration, which contains the three distinctive elements of the type: a distinctive audience of 'professionals'; experts whose legitimacy is a matter of acceptance by these professionals, but who are not accepted as experts by the public (and ordinarily are not even known to the public); and whose audience of 'professionals' is itself not (or, at most, partially) recognized as possessing 'expertise' by the public.

It would be useful to survey the major national administrative traditions to better understand the rôle of this kind of expert knowledge in each of them. Doing so, I suspect, would point to sharp differences with deep historical roots. But there are some commonalities as well, that result in part from the historical fact that public administration itself was the product of a strategy for the creation of expertise which had American roots. In what follows I will simply describe the strategy and its origins, and consider the political meaning of the strategy in terms of the elementary political theory problems with which I began.

Public administration was a major target of the reformers of the early part of the 20th century—corrupt and incompetent city officials, given jobs as part of a system of patronage appointments, were major obstacles to the correction of the conditions the reformers objected to. But political reformers—reform Mayors, for example—came and went, and the underlying problem of ineptitude and corruption remained. The movement for the professionalization of public administration, sponsored in large part by the Rockefellers (who had previously invested heavily in the professionalization of social work as well), changed this.

The professionalization strategy was rooted in the successful experience of Abraham Flexner in the reform of medical education, and in the Rockefeller efforts in creating a medical profession in China. It targeted practitioners and sought to turn them into an audience for expertise. One of the pillars of the reform of medical education was to make it 'scientific', and this meant in part the creation of a sharp distinction between medicine as a craft skill to be conveyed from one practitioner to another and medicine taught and validated by medical scientists, and the elimination of the former. One of the major goals of the reform of medical education was the elimination of part-time clinical faculty: this was made a condition of grants for improvement (Brown, 1979: xv).

The professionalizing strategy employed by the Rockefeller philanthropists in this and other domains ignored, for the most part, the 'general public', except to educate the general public in the differences between professional and non-professional workers. This education was supplemented by legal requirements and schemes of certification designed to drive non-professionals from occupations that had previously been weakly professionalized. The strategy, by the time it was applied to public administration, was well-tested and mature, and the machinery for implementing it was already in the hands of the Rockefeller founders. The Rockefeller philanthropies already had a well-established relationship with the social sciences, particularly through such individuals as Robert Merriam and such organizations as the Social Science Research Council, as well as longstanding relationships with certain major universities—some of which were 'major' largely as a consequence of Rockefeller largesse, and one of which, the University of Chicago, was a Rockefeller creation. During the 1930s, at a time when Rockefeller funding was being redirected away from 'pure' social research—the professionalization of the social sciences themselves was a Rockefeller project of the 1920s, and social science institutions were still dependent on Rockefeller funds—and many universities were in dire financial straits, the Rockefeller philanthropists induced, through the use of their financial muscle, several key universities, such as the University of North Carolina, to establish training programmes in public administration (Johnson & Johnson, 1980: 111–12).

The 1930s saw the creation of a number of schools of public administration, of professional organizations of public administrators, and the gradual creation of a class of specially trained public administrators. The remnants of these original Rockefeller efforts still persist, in such forms as various schools of Public Administration and the professional associations of public administrators. The training, by experts, of municipal workers who had traditionally been 'amateurs' appointed as political favours led to the creation of a distinction between trained and untrained administrators, and between political and professional administrators. The expertise of the teachers of public administrators was no different than the expertise of municipal research bureau researchers. The institutional structures were novel and took the form not of training schools but of university departments which eventually produced professional academic public administrators. These then became the experts, and their audience became the professional public administrators.

The striking feature of this development is that it solves the problem of the audience of the expert by creating an audience for the expert and assuring indirectly that this audience is in a position to compete successfully with amateurs. A similar kind of development took place during and after World War II with respect to foreign policy, area studies, and similar domains related to the postwar American imperium. Such organizations as the Russian Research Center at Harvard, for example, were the product of the same strategy, and involved some of the same players—previous recipients of Rockefeller funds. Later the newly created Ford Foundation played a significant role in the creation of foreign policy experts. In this case the primary consumer of professional employees was the federal government, often indirectly: training of foreign service officers, military officers, and the like, was a major task of these experts. Indeed, the Harvard investment in regional studies began with contracts during World War II for the training of occupation army officers (Buxton & Turner, 1992).

BUREAUCRATIC DISCRETION AND SECTARIAN EXPERTISE

It is with this step that the problem of democracy and expertise becomes salient. The experts whose expertise is employed are experts in the sense that they have an audience that recognizes their expertise by virtue of being trained by these experts. The audience, in a sense, is the creation of the experts. In this respect the expert more closely resembles the theologian whose expertise is recognized by the sect he successfully persuades of his theological expertise. In the case of theologians, however, liberal governments withdrew (or were based on the withdrawal of) public recognition of expertise from such sectarian 'experts'. In the case of the kinds of experts I have been discussing here, there is, in contrast, a discrepancy between the sectarian character of their audience and their rôle in relation to political authority. Since a great deal of political authority in modern democratic regimes resides in discretionary actions of bureaucrats, the control of the bureaucracy by a sect can amount to the denial of the original premises of liberal regimes.

Analogues to these 'sects', as I have characterized them, exist in all modern bureaucratic traditions: the élite Civil Servant in Britain, the graduate of the

Grandes Écoles in France, and in Germany the bureaucracy with its own distinctive internal culture. The German case perhaps does fit Habermas's category of 'expert cultures'. To the extent that these groups exercise power in terms of a distinctive 'culture' that is neither understood nor accountable, they violate equality and neutrality. But one can also claim that there is a kind of tacit consent to their expertise.

In the case of physics, with which we began, there was a kind of generalized approbation and acceptance on the grounds of indirect evidence of the physicist's claim to expertise, and the claim to exercise powers of self-regulation and certification that should be honoured by the public at large. In the case of professional bureaucrats and administrators there is perhaps something analogous. In the course of creating an audience for public administrators and area studies experts, there was indeed a moment in which the offer of 'professionally trained' workers could have been resisted, and the amateurism of the past been allowed to persist. Similarly, there might have been, in the United States, a strong civil service core that exercised some sort of generalized quasi-representative functions for the nation, as arguably is the case in, for example, France and Britain. There, professional administrators did displace 'amateurs', and this occurred with democratic consent of a sort.

Professionalization was a mechanism of reform that was appealing to reformers who lacked a sufficient body of amateur political friends to fill the jobs that existed, or the needs for personnel that arose. War, here as elsewhere, was a significant catalyst of these changes, especially in the realm of foreign policy, where the need for occupation army expertise was soon followed by a need for expertise in dealing with foreign aid. It should be obvious that this kind of expertise more closely resembles sectarian expertise than the expertise of physics. The distinction is not that the pronouncements of ideologists and theologians are ideological, and those of physicists are not. The distinction is between what might best be described as generalized public validating audiences and specialized validating audiences that do not correspond with the general public. No foreign policy expert is obligated to demonstrate the validity of his views on foreign policy by producing an unambiguous success, like curing cancer or constructing atomic weaponry. Indeed, there has often been a large disparity between the views of experts and the kinds of facts upon which these views are alleged to be based, on the one hand, and, on the other, the views of politicians and the kinds of facts and results on which their acceptance or validation by the general public is based.[4]

In the case of foreign policy, opinions based on secret information gain a certain prestige, and a foreign policy analyst who does not have access to information that the public does not have is diminished in his credibility in the eyes of the target audience of the expert—namely, government officials who themselves operate on the basis of information that the general public does not possess. The implications of this discrepancy are obvious. Conflicts between democratic and expert opinion are inevitable, not so much because the expert invariably possesses secret information (though that may be the case with respect to foreign policy, and in practice is the case with respect to bureaucratic secrets generally), but is a simple consequence of the fact that the processes by which knowledge is validated by audiences are separate, just as the processes of validation of theological expertise by sect are distinct from the processes by which public validation is achieved.

Conflicts between expert knowledge of this special 'sectarian' kind and democratic opinion are thus, if not inevitable, systematic and systematically produced by the very processes by which expertise itself is validated. The liberal ideal of a state that refuses to decide sectarian questions does not work very well when the sects are, so to speak, within the bureaucracy and their sectarian beliefs have their main effects on the discretionary activities of bureaucrats.[5] It is this peculiar combination of circumstances that allows for this kind of conflict, and it cannot be remedied by ordinary means. The whole process of bureaucratic selection and training is the means by which this kind of influence is exercised: to root out sectarianism would involve rooting out the existing system of bureaucratic professionalization. Whether there is a practical alternative to this system of professional governance is a question I leave as an exercise to the reader. In specific cases, of course, government is 'deprofessionalized': professional diplomats are replaced by individuals from other backgrounds, formerly bureaucratic positions are made into 'political appointments' or elected positions, government is run 'like a business' by businessmen, and the like.

I have suggested here that the difficulties that have concerned theorists of democracy about the role of expert knowledge must be understood as arising not from the character of expert knowledge itself (and its supposed inaccessibility to the masses), but from the sectarian character of the kinds of expert knowledge that bear on bureaucratic decision-making.[6] There is, in the case of science, an important check on claims of expert knowledge that is lacking in the case of experts of the kind who threaten or compete with democratic decision processes: scientists need to legitimate themselves to the public at large. The expert who is a threat is the expert who exerts influence through the back door of training and validating the confidence of professionals, and whose advice is regarded as authoritative by other bureaucrats but not by the public at large. The authority of the expert whose expertise is not validated by public achievements is the authority that comes into conflict with democratic processes. Of course, there is, in a sense, a check: governments that fail to deliver on promises may earn the contempt of their citizenry. But this is not the same as the check on science, for it is quite indirect. If we know that the juvenile justice system is failing, this is not the same as knowing who in the system is to blame, or which of its various 'professions' with claims to expertise ought not to be regarded as expert. The 'public' may be dissatisfied, and find outlets for its dissatisfaction, but the very fact that the bureaucrats themselves are not directly elected and do not appeal to the general public for legitimation means that there is no direct relationship.

RECONCILING EXPERTISE AND LIBERAL DEMOCRACY

The discussion so far has distinguished five kinds of experts: experts who are members of groups whose expertise is generally acknowledged, such as physicists; experts whose personal expertise is tested and accepted by individuals, such as the authors of self-help books; members of groups whose expertise is accepted only by particular groups, like theologians whose authority is accepted only by their sect; experts whose audience is the public but who derive their support from subsidies from parties interested in the acceptance of their opinions as authoritative; and experts whose audience is bureaucrats with discretionary power, such as experts in public administration whose views are accepted as authoritative by public administrators. The first two do not present any real problem for either democracy or liberalism: physicists are experts by general consent, and their authority is legitimated by rational beliefs in the efficacy of the knowledge they possess. The expertise of self-help authors is private, and the state need not involve itself in the relation between sellers of advice and buyers. Theologians and public administrators present a different problem. Neutrality is the proper liberal state's stance toward theologians, because the audience that grants them legitimacy is sectarian. The state ought not to subsidize them or to give one sect preferential treatment over another. The fourth and fifth type present more serious problems. Both typically are subsidized by the state, indirectly—foundations derive some of their money from tax expenditures. Had the Rockefeller fortune been taxed as an estate, there would have been no Foundation, or if there were it would have been smaller.

What does this kind of laundry list establish? It makes no claims to completeness as a taxonomy of expertise. But it does contain all of the types of experts that figure in the problem as traditionally conceived. Habermas's shadowy expert cultures are there, but not in the same form: they appear not merely because there is expert consensus, but also because there are bureaucrats with discretionary powers who share in this consensus and are guided by it. Ian Hacking's classic paper on child abuse (1991) is an example of this kind of argument involving expertise in this sense: here the triumph of an expanded concept of child abuse is seen as the successful imposition of a definition which serves the interests of certain professional groups. The experts in question here are perfect embodiments of what I have called the fifth type of expertise. The political issue here is not expert knowledge as such, but discretionary power: the reason child abuse is a problematic category is because social workers and physicians acting in the name of the state employ this concept and operate in terms of a consensus about it.

If we reconsider the traditional problems—and the Fish/Foucault 'cultural studies' form of the problem as well—in the light of the list, some of the difficulties vanish or are greatly modified, and some features of the problem stand out more sharply. Begin with the expert whose racist biases are passed off as science and become part of the culture through repetition and through presentations of 'facts' in which the prejudices are concealed but nevertheless imposed on the public recipient. Two things become apparent about such experts: (a) that their expertise was not simply given, but somehow had to be earned or created—in the language of the present discussion, legitimated; and (b) that their expertise typically operates in the penumbral regions of science—that is to say, topics on which there is neither agreement on conclusions, on appropriate methods, or on whether the topic is itself entirely 'scientific'. To be sure many things may pass, in the eye of the public, for science. Scientific views, and scientific consensuses, may of course change, and the public may well legitimate and accept scientific communities whose views later appear to be wrong. The 'public' is not merely the passive recipient of science and the prejudices and errors of scientists, but plays a rôle in their legitimation. The hard road that Darwinism had to acceptance should suffice to remind us that, although it may be easier to get public acceptance for views that flatter the self-image or prejudices of the public, the public is not a passive receptor. The legitimating done by the public may lag the legitimating done by the professional community by decades. And the public is not very adept at distinguishing the core of expert knowledge from the penumbra: this is a distinction made within the community of experts, and it may be erroneously made by them—in the sense that retrospectively the community may come to conclude that only a fragment of what was formerly held to be true was in fact true. So experts are fallible, and the public is fallible in judging claims to expertise. But this does not mean that the public is powerless to make judgements.

More importantly, however, 'expert' claims of this sort do not permanently or inherently occupy the status of 'scientific'. When issues arise in which there are grounds for questioning the legitimacy of expert claims, they may, along with the legitimacy of the experts, come under public scrutiny and lose legitimacy. Indeed, this process, call it 'politicization', is a normal fact of political life, and it goes both ways. That which is taken to be a matter of expert truth, or taken for granted as true by the public and therefore regarded by the liberal state as itself neutral, may cease to be taken as neutral truth. That which has hitherto seemed to be an appropriate matter for 'politics', or for the negotiation of interest groups, or for 'public' discussion, may come to be regarded as a 'professional' matter that only experts certified by their appropriate communities have any real business discussing authoritatively. We do not regard legislators as experts on physics, and for one to oppose a particular physical theory or its teaching we would regard as an inappropriate act—as 'politicization' of that toward which the state should be neutral.

This is typically an 'academic' issue—academic in the pejorative sense. But sometimes it is not academic. Illiberal regimes, notoriously, do not accept these distinctions, leading to such things as Islamic science, Aryan physics and socialist genetics, each sponsored by the state in opposition to expertise that has been defined as enemy ideology. But these are obviously very different cases than the ordinary expertise of, for example, physicians promoting CPR. It is nevertheless still common to collapse all of these cases of expertise, and the case of science itself, into the category of ideology. The term 'ideology' itself is a good place to start with this issue, for it figures in Fish's attack on liberalism. Fish regards liberalism as a sham because it rests on a bogus notion of reason— that is on the assumption that there is such a thing as neutral 'reason', that is reason that is outside of the battle between world-views (Fish, 1994: 135). So Fish thinks of liberalism as founded on an ideology it takes for granted, an ideology that is not neutral, and thus, paradoxically, liberalism cannot exist, for the idea of liberalism as neutrality represents a kind of self-contradiction. It can exist only by hiding the untruth of its foundations.

There is bite to this criticism, bite that derives from the naturalism of natural right thinking out of which liberalism historically grew. The liberalism of the American founding tended to regard the truths relevant to politics as immutable and self-evident, and accordingly could regard them as neutral facts. It seems that the same kinds of claims arise in connection with scientific expertise. It is at this point that science studies, particularly controversy studies and such topics as the law's construction of science, become relevant as well. One might read these studies to be making a point that is similar to Fish's: the self-descriptions of scientists are applications of an ideology of precisely the same kind, an ideology of neutral reason. A closer grained description of the activities of science of the sort that science studies offers conflicts with these self-descriptions. Thus these studies undermine the distinction between politics and science by showing them to be constructed, historical, and so forth, meaning that science is really ideology.

Or does it mean this? To be sure, 'controversy studies' in the literature on science studies *have* sometimes focused on the problem of expertise in order to problematize the claims of experts (e.g. Timmermans, 1999, on CPR), to show that the public construction of science is wrong (e.g. Collins & Pinch, 1993), or that the law's construction of science is wrong, or at least arbitrary and misguided (e.g. Jasanoff, 1995: 60). But are the implications of these studies that expertise is ideological and therefore non-neutral? Or can they be taken differently?

When Collins and Pinch (1993: 145) discuss the contribution of science studies to citizenship, they wrestle inconclusively with the question of what is implied by their redescriptions of science. Their message is that 'scientists . . . are merely experts, like every expert on the political stage', like plumbers, as they put it, but experts who happen to have an immaculate conception of themselves, unlike plumbers. They suggest that both scientists and the public would be better off without this conception. The same point is made by Jasanoff with respect to judges. But in both cases it is not clear what the end game of this argument is. Professional ideologies of the sorts that scientists have and plumbers do not, can of course be exposed as false, or, to put the point in Schmittian terms, to be political or non-neutral. But there are at least two possible ways of reasoning from this kind of argument, which need to be carefully distinguished.

The differences between the two ways is parallel to the difference between the ways in which one might reason from Fish's account of liberalism as a faith which pretends it is neutral to faith. If we reason as Fish does, moving the givens of liberal theory into the category of faith (that is, outside the category of reason), we get a contradiction at the base of liberalism. Moving science and expertise generally from the category of neutral to ideological in the manner of Feyerabend begets a conflict with the practice of treating science as unproblematically a source of truth, or when the self-descriptions of scientists involve appeals which are absolute, such as the metaphysical claim that present science has succeeded in establishing the truth about the universe, or has a method for doing so—which I take it is part of what Collins and Pinch have in mind with the phrase 'immaculate conception'. But there is an alternative: we can take all of these presumptively absolute conceptions in a different way. Schmitt has a slogan that bears on this problem, the saying that what is political is a political question—NB a political rather than a scientific or philosophical question, and for Schmitt this means a matter of decision, not truth. Terms like faith and reason, and also science, can be thought of as political in this sense as well, and indeed this is precisely how Schmitt thinks of such terms (Schmitt, [1932] 1996: 31). But treating these terms as political at the 'meta' level is not the same as discrediting them, eliminating them, or collapsing them into the category of 'ideological'. On the contrary, we can recognize them as political, recognize the foundations of liberalism as non-absolute, and still accept them in practice as a political necessity. To put this somewhat differently, consider the arguments made by Collins, Pinch and Jasanoff. One end game is to treat such things as the law's construction of science as ideological; another is to treat both categories as themselves political. If calling something a matter of expertise is a political decision, so is calling something 'ideological'. None of these are, or need be, natural or absolute categories.

CONSTRUCTIONISM AS LIBERALISM

The answer to Fish is to treat the liberal principle of neutrality not as an absolute assertion about the nature of beliefs, but as a core rule, whose application varies historically, whose main point is to establish a means of organizing the discussion of political matters, that is to say the discussion of political decisions. We can apply this to the problem of expertise as follows: it is no surprise that, in order for there to be genuine discussion in Schmitt's sense, some things would be temporarily taken for fact, or, alternatively, some things would be left to the experts to settle. 'Politicizing' everything, making everything into the subject of political decision-making (or treating it as an analogue to political decision-making), would lose the advantages of the intellectual division of labour and make reasoned persuasion impossible. Some facts need to be taken for granted in order for there to be genuine political discussion, and some of the work of establishing the facts is, properly, delegated to experts. Indeed, to imagine a world in which such delegation did not occur would be to imagine a simpler society, at best a society of Jeffersonian yeomen, in which everyone knew pretty much what everyone else knew that was relevant to public decision-making.

To preserve the possibility of political discussion that such societies established, it is essential to delegate to experts and grant them cognitive authority. But granting them cognitive authority is not the same as granting them some sort of absolute and unquestionable power over us. The fact that expertise goes through a process of legitimation also means that legitimacy may be withdrawn and the cognitive authority of experts may collapse, and this suggests something quite different than the idea that liberalism is a kind of self-contradiction, and also something much more interesting. We, the non-experts, decide whether claims to cognitive authority, which in political terms are requests to have their conclusions treated as neutral fact, are to be honoured. And we have, historically, changed our minds about who is 'expert', and what is to be treated as neutral fact.

This is, so to speak, a 'liberal' argument about expertise. It grants that cognitive authority and the acceptance of expertise, in modern conditions, is a condition of genuine public discourse. Liberalism, in the form of the principle of neutrality, is a means to the end of the creation of the conditions for public discourse. It is a means, however, that is not given by God, or the courts, or 'reason', but lives in the political decisions we make to regard assertions as open to public discussion or not. Historically, liberalism established the space for public discussion by expelling religious sectarian 'expertise'. The challenge of the present is, in part, to deal with the claims of non-religious experts to cognitive authority. There is no formula for meeting this challenge. But there is a process of legitimation and delegitimation. And it should be no surprise that this process has come to occupy more of public discourse than ever before. But the very vigour of discussion, and the ability of the public to make decisions about what claims are legitimate, belies the image of the liberal public as victim.

Is this enough? Or is there a higher standard of proper public deliberation to which public acceptance of expert claims ought to be held? Antiliberals, following the arguments of Habermas and Foucault, have generally said that it is not enough. For them, it is precisely the point of the critique of expertise to show how our forms of reasoning in public deliberation are preconditioned by unchallenged and, practically speaking, unchallengeable forming assumptions that derive from experts.[7]

The kind of social constructionism that has been practised in much of science studies is different in character, and has different implications, for it is concerned not with showing that some forms of discussion involve social construction and others do not, but with showing that even science has this character. As I have suggested, to the extent that it has been concerned with establishing the conventional and mutable character of many of the distinctions that philosophers of science have attempted to absolutize, that is to say to make scientists less immaculate and more like plumbers, social constructionism parallels a moment in liberal theory. The moment is the one at which it was recognized that the history of liberalism is a matter of 'continuation by other means', in which the 'foundations' of actual liberal democracies are conventions, custom, flexibly applied and typically somewhat vague 'principles' rather than rigid doctrines or acts of faith. A corollary recognition to this political realization is that despite being mutable and shifting, conventions have sufficed to preserve what Schmitt ([1926] 1985: 5) characterized as the real possibility of 'persuading one's opponents through argument of the truth or justice of something, or allowing oneself to be persuaded of something as true or just'.

The parallel claim that what counts as 'expert' is conventional, mutable and shifting, and that people are persuaded of claims to expertise through mutable, shifting conventions does not make the decisions to accept or reject the authority of experts less than reasonable in the sense appropriate to liberal discussion. To grant a rôle to expert knowledge does not require us to accept the immaculate conception of expertise. The lesson of the second kind of social constructionism is that these conditions, the conditions of mutability—and not some sort of analogue to Habermas's ideal-speech situation—are the conditions under which scientific consensus itself occurs, and that there is no alternative. This is a negative message, but nevertheless an important one, in that it excludes a certain kind of utopianism about expertise and its 'control' by some sort of higher reason. Excluding this kind of utopianism is a kind of answer to the issues with which we began. Expertise is a deep problem for liberal theory only if we imagine that there is some sort of standard of higher reason against which the banal process of judging experts as plumbers can be held, and if there is not, it is a deep problem for democratic theory only if this banal process is beyond the capacity of ordinary people.

NOTES

This paper forms a part of a larger project, tentatively titled *Liberal Democracy 3.0: Politics in the Age of Expertise*, to be published by Sage in its *Theory, Culture & Society* series. Research for this project has been supported by the National Science Foundation Ethics and Value Studies and the Science and Technology Studies Programs.

1. There are exceptions to this, such as Steve Fuller's *The Governance of Science* (2000). In political science itself, the writings of Aaron Wildavsky (1995) and Charles Lindblom and Edward Woodhouse (Lindblom & Woodhouse, 1993) should also be mentioned. Each reflects the concerns of Robert Dahl (1993) with the competence of citizens.

2. The American system of 'extension education' in agriculture is an example of this. It soon created a new kind of inequality, between early adoptees and laggards.

3. The claims about the nature of intelligence to which the letter-writer to *Newsweek* objected (Jaffe, 1994), curiously, produced a similar kind of collective letter signed by a large number of prominent psychologists, designed to correct what they saw to be the alarming disparity between what was presented by journalists and commentators as the accepted findings of psychological research on intelligence and what psychologists in fact accepted. Here the issues were different: the accepted facts were simply not known to the journalists, who seemed to assume that the facts fit with their prejudices.

4. This is a large theme of the literature that inaugurated 'professional' diplomacy and foreign policy analysis. Hans Morgenthau (1946), for example, stressed the idea that it was often a necessity for the leader to act against the democratic consensus with respect to foreign policy.

5. For a detailed theoretical account of the notion of discretionary power, see the discussion of the notion of decision in the work of Schmitt. Schmitt focuses on the puzzle of declarations of states of exception (or states of siege) which are not, by definition, governed by rules that fully define the conditions under which the decision-maker can act authoritatively. This is of course the same phenomenon as administrative discretion: the law is not, and perhaps cannot be, written to cover every contingency, so the bureaucrat, or judge, is given power to apply the law as he or she sees fit (Schmitt, [1922] 1985: 31–34).

6. Elsewhere, I have discussed some other aspects of the problem of expert knowledge in relation to power. In 'Forms of Patronage' (1990), I discussed the problem faced both by scientists and by governmental patrons in deciding whether to patronize scientists, and I suggested that there was a generic problem that arose from the fact that politicians and bureaucrats were not trained in a way that enabled them to judge the promises made to them by scientists. It is questionable whether scientists are able adequately to judge such promises, as they do, for example, in peer review decisions on grant applications. I pointed out in

that paper that the knowledge possessed by scientific experts was so specialized and fragmented that there was no general threat of scientists or experts as a group supplanting democracy. In 'Truth and Decision' (1989), I discussed the issue of the limitations of specialist knowledge in the face of ill-structured decisions of the sort that policy makers and politicians actually face. I noted that typically experts with different backgrounds framed issues in ways that conflicted, and that consequently there was no univocal expert opinion in such decisions. This speaks to the notion that 'expert culture' is some sort of unified whole: clearly it is not.

7. Thinkers like Foucault and Habermas present a more serious challenge than Fish does when they attack the power of the public to judge, because this undermines the notion of democratic or liberal legitimacy itself. For Habermas, for example, the communication on which the legitimacy of uncontested as well as contested viewpoints is based may be 'distorted', and its results therefore bogus. Foucault is even more direct. The beliefs that we share or accept widely as true as well as the (for him) small domain in which political contests occur are all essentially the product of non-consensual manipulation, or rather a kind of hegemonic intellectual influence which does not require conscious manipulators but which prevents the ordinary citizen from, to put it in somewhat different language, giving 'informed consent' to the arrangements under which he or she is compelled to live.

For Foucault, the condition of religious believer, that is to say the voluntary acceptance of the authoritative character of that which cannot be understood, is realized in an involuntary way by the citizen: the religious believer voluntarily accepts mystical authority; the ordinary citizen is mystified into the acceptance of uncontested givens through which he or she is deprived of the volitional and cognitive powers necessary for citizenship. Foucault holds out no hope that there can be any escape from this kind of 'control' and provides no exemptions from its effects, except perhaps to intellectuals who can recognize and protest against their fate, but who are politically irrelevant because they have no alternative to this fate. In Foucault, the experts and the public disappear simultaneously into the thrall of forms of discourse which is constitutive of their mental world. In Habermas, in contrast, there is an exemption for experts, of a sort. The people who do the steering are not trapped within the limitations of the life-world that they steer. This is not to say that they are not limited, however, by the effects of distorted communication. But their limitations are different from the limitations of those they administer over. Their control cannot be truly legitimate because the consent that they depend upon is not genuinely 'informed'. Those who assent are governed by myths that preclude their being truly informed or informable.

As I suggest in the conclusion, both lines of argument depend on a kind of utopianism about the character of knowledge that social constructionism undermines.

REFERENCES

Brown (1979). Richard E. Brown, *Rockefeller Medicine Men: Medicine and Capitalism in America* (Berkeley: University of California Press).

Buxton & Turner (1992). William Buxton and Stephen Turner, 'From Education to Expertise: Sociology as a "Profession"', in Terence C. Halliday and Morris Janowitz (eds), *Sociology and its Publics* (Chicago, IL: The University of Chicago Press): 373–407.

Collins & Pinch (1993). Harry Collins and Trevor Pinch, *The Golem: What Everyone Should Know About Science* (Cambridge & New York: Cambridge University Press).

Dahl (1993). Robert A. Dahl, 'Finding Competent Citizens: Improving Democracy', *Current*, No. 351: 23–30.

Feyerabend (1978). Paul Feyerabend, *Science in a Free Society* (London: New Left Books).

Fish (1994). Stanley Fish, *There's No Such Thing as Free Speech and It's a Good Thing, Too* (New York: Oxford University Press).

Fuller (2000). Steve Fuller, *The Governance of Science* (Buckingham, UK & Philadelphia, PA: Open University Press).

Gieryn (1994). Thomas F. Gieryn, 'Boundaries of Science', in Sheila Jasanoff, Gerald E. Markle, James C. Petersen and Trevor Pinch (eds), *Handbook of Science and Technology Studies* (Newbury Park, CA: Sage/4S): 393–443.

Gieryn & Figert (1986). Thomas F. Gieryn and Anne Figert, 'Scientists Protect their Cognitive Authority: The Status Degradation Ceremony of Sir Cyril Burt', in Gernot Bühme and Nico Stehr (eds), *The Knowledge Society* (Dordrecht, Holland: Reidel): 67–86.

Habermas ([1985] 1987). Jürgen Habermas, trans. Thomas McCarthy, *The Theory of Communicative Action*, Vol. 2 (Boston, MA: Beacon).

Hacking (1991). Ian Hacking, 'The Making and Molding of Child Abuse', *Critical Inquiry* 17: 253–88.

Haraway (1984–85). Donna Haraway, 'Teddy Bear Patriarchy: Taxidermy in the Garden of Eden, New York City, 1908–1936', *Social Text* 11: 20–64. Reprinted in Haraway, *Primate Visions: Gender, Race, and Nature in the World of Modern Science* (New York: Routledge, 1989): 26–59.

Jaffe (1994). Naomi Jaffe, *Newsweek* (21 November): 26.

Jasanoff (1995). Sheila Jasanoff, *Science at the Bar: Law, Science, and Technology in America* (Cambridge, MA & London: Harvard University Press).

Johnson & Johnson (1980). Guy Johnson and Guion Johnson, *Research in Service to Society: The First Fifty Years of the Institute for Research in Social Science at the University of North Carolina* (Chapel Hill: University of North Carolina Press).

Kellogg ([1912] 1985). Paul U. Kellogg, 'The Spread of the Survey Idea', in *The Social Survey: Papers by Paul Kellogg, Shelby M. Harrison and George T. Palmer* (New York: Russell Sage Foundation, 2nd edn). Reprinted from *The Proceedings of the Academy of Political Science*, Vol. II, No. 4 (July 1912): 1–17.

Lindblom & Woodhouse (1993). Charles E. Lindblom and Edward J. Woodhouse, *The Policy-Making Process* (Upper Saddle River, NJ: Prentice Hall, 3rd edn).

Merton (1976). Robert K. Merton, *Sociological Ambivalence and Other Essays* (New York: Free Press).

Morgenthau (1946). Hans J. Morgenthau, *Scientific Man vs Power Politics* (Chicago, IL: The University of Chicago Press).

Polanyi (1946). Michael Polanyi, *Faith and Society* (London: Oxford University Press).

Price (1965). Don K. Price, *The Scientific Estate* (London & New York: Oxford University Press).

Schmitt ([1922] 1985). Carl Schmitt, trans. George Schwab, *Political Theology: Four Chapters on the Concept of Sovereignty* (Cambridge, MA: MIT Press).

Schmitt ([1926] 1985). Carl Schmitt, trans. Ellen Kennedy, *The Crisis of Parliamentary Democracy* (Cambridge, MA & London: MIT Press).

Schmitt ([1932] 1996). Carl Schmitt, trans. George Schwab, *The Concept of the Political* (Chicago, IL: The University of Chicago Press).

Schudson (1997). Michael Schudson, 'Cultural Studies and the Social Construction of "Social Construction": Notes on "Teddy Bear Patriarchy"', in Elizabeth Long (ed), *From Sociology to Cultural Studies: New Perspectives* (Oxford: Blackwell): 379–99.

Schumpeter ([1942] 1950). Joseph Schumpeter, *Capitalism, Socialism and Democracy* (New York: Harper & Row).

Timmermans (1999). Stefan Timmermans, *Sudden Death and the Myth of CPR* (Philadelphia, PA: Temple University Press).

Turner (1989). Stephen Turner, 'Truth and Decision', in Daryl Chubin and Ellen W. Chu (eds), *Science off the Pedestal: Social Perspectives on Science and Technology* (Belmont, CA: Wadsworth): 175–88.

Turner (1990). Stephen Turner, 'Forms of Patronage', in Susan Cozzens and Thomas F. Gieryn (eds), *Theories of Science in Society* (Bloomington: Indiana University Press): 185–211.

Wildavsky (1995). Aaron Wildavsky, *But Is It True?: A Citizen's Guide to Environmental Health and Safety Issues* (Cambridge, MA & London: Harvard University Press).

5. MORAL EXPERTS

PETER SINGER

THE FOLLOWING POSITION has been influential in recent moral philosophy: there is no such thing as moral expertise; in particular, moral philosophers are not moral experts. Leading philosophers have tended to say things like this:

> It is silly, as well as presumptuous, for any one type of philosopher to pose as the champion of virtue. And it is also one reason why many people find moral philosophy an unsatisfactory subject. For they mistakenly look to the moral philosopher for guidance.
>
> (A. J. Ayer, 'The Analysis of Moral Judgments', in *Philosophical Essays.*)

or like this:

> It is no part of the professional business of moral philosophers to tell people what they ought or ought not to do. . . . Moral philosophers, as such, have no special information not available to the general public, about what is right and what is wrong; nor have they any call to undertake those hortatory functions which are so adequately performed by clergymen, politicians, leader-writers . . .
>
> (C. D. Broad, *Ethics and the History of Philosophy.*)

Assertions like these are common; arguments in support of them less so. The role of the moral philosopher is not the role of the preacher, we are told. But why not? The reason surely cannot be, as Broad seems to suggest, that the preacher is doing the job 'so adequately'. It is because those people who are regarded by the public as "moral leaders of the community" have done so badly that 'morality', in the public mind, has come to mean a system of prohibitions against certain forms of sexual enjoyment.

Another possible reason for insisting that moral philosophers are not moral experts is the idea that moral judgments are purely emotive, and that reason has no part to play in their formation. Historically, this theory may have been important in shaping the conception of moral philosophy that we have today. Obviously, if anyone's moral views are as good as anyone else's, there can be no moral experts. Such a crude version of emotivism, however, is held by few philosophers now, if indeed it was ever widely held. Even the views of C. L. Stevenson do not imply that anyone's moral views are as good as anyone else's.

A more plausible argument against the possibility of moral expertise is to be found in Ryle's essay "On Forgetting the Difference between Right and Wrong," which appeared in *Essays in Moral Philosophy*, edited by A. Melden. Ryle's point is that knowing the difference between right and wrong involves caring about it, so that it is not, in fact, really a case of knowing. One cannot, for instance, forget the difference between right and wrong. One can only cease to care about it. Therefore, according to Ryle, the honest man is not 'even a bit of an expert at anything' (p. 157).

It is significant that Ryle says that 'the honest man' is not an expert, and later he says the same of 'the charitable man.' His conclusion would have had less initial plausibility if he had said 'the morally good man.' Being honest and being charitable are often—though perhaps not as often as Ryle seems to think—comparatively simple matters, which we all can do, if we care about them. It is when, say, honesty clashes with charity (If a wealthy man overpays me, should I tell him, or give the money to famine relief?) that there is need for thought and argument. The morally good man must know how to resolve these conflicts of values. Caring about doing what is right is, of course, essential, but it is not enough, as the numerous historical examples of well-meaning but misguided men indicate.

Only if the moral code of one's society were perfect and undisputed, both in general principles and in their application to particular cases, would there be no need for the morally good man to be a thinking man. Then he could just live by the code, unreflectively. If, however, there is reason to believe that one's society does not have perfect norms, or if there are no agreed norms on a whole range of issues, the morally good man must try to think out for himself the question of what he ought to do. This 'thinking out' is a difficult task. It requires, first, information. I may, for instance, be wondering whether it is right to eat meat. I would have a better chance of reaching the right decision, or at least, a soundly based decision, if I knew a number of facts about the capacities of animals for suffering, and about the methods of rearing and slaughtering animals now being used. I might also want to know about the effect of a vegetarian diet on human health, and, considering the world food shortage, whether more or less food would be produced by giving up meat production. Once I have got evidence on these questions, I must assess it and bring it together with whatever moral views I hold. Depending on what method of moral reasoning I use, this may involve a calculation of which course of action produces greater happiness and less suffering; or it may mean an attempt to place myself in the positions of those affected by my decision; or it may lead me to attempt to "weigh up" conflicting duties and interests. Whatever method I employ, I must be aware of the possibility that my own desire to eat meat may lead to bias in my deliberations.

None of this procedure is easy—neither the gathering of information, nor the selection of what information is relevant, nor its combination with a basic moral position, nor the elimination of bias. Someone familiar with moral concepts and with moral arguments, who has ample time to gather information and think about it, may reasonably be expected to reach a soundly based conclusion more often than someone who is unfamiliar with moral concepts and moral arguments and has little time. So moral expertise would seem to be possible. The problem is not so much to know 'the difference between right and wrong' as to decide what is right and what wrong.

If moral expertise is possible, have moral philosophers been right to disclaim it? Is the ordinary man just as likely to be expert in moral matters as the moral philosopher? On the basis of what has just been said, it would seem that the moral philosopher does have some important advantages over the ordinary man. First, his general training as a philosopher should make him more than ordinarily competent in argument and in the detection of invalid inferences. Next, his specific experience in moral philosophy gives him an understanding of moral concepts and of the logic of moral argument. The possibility of serious confusion arising if one engages in moral argument without a clear understanding of the concepts employed has been sufficiently emphasized in recent moral philosophy and does not need to be demonstrated here. Clarity is not an end in itself, but it is an aid to sound argument, and the need for clarity is something which moral philosophers have recognized. Finally, there is the simple fact that the moral philosopher can, if he wants, think full-time about moral issues, while most other people have some occupation to pursue which interferes with such reflection. It may sound silly to place much weight on this, but it is, I think very important. If we are to make moral judgments on some basis other than our unreflective intuitions, we need time, both for collecting facts and for thinking about them.

Moral philosophers have, then, certain advantages which could make them, relative to those who lack these advantages, experts in matters of morals. Of course, to be moral experts, it would be necessary for moral philosophers to do some fact-finding on whatever issue they were considering. Given a readiness to tackle normative issues, and to look at the relevant facts, it would be surprising if moral philosophers were not, in general, better suited to arrive at the right, or soundly based, moral conclusions than non-philosophers. Indeed, if this were not the case, one might wonder whether moral philosophy was worthwhile.

PART II
EXPERTISE AND PRACTICAL KNOWLEDGE

THE FIVE ARTICLES in this section represent different approaches to theorizing the unique kind of practical knowledge that experts exhibit. In "Moral Knowledge as Practical Knowledge," Julia Annas reminds us that this subject is embedded firmly within the history of philosophy. She claims that Plato establishes a provocative relation between the specific features of practical expertise (*techné*) and more general epistemic criteria. "When Socrates seeks moral knowledge," Annas writes, "then, it is only to be expected that this will be seen on the model of practical expertise, since this is the model for knowledge in general." By explaining why practical expertise was understood as a model for knowledge in general, Annas intervenes critically into debates concerning how to best characterize the insights expressed in different periods of philosophical history. Not only does she aim to correct some of the dominant modern misconceptions concerning ancient philosophy, but, in doing so, she hopes to expose "a weakness in the modern approaches to moral epistemology" and reveal "a point where we might actually learn from the ancients."

Hubert Dreyfus's contribution, "How Far Is Distance Learning from Education?" represents one of his most recent attempts to use phenomenological description to: (1) classify the developmental stages of apprenticeship that typical novice learners necessarily pass through in order to become experts, and (2) use this classification as the basis for criticizing those who distort the proper understanding of expertise by characterizing it in disembodied terms. In the past, Dreyfus (sometimes in collaboration with others) has appealed to his model of expert skill acquisition in order to: (1) demythologize extravagant claims associated with artificial intelligence projects, in particular, "expert computer systems" designed to simulate human expertise; (2) assess social biases that "endanger" professional experts (such as nurses, doctors, teachers, pilots, and scientists) by imposing "rationalization" constraints; (3) explain what is wrong with dominant tendencies in American styles of business management; and (4) detail the expertise of political action groups. In the article included here, Dreyfus investigates the educational potential of "distance learning" programs by assessing the proposal that the traditional classroom experience can be replaced effectively through automation, using the Internet's technological capabilities. This proposal is embedded in a number of contexts, each of which suggests a range of binary oppositions:

- technical control versus traditional methodology
- face-to-face interaction versus depersonalization

- real versus virtual experience
- body-oriented experience versus mind-oriented experience
- profit-driven motives versus education-driven motives
- nostalgic aspirations versus visionary aspirations
- educational goals versus training goals
- the shape of society versus the shape of higher education

Dreyfus insists that what this proposal concerns, ultimately, is expertise, and that the phenomenological approach to expertise is sufficient for presenting a deep evaluation of it. He thus assesses the educational potential of distance learning by raising the following two questions: Can the typical learner become an expert through distance learning? And can an instructor's teaching expertise be transmitted through distance learning? Dreyfus answers both of these questions in the negative, claiming that "expertise cannot be acquired in disembodied cyberspace."

The next contribution, "Dreyfus on Expertise: The Limits of Phenomenological Analysis," is our attempt to assess Dreyfus's model of expert skill acquisition. We argue that his model is philosophically important because it shifts the focus on expertise away from its social and technical externalization in science and technology studies, and its relegation to the historical and psychological context of discovery in the classical philosophy of science, to universal structures of embodied cognition and affect. In doing so, we believe that Dreyfus provides a compelling explanation for why expert authority is not reducible to ideology or skillful networking. Furthermore, we demonstrate that because Dreyfus analyzes expertise from a first-person perspective, his phenomenology can be appealed to in order to reveal the limitations of and sometimes superficial treatment that comes from investigating expertise from a third-person perspective. However, we insist that both Dreyfus's descriptive model and his normative claims are flawed due to lack of hermeneutical sensitivity. He assumes an expert's knowledge has crystallized out of contextual sensitivity plus experience, and that an expert has shed, during the training process, whatever prejudices, ideologies, hidden agendas, or other forms of cultural embeddedness that person might have possessed initially. One would never imagine from Dreyfus's account that society could possibly be endangered by experts, only how society's expectations and actions could endanger experts. We claim that fictional stories and historical accounts of controversies involving experts demonstrate that things do not work the way Dreyfus claims, and would, in fact, be less salutary if they did.

The critical assessment of Dreyfus continues in Evan Selinger and John Mix's "On Interactional Expertise: Pragmatic and Ontological Considerations." In the context of evaluating Harry Collins's investigation into a third form of knowledge, "interactional expertise," it is claimed that while Collins corrects Dreyfus's untenable analysis of "authentic" expert language, he generates new problems for theorizing expertise and practical knowledge. According to Selinger and Mix, three deficiencies diminish Collins's account: (1) the "contributory" potential of interactional expertise is unduly limited; (2) the nature of embodiment is misunderstood; and (3) the proper relation between embodiment and interactional expertise is distorted. Since Collins attempts to discern the proper social role for all kinds of interactional experts—activists, critics, sociologists, journalists, and some science administrators—the reader should consider whether Selinger and Mix are right to insist that Collins potentially undermines the value that a more rigorously construed concept of interactional expertise might have.

The themes of embodiment and practical knowledge continue to be explored in Hélène Mialet's article, "Do Angels Have Bodies? The Cases of William X and Mr. Hawking." Mialet uses some of the analytic tools developed by anthropologists and sociologists of science to present a concrete account of how expertise is formed and maintained. On the basis of analyzing two case studies—one involving William X, a researcher working at France's largest petroleum company, the other involving Stephen Hawking, the famous physicist— she poses a variation of the traditional demarcation problem: How do these scientists come to distinguish themselves as creative geniuses? Her provocative answer is part of a larger argument: she proposes that the expert knower needs to be conceptualized in terms of a distributed identity—a "knowing subject" who is tied materially to specific tools, practices, and social networks. She thus seeks to show how individual experts accommodate themselves to others and how collective operations are constitutive of the singularity of an expert's character. In trying to understand how innovation is possible within an institutional context, one that contains pervasive technical, economic, and organizational constraints, she investigates practical knowledge in a way that differs greatly from the majority of phenomenologists and sociologists.

6. HOW FAR IS DISTANCE LEARNING FROM EDUCATION?

HUBERT DREYFUS

With knowledge doubling every year or so, "expertise" now has a shelf life measured in days; everyone must be both learner and teacher; and the sheer challenge of learning can be managed only through a globe-girdling network that links all minds and all knowledge. I call this new wave of technology hyperlearning. . . . It is not a single device or process, but a universe of new technologies that both possess and enhance intelligence. The hyper in hyperlearning refers not merely to the extraordinary speed and scope of new information technology, but to an unprecedented degree of connectedness of knowledge, experience, media, and brains—both human and nonhuman. . . . We have the technology today to enable virtually anyone who is not severely handicapped to learn anything, at a "grade A" level, anywhere, anytime.
—Lewis J. Perelman, *School's Out* (Avon/Education, 1993), 22–23.

Do not spend vast sums of money to buy machinery that you are going to set down on top of existing dysfunctional institutions. The Internet, for example, will not fix your schools. Perhaps the Internet can be part of a much larger and more complicated plan for fixing your schools, but simply installing an Internet connection will almost surely be a waste of money.
—Phil Agre, *Telematics and Informatics* 15(3) (1998): 231–234.

IN 1922 THOMAS EDISON predicted that "the motion picture is destined to revolutionize our educational system and . . . in a few years it will supplant largely, if not entirely, the use of textbooks." Twenty-three years later, in 1945, William Levenson, the director of the Cleveland public schools' radio station, claimed that "the time may come when a portable radio receiver will be as common in the classroom as the blackboard." Forty years after that the noted psychologist B. F. Skinner, referring to the first days of his "teaching machines," in the late 1950s and early 1960s, wrote, "I was soon saying that, with the help of teaching machines and programmed instruction students could learn twice as much in the same time and with the same effort as in a standard classroom."[1]

For two decades now computers have been touted as a new technology that will revitalize education. In the eighties they were proposed as tutors, tutees, and drillmasters but none of those ideas seem to have taken hold.[2] Now the latest proposal is that somehow the power of the World Wide Web will make possible a new approach to education for the twenty-first century in which each student will be able to stay at home and yet be taught by great teachers from all over the world.

Many influential people in the United States, including Al Gore, believe that the development of the Internet will solve the problems of our current educational system.[3] At the secondary school level, we will no longer have to worry about crammed classes, a deficient infrastructure, or the lowering of standards, and at the college level, we will be able to leave behind the demographic difficulties posed by too many students, limited access to the most expensive universities, and the need for constant retraining as skill requirements change. If the new technology is put to use in the right way, they maintain, a first class education will be available to everyone, everywhere in so far as they master the relevant information technology.

The implementation of this vision is well underway. Reed Hundt, who was from 1993 to 1997 Chairman of the Federal Communications Commission presiding over the implementation of the Telecommunications Act of 1996 and helping to negotiate the World Trade Organization Telecommunications Agreement, has no doubts or reservations about the power of the Net to transform education. Indeed, he is euphoric. He boasts that under his guidance

> the nation began the largest single national program ever to better education from K through 12—the Snowe-Rockefeller Amendment to the 1996 Telecommunications Act, which at this very moment is causing $4 billion dollars in new money to be spent to put the Internet in every classroom in the country.

And he goes on triumphantly:

> The mayor of Philadelphia told me that this particular feature was the most important thing done by the federal government for education in his lifetime. Rudy Giuliani told me that this would transform education in New York. I have been told the same thing by all the urban city mayors. In rural areas, the same message is coming forth.

It's a bit hard to see what specific vision of the Net's educational power could generate all this excitement, and Hundt's explanation of what the new contentedness will enable teachers and students to do only adds to the puzzle:

It is a transformation for education, K through 12. We have always followed the following view. Teachers should be isolated with students. . . . We should never have up-to-date information on any child. . . . It should be impossible to have a dialogue between parents and teachers on how kids are doing. Information should be hoarded, concealed, or destroyed. It should not be created, shared, developed, or learned from. We should make distance learning extremely expensive and hard to accomplish, awkward technologically, and economically impossible to implement. . . .

No doubt the Net can change all that, which means that, as far as the public schools are concerned, all one will get for their 4 billion dollars is an efficient e-mail system linking teachers, administrators and parents, and, for students, access to a lot of on-line information. (Also some kind of distance learning is mentioned but not explained.) But it's hard to see what this transformation in the method of communication of those involved in secondary education has to do with what goes on in the classroom. What proposed change in the method of education generates all the excitement?

The claims for universities are a lot more specific but, as we shall see, equally irrelevant. Hundt goes on:

I went back to my old school, Yale, and the dean of one of the professional schools told me: "Number one, the historic, primary purpose of the university was to have a library so that scholars could gather around it. And second, scholars could meet other scholars and work and talk to them. And third, there would be a validation system so that smart people would be stamped: grade A, Yale; grade A, University of Wisconsin at Madison— whatever. And fourth, it's a place of quiet contemplation.

All four of these purposes of a university are not just jeopardized but are probably invalid in the information age. No particular reason to go anywhere to have a library when the libraries of the world are available at your fingertips. No particular reason for scholars to actually physically meet with scholars. When you look at the reality in higher education today, the communities of scholars that interact with each other are on the Net; they're not in person anymore. . . . In terms of validation, how long will a validation system last when fundamentally the Internet disintermediates those systems? And last, in terms of quiet contemplation, it doesn't get any quieter than if you live exactly where you want to live.

So, he said, as far as he could tell, the whole idea of, in his case, Yale University, was threatened. . . .

Now what is striking about this Dean's four points as reported by Hundt is that there isn't a word about the role of the university in educating students.

Once a university education is defined as a way for scholars to collect information and talk to each other or be left alone, they don't need to be bodily present to each other, and so it looks like the Net could easily replace that vision of the University.

When Hundt attempts to include the students whom the Dean seems to have overlooked, he sees them as consumers of information.

> The Internet . . . disintermediates everything. And insofar as the university itself is the retailer of knowledge to the consumer, the student, it is disintermediated. Now all that's necessary is for people to be able to trust the new Internet system of education to bring down that old system.[4]

Granted that, insofar as education consists in sending facts from someone who has a lot of information to those who don't have it, the Web works well, but so would videotapes or any recording medium. There must be something more than information-consumption going on in distance learning or there is no point in adding the Internet to the canned lecture. Hundt goes on about empowering the individual scholar, education for all, getting rid of elite universities, but he adds nothing helpful about what education is supposed to be once the Net disintermediates everything.[5]

Of course, many educators hold the opposite view—viz. that colleges and universities are engaged in education, and that education requires face to face interaction between teachers and students. For example, Nancy Dye, President of Oberlin College, is sure that "learning is a deeply social process that requires time and face-to-face contact. That means professors interacting with students."[6] Likewise, *The New York Times* reports that "the American Federation of Teachers, . . . critical of the sterility of distance learning, noted, 'All our experience as educators tells us that teaching and learning in the shared human spaces of a campus are essential to the undergraduate experience.'"[7]

But neither side gives us any reason to accept their pronouncements. In the face of this stand-off with no arguments on either side, we have to take a careful look at education in the light of the new possibilities for distance learning and ask: Can distance learning enable students to acquire the skills they need in order to be good citizens skilled in various domains? Or, does learning really require face to face engagement, and, if so, why? Just what goes on in classrooms, lecture halls, seminar rooms, and wherever skills are learned?

First, we need to get clear about what skills are and how they are acquired.[8] So, before seeking to evaluate the conflicting claims concerning distance learning, I'll lay out briefly what seem to be the stages in which a student learns by means of instruction, practice, and, finally, apprenticeship, to become an expert in some particular domain and in everyday life. The question then becomes: can these stages be implemented and encouraged on the Web?

STAGE 1: NOVICE

Normally, the instruction process begins with the instructor decomposing the task environment into context-free features that the beginner can recognize without the desired skill. The beginner is then given rules for determining actions on the basis of these features, like a computer following a program.

For purposes of illustration, I'll consider three variations: a motor skill, an intellectual skill, and what takes place in the lecture hall. The student automobile driver learns to recognize such domain-independent features as speed (indicated by the speedometer) and is given rules such as shift to second when the speedometer needle points to ten. The novice chess player learns a numerical value for each type of piece regardless of its position, and the rule: "Always exchange if the total value of pieces captured exceeds the value of pieces lost." The player also learns to seek center control when no advantageous exchanges can be found, and is given a rule defining center squares and one for calculating extent of control.

In the classroom and lecture hall, the teacher supplies the facts and procedures that need to be learned in order for the student to begin to develop an understanding of some particular domain. The student learns to recognize the features and follow the procedures by drill and practice. Hundt is right that, as long as students are merely consumers of information, as they are at this stage, they don't need to be in a classroom with each other and a teacher at all. Each can learn at his own terminal, wherever and whenever is convenient. Clearly, in this way the Internet can offer an improved version of the correspondence course, but this can't be what the enthusiasts are shouting about.

In any case, merely following rules will produce poor performance in the real world. A car stalls if one shifts too soon on a hill or when the car is heavily loaded; a chess player who always exchanges to gain points is sure to be the victim of a sacrifice by the opponent who gives up valuable pieces to gain a tactical advantage. Understanding a language or a science is much more than memorizing the elements and the rules relating them. The student needs not only the facts but also an understanding of the context in which that information makes sense.

STAGE 2: ADVANCED BEGINNER

As the novice gains experience actually coping with real situations and begins to develop an understanding of the relevant context, he or she begins to note, or an instructor points out, perspicuous examples of meaningful additional aspects of the situation or domain. After seeing a sufficient number of examples, the student learns to recognize these new aspects. Instructional maxims can then refer to these new situational aspects, recognized on the basis of experience, as well as to the objectively defined non-situational features recognizable by the novice.

The advanced beginner driver uses (situational) engine sounds as well as (non-situational) speed in deciding when to shift. He learns the maxim: Shift up when the motor sounds like it's racing and down when it sounds like it's straining. Engine sounds cannot be adequately captured by a list of features, so features cannot take the place of a few choice examples in learning the relevant distinctions.

With experience, the chess beginner learns to recognize overextended positions and how to avoid them. Similarly, she begins to recognize such situational aspects of positions as a weakened king's side or a strong pawn structure, despite the lack of precise and situation-free definitions. The player can then follow maxims such as: attack a weakened king's side. Unlike a rule, a maxim requires that one already have some understanding of the domain to which the maxim applies.[9]

At school, mere information is contextualized so that the student can begin to develop an understanding of its significance. The instructor takes on the role of a coach who helps the student pick out and recognize the relevant aspects that organize and make sense of the material. Though aspects can be presented to passive students in front of their terminals, it is more efficient for the students to attempt to use the maxims that have been given them, whereas the instructor points out aspects of the current situation to the student as the student encounters them. Here the teacher needs to be present with the student in the actual situation of thought or action.

Still, at this stage, learning, whether it takes place at a distance or face to face, can be carried on in a detached, analytic frame of mind, as the student follows instructions and is given examples. But to progress further requires a special kind of involvement.

STAGE 3: COMPETENCE

With more experience, the number of potentially relevant elements and procedures that the learner is able to recognize and follow becomes overwhelming. At this point, since a sense of what is important in any particular situation is missing, performance becomes nerve-wracking and exhausting, and the student might well wonder how anybody ever masters the skill.

To cope with this overload and to achieve competence, people learn, through instruction or experience, to devise a plan, or choose a perspective, that then determines which elements of the situation or domain must be treated as important and which ones can be ignored. As students learn to restrict themselves to only a few of the vast number of possibly relevant features and aspects, understanding and decision making becomes easier.

Naturally, to avoid mistakes, the competent performer seeks rules and reasoning procedures to decide which plan or perspective to adopt. But such rules are not as easy to come by as are the rules and maxims given beginners in manuals and lectures. Indeed, in any skill domain the performer encounters a vast number of situations differing from each other in subtle ways. There are, in fact, more situations than can be named or precisely defined, so no one can prepare for the learner a list of types of possible situations and what to do or look for in each. Students, therefore, must decide for themselves in each situation what plan or perspective to adopt without being sure that it will turn out to be appropriate.

Given this uncertainty, coping becomes frightening rather than merely exhausting. Prior to this stage, if the rules don't work, the performer, rather than feeling remorse for his mistakes, can rationalize that he hadn't been given adequate rules. But, since at this stage, the result depends on the learner's choice of perspective, the learner feels responsible for his or her choice. Often, the choice leads to confusion and failure. But sometimes things work out well, and the competent student then experiences a kind of elation unknown to the beginner.

A competent driver leaving the freeway on an off-ramp curve, learns to pay attention to the speed of the car, not whether to shift gears. After taking into account speed, surface condition, criticality of time, and so on, he may decide he is going too fast. He then has to decide whether to let up on the accelerator, remove his foot altogether, or step on the brake, and precisely when to perform any of these actions. He is relieved if he gets through the curve without mishap, and shaken if he begins to go into a skid.

The class A chess player, here classed as competent, may decide after studying a position that her opponent has weakened his king's defenses so that an attack against the king is a viable goal. If she chooses to attack, she ignores weaknesses in her own position created by the attack, as well as the loss of pieces not essential to the attack. Pieces defending the enemy king become salient. Since pieces not involved in the attack are being lost, the timing of the attack is critical. If she attacks too soon or too late, her pieces will have been lost in vain and she will almost surely lose the game. Successful attacks induce euphoria, while mistakes are felt in the pit of the stomach.

If we were disembodied beings, pure minds free of our messy emotions, our responses to our successes and failures would lack this seriousness and excitement. Like a computer we would have goals and succeed or fail to achieve them but, as John Haugeland once said of chess machines that have been programmed to win, they seek their goal, but when it comes to winning, they don't give a damn. For embodied, emotional beings like us, however, success and failure do matter. So the learner is naturally frightened, elated, disappointed, or discouraged by the results of his or her choice of perspective. And, as the competent student becomes more and more emotionally involved in his task, it becomes increasingly difficult for him to draw back and adopt the detached maxim-following stance of the advanced beginner.

But why let learning be infected with all that emotional stress? Haven't we in the West, since the Stoics, and especially since Descartes, learned to make progress by mastering our emotions and being as detached and objective as possible? Wouldn't rational motivations, objective detachment, honest evaluation, and hard work be the best way to acquire expertise?

While it might seem that involvement could only interfere with detached rule-testing, and so would inevitably lead to irrational decisions and inhibit further skill development, in fact, just the opposite seems to be the case. Patricia Benner has studied nurses at each stage of skill acquisition. She finds that, unless the trainee stays emotionally involved and accepts the joy of a job well done, as well as the remorse of mistakes, he or she will not develop further, and will eventually burn out trying to keep track of all the features and aspects, rules and maxims that modern medicine requires. In general, resistance to involvement and risk leads to stagnation and ultimately to boredom and regression.[10]

Since students tend to imitate the teacher as model, teachers can play a crucial role in whether students will withdraw into being disembodied minds or become more and more emotionally involved in the learning situation. If the teacher is detached and computer-like, the students will be too. Conversely, if the teacher shows his involvement in the way he pursues the truth, considers daring hypotheses and interpretations, is open to student's suggestions and objections, and emotionally dwells on the choices that have lead him to his conclusions and actions, the students will be more likely to let their own successes and failures matter to them, and rerun the choices that lead to these outcomes.

In the classroom and lecture hall the stakes are less dramatic than the risk of having a car accident while driving or of losing an important game of chess. Still, there is the possibility of taking the risk of proposing and defending an idea and finding out whether it fails or flies. If each student is at home in front of his or her terminal, there is no place for such emotional involvement. On the contrary, the correspondence-course model of anonymous information consumers that promoters of distance learning seem to have in mind when they say that the course material will be available to anyone, anywhere, anytime, makes such involvement impossible. But, even if we drop the anytime, and suppose that the students are all watching the professor at the same time, as with interactive video, and everyone watching hears each student's question, each student is still anonymous and there is still no class before which the student can shine and also risk making a fool of himself. The professor's approving or disapproving response might carry some emotional weight, but it would be much less intimidating to offer a comment and get a reaction from the professor if one had never met the professor and was not in her presence. Thus, those who think like President Dye and The American Federation of Teachers may well be right. The Net's limitations where embodiment is concerned—the absence of face to face learning—may well leave students stuck at competence.

STAGE 4: PROFICIENCY

Only if the detached, information-consuming stance of the novice, advanced beginner, and distance learner is replaced by involvement, is the student set for further advancement. Then, the resulting positive and negative emotional experiences will strengthen successful responses and inhibit unsuccessful ones, and the performer's theory of the skill, as represented by rules and principles, will gradually be replaced by situational discriminations, accompanied by associated responses. Proficiency seems to develop if, and only if, experience is assimilated in this embodied, atheoretical way. Only then do intuitive reactions replace reasoned responses.

As usual, this can be seen most clearly in cases of action. As the performer acquires the ability to discriminate among a variety of situations, each entered into with involvement, plans are evoked and certain aspects stand out as important without the learner standing back and choosing those plans or deciding to adopt that perspective. Action becomes easier and less stressful as the learner simply sees what needs to be done rather than using a calculative procedure to select one of several possible alternatives. When the goal is simply obvious, rather than the winner of a complex competition, there is less doubt as to whether what one is trying to accomplish is appropriate.

To understand this stage of skill acquisition we must remember that the involved, experienced performer sees goals and salient aspects, but not what to do to achieve these goals. This is inevitable since there are far fewer ways of seeing what is going on than there are ways of reacting. The proficient performer simply has not yet had enough experience with the outcomes of the wide variety of possible responses to each of the situations he can now discriminate, to react automatically. Thus, the proficient performer, after spontaneously seeing the point and the important aspects of the current situation, must still *decide* what to do. And to decide, he must fall back on detached rule and maxim following.

The proficient driver, approaching a curve on a rainy day, may *feel in the seat of his pants* that he is going dangerously fast. He must then *decide* whether to apply the brakes or merely to reduce pressure by some specific amount on the accelerator. Valuable time may be lost while making a decision, but the proficient driver is certainly more likely to negotiate the curve safely than the competent driver who spends additional time *considering* the speed, angle of bank, and felt gravitational forces, in order to *decide* whether the car's speed is excessive.

The proficient chess player, who is classed a master, can recognize almost immediately a large repertoire of types of positions. She then deliberates to determine which move will best achieve her goal. She may know, for example, that she should attack, but she must calculate how best to do so.

A student at this level sees the question that needs to be answered but has to figure out what the answer is.

STAGE 5: EXPERTISE

The *proficient performer*, immersed in the world of his skillful activity, sees what needs to be done, but *decides* how to do it. The *expert* not only sees what needs to be achieved; thanks to his vast repertoire of situational discriminations, he also sees immediately how to achieve his goal. Thus, the ability to make more subtle and refined discriminations is what distinguishes the expert from the proficient performer. Among many situations, all seen as similar with respect to plan or perspective, the expert has learned to distinguish those situations requiring one reaction from those demanding another. That is, with enough experience in a variety of situations, all seen from the same perspective but requiring different tactical decisions, the brain of the expert gradually decomposes this class of situations into subclasses, each of which requires a specific response. This allows the immediate intuitive situational response that is characteristic of expertise.

The chess Grandmaster experiences a compelling sense of the issue and the best move. Excellent chess players can play at the rate of 5 to 10 seconds a move and even faster without any serious degradation in performance. At this speed they must depend almost entirely on intuition and hardly at all on analysis and comparison of alternatives. It has been estimated that an expert chess player can distinguish roughly 50,000 types of positions. For much expert performance, the number of classes of discriminable situations, built up on the basis of experience, must be comparatively large.

The expert driver not only feels in the seat of his pants when speed is the issue; he knows how to perform the appropriate action without calculating and comparing alternatives. On the off-ramp, his foot simply lifts off the accelerator and applies the appropriate pressure to the brake. What must be done, simply is done. As Aristotle says, the expert "straightway" does "the appropriate thing, at the appropriate time, in the appropriate way."

The student, who has mastered the material, immediately sees the solution to the current problem.

What is the role of the teacher at this stage? A student learns by small random variations on what he is doing, and then checking to see whether or not his performance has improved. Of course, it would be better for learning if these small random variations were not random—if they were sensible deviations. If the learner watches someone good at doing something, that could limit the learner's random trials to the more promising ones. So observation and imitation of the activity of an expert can replace a random search for better ways to act. In general, this is the advantage of being an apprentice. Its importance is particularly clear in professional schools.

One thing that professional schools must teach is the way the theory the student has learned can be applied in the real world. One way to accomplish this without apprenticeship is for the school to simulate the surroundings that the students are to function in at a later point in their careers. Business schools provide an instructive example. At American schools of business administration two different modes of thought dominate. One is to be found in the so-called analytical school where most teaching focuses on theory. This type of school rarely produces capable business people who are intuitive experts. The other tradition is based on case studies, where real-life situations are described to the students and discussed. This produces better results.

To become an expert, however, it is not sufficient to have worked through a lot of cases. As we have already seen in discussing the move from competence to proficiency, the cases must matter to the learner. Just as flight simulators work only if the trainee feels the stress and risk of the situation and does not just sit back and try to figure out what to do, for the case method to work, the students must become emotionally involved. So, in a business school case study, the student should not be confronted with objective descriptions of situations, but rather be led to identify with the situation of the senior manager and experience his agonized choices and subsequent joys and disappointments. Provided that they draw in the embodied, emotional student, not just his mind, simulations—especially computer simulations—can be useful. The most reliable way to produce involvement, however, is to require that the student work in the relevant skill domain. So we are back at apprenticeship.

Even where the subject matter is purely theoretical, apprenticeship is necessary. Thus, in the sciences, post-doctoral students work in the laboratory of a successful scientist to learn how their disembodied, theoretical understanding can be brought to bear on the real world. By imitating the master, they learn abilities for which there are no rules, such as how long to persist when the work does not seem to be going well, just how much precision should be sought in each different kind of research situation, and so forth. In order to bring theory into relation with practice, this sort of apprenticeship turns out to be essential.

Even in the humanities where there are no agreed upon theories, the graduate student needs personal guidance. Thus, she normally becomes a teaching assistant where she can interact with a practicing researcher and teacher. The teacher can't help but exhibit a certain style of approaching texts and problems and of asking questions. For example, he may manifest an aggressive style of never admitting he is wrong or a receptive style of soliciting objections and learning from his mistakes. It is their teacher's style more than anything else that the teaching assistants pick up and imitate, even though they usually don't realize that they are doing so. An inspiring teacher like Wittgenstein left several succeeding generations of students not only imitating his style of questioning but even his gestures of puzzlement and desperation.

STAGE 6: MASTERY

For passing on a style, apprenticeship is the only technique available. However, if what the expert produced were clones of his or her own style, apprenticeship would be stultifying. Taking the notion of apprenticeship seriously, one has to ask how, within this framework, new styles and innovative ability can be developed? The training of musicians provides a clue. If you are training to become a performing musician, you have to work with an already recognized master. The apprentice cannot help but imitate the master, because when you admire someone and spend time with her, her style becomes your style. But then the danger is that the apprentice will become merely a copy of the master, while being a virtuoso performing artist requires developing a style of one's own.

Musicians have learned from experience that those who follow one master are not as creative performers as those who have worked sequentially with several.[11] The apprentice, therefore, needs to leave his first master and work with a master with a different style. In fact, he needs to study with several such masters. Journeymen in medieval times, and performing artists even now, when they become good enough to develop a style of their own, travel around and work in various communities of practice. In music, the teachers encourage their students to work with them for a while and then go on to other teachers. Likewise, graduate students usually assist several professors, and young scientists may work in several laboratories.

It is easy for us moderns to misunderstand this need for apprenticeship to several teachers. We tend to think that the music apprentice needs to go to one master, for example, because she is good at fingering, to another because she is good at phrasing, and yet another because she is good at dynamics. That would suggest one could divide a skill into components, which is the wrong way to look at it. Rather, one master has one whole style and another has a wholly different style.[12] Working with several masters destabilizes and confuses the apprentice so that he can no longer simply copy any one master's style and so is forced to begin to develop a style of his own. In so doing he achieves the highest level of skill. Let us call it *mastery*. Such mastery would seem to be out of reach of the distance learner.

STAGE 7: PRACTICAL WISDOM

Not only do people have to acquire skills by imitating the style of experts in specific domains; they have to acquire the style of their culture in order to gain what Aristotle calls practical wisdom. Children begin to learn to be experts in their culture's practices from the moment they come into the world. In this task, they are apprenticed to their parents from the word go.

Our cultural style is so embodied and pervasive that it is generally invisible to us, so it is helpful to contrast our style with some other cultural style and how it is learned. Sociologists point out that mothers in different cultures handle their babies in different ways.[13] For example, American mothers tend to place babies in their cribs by putting them on their stomachs, which encourages the babies to move around more. Japanese mothers, contrariwise, put their babies on their backs so they will lie still, lulled by whatever they hear and see. American mothers encourage passionate gesturing and vocalizing, while Japanese mothers are much more soothing and mollifying. In general American mothers situate the baby's body and respond to the baby's actions in such a way as to promote an active and aggressive style of behavior. Japanese mothers, in contrast, promote a greater passivity and sensitivity to harmony. Thus, what constitutes the American baby as an *American* baby is its style, and what constitutes the Japanese baby as a *Japanese* baby is its quite different style.

The general cultural style determines how the baby encounters himself or herself, other people, and things. Starting with a style, various practices will make sense and become dominant and others will either become subordinate or will be ignored altogether. So, for example, babies never encounter a bare rattle. For an American baby a rattle-thing is encountered as an object to make lots of expressive noise with and to throw on the floor in a willful way in order to get a parent to pick it up. A Japanese baby may treat a rattle-thing this way more or less by accident, but generally, I suspect, a rattle-thing is encountered as serving a soothing, pacifying function like a Native American rainstick.

Once we see that a style governs how anything can show up as anything, we can see that the style of a culture governs not only the babies. The adults in each culture are completely shaped by it too. For example, it should come as no surprise to us, given the sketch of Japanese and American culture already presented, that Japanese adults seek contented, social integration, while American adults are still striving willfully to satisfy their individual desires. Likewise, the style of enterprises and of political organizations in Japan aims at producing and reinforcing cohesion, loyalty, and consensus, while what is admired by Americans in business and politics is the aggressive energy of a *laissez-faire* system in which everyone strives to express his or her own individuality, and where the state, businesses, or other organizations function to maximize the number of desires that can be satisfied without destructive instability.

Like embodied commonsense understanding, cultural style is too embodied to be captured in a theory, and passed on in courses. It is simply passed on silently from body to body, yet it is what makes us human beings and provides the background against which all other learning is possible. It is only by being an apprentice to one's parents and teachers that one gains what Aristotle calls practical wisdom—the general ability to do the appropriate thing, at the appropriate time, in the appropriate way. To the extent that we were able to leave our bodies behind and live in cyberspace and chose to do so, nurturing children and passing on one's variation of one's cultural style to them would become impossible.

CONCLUSION

At every stage of skill acquisition beyond the first three, involvement and mattering are essential to the acquisition of skills. Like expert systems following rules and procedures, the immortal pure minds envisaged by futurists like Moravec would at best be competent.[14]

Distance learning enthusiasts like Hundt need to realize that only emotional, involved, embodied human beings can become proficient and expert. So, while they are teaching specific skills, teachers must also be incarnating and encouraging involvement. Moreover, learning through apprenticeship requires the presence of experts, and picking up the style of life that we share with others in our culture requires being in the presence of our elders. On this basic level, as Yeats said, "Man can embody the truth, but he cannot know it."[15]

When one looks at education in detail—from coaching, to manifesting the necessary involvement, to showing how the theory of the domain can be brought to bear on real situations, to developing one's own style—one can see why the university can't be disintermediated. While the Dean Hundt quotes may well turn out to be superfluous, there is plenty of work left for universities like Yale.

Thus, in so far as we want to teach expertise in particular domains and practical wisdom in life, which we certainly want to do, we finally run up against the most important question a philosopher can ask those who believe in the educational promise of the World Wide Web: can the bodily presence required for acquiring skills in various domains and for acquiring mastery of one's culture be delivered by means of the Internet?

The promise of telepresence holds out hope for a positive answer to this question. If telepresence could enable human beings to be present at a distance in a way that captures all that is essential about bodily presence, then the dream of distance learning at all levels could, in principle, be achieved. But if telepresence cannot deliver the classroom coaching and the lecture hall presence through which involvement is fostered by committed teachers, as well as the presence to apprentices of masters whose style is manifest on a day to day basis so that it can be imitated, distance learning will produce only competence, while expertise and practical wisdom will remain completely out of reach. Hyper-learning would then turn out to be mere hype. So our question becomes: How much presence can telepresence deliver?

NOTES

1. Todd Oppenheimer, "The Computer Delusion," *The Atlantic Monthly*, July 1997.

2. See Dreyfus and Dreyfus, *Mind over Machine* (Free Press, 1988), Chapter 5.

3. It seems that this optimism is shared in China. Reuters reports on August 22, 2000:— Chinese President Jiang Zemin offered a ringing endorsement of the Internet on Monday, saying e-mail, e-commerce, distance learning and medicine would transform China.

4. This speech was given on April 29, 1999, at the Networking '99 conference in Washington, D.C. It was published in *Educom Review*, Volume 34, Number 6, 1999, and can be found on the web at: http://www.educause.edu/ir/library/html/erm9963.html.

5. Here are Hundt's remarks on this last point: "The Internet is also an assault on elites. One of the top three Ivy Schools reported the other day it was a little worried about the skewing of students with respect to the income base. . . . Here was the statistic. Eighty-five percent of the students at this top Ivy League school, . . . were from upper-income families or higher. . . . It's high time for the highest level of education to be democratized in this country."

6. "The Paula Gordon Show," Broadcast Feb. 19, 2000, on WGUN.

7. Trip Gabriel, "Computers Can Unify Campuses, But Also Drive Students Apart," *The New York Times*, November 11, 1996.

8. For more details, see *Mind over Machine*.

9. See Michael Polanyi, *Personal Knowledge* (Routledge and Kegan Paul, 1958).

10. Patricia Benner has described this phenomenon in *From Novice to Expert: Excellence and Power in Clinical Nursing Practice* (Addison-Wesley, 1984), 164. Furthermore, failure to take risks leads to rigidity rather than the flexibility we associate with expertise. When a risk-averse person makes an inappropriate decision and consequently finds himself in trouble, he tries to characterize his mistake by describing a certain class of dangerous situations and then makes a rule to avoid them in the future. To take an extreme example, if a driver, hastily pulling out of a parking space, is side-swiped by an oncoming car he mistakenly took to be approaching too slowly to be a danger, he may resolve to follow the rule, never pull out if there is a car approaching. Such a rigid response will make for safe driving in a certain class

of cases, but it will block further skill refinement. In this case, it will prevent acquiring the skill of flexibly pulling out of parking places. In general, if one seeks to follow general rules one will not get beyond competence. Progress is only possible if, responding quite differently, the driver accepts the deeply felt consequences of his action without detachedly asking himself what went wrong and why. If he does this, he is less likely to pull out too quickly in the future, but he has a much better chance of ultimately becoming, with enough frightening or, preferably, rewarding experiences, a flexible, skilled driver.

One might object that this account has the role of emotions reversed; that the more the beginner is emotionally committed to learning the better, while an expert could be, and, indeed, often should be, coldly detached and rational in his practice. This is no doubt true, but the beginner's job is to follow the rules and gain experience, and it is merely a question of motivation whether he is involved or not. Furthermore, the novice is not emotionally involved in choosing an action, even if he is involved in its outcome. Only at the level of competence is there an emotional investment in the choice of action. Then emotional involvement seems to play an essential role in switching over from what one might roughly think of as a left-hemisphere analytic approach to a right-hemisphere holistic one. Of course, not just any emotional reaction such as enthusiasm, or fear of making a fool of oneself, or the exultation of victory, will do. What matters is taking responsibility for one's successful and unsuccessful choices, even brooding over them; not just feeling good or bad about winning or losing, but replaying one's performance in one's mind step by step or move by move. The point, however, is not to analyze one's mistakes and insights, but just to let them sink in. Experience shows that only then will one become an expert. After one becomes an expert one can rest on one's laurels and stop this kind of obsessing, but if one is to be the kind of expert that goes on learning, one has to go on dwelling emotionally on what critical choices one has made and how they affected the outcome.

11. Klaus Nielsen, "Musical Apprenticeship, Learning at the Academy of Music as Socially Situated," *Nordic Journal of Educational Research*, Vol. 3, 1997.

12. If we take a closer look at apprenticeship, we find that this kind of training contains important insights for testing as well as teaching. The apprentice becomes a master by imitating a master. He gradually learns how to do the whole task. Since skills are not learned by components but, rather, by small holistic improvements, there is thus no way to test the student in each component of the relevant skill. Where mastery is at stake, the kind of examinations used in most universities and necessarily on the Internet is not useful, and even counter-productive. Rather, instead of giving the apprentice periodic examinations to see if he has mastered the components that are normally mastered by students at his stage, when it seems to the master that an apprentice has learned his craft, he is asked to do what is normally done by an expert in his domain of expertise. For example, if he is learning to make a musical instrument, he may be asked to make, say a violin. But without an examination scored on a normal curve, who is to decide whether or not the apprentice has made a good violin? Only an expert can tell. So the masters gather around and play the apprentice's violin to test it. If the apprentice has made a good violin, he is sent to another master. Otherwise, he is put back to work to gain more experience.

13. To get at the gist of the way style works, I've simplified the specific sociological claims. For more precise details, see, for example, W. Caudill and H. Weinstein, "Maternal Care and Infant Behavior in Japan and America," *Readings in Child Behavior and Development*, eds. C. S. Lavatelli and F. Stendler (New York: Harcourt Brace, 1972), 78ff.

14. Deep Blue, the program that is currently world chess champion, is not an expert system operating with rules obtained from experts. Experts look at at most 200 possible moves, while Deep Blue uses brute force to look at a billion moves a second and so can look at all moves seven moves into the future without needing to understand anything.

15. Yeats's last letter, in *The Letters of W. B. Yeats*, ed. Allen Wade (Macmillan: New York, 1995), 922, written, just before his death, to Lady Elizabeth Pelham.

EVAN SELINGER AND ROBERT P. CREASE

INTRODUCTION

Expertise is of central importance to contemporary life, in which many economic, political, scientific, and technological decisions are routinely delegated to experts (Barbour, 1993, pp. 213–223). Citizens defer to the authority of experts not only in circumstances involving technical dimensions, but also in "all sorts of common decisions" (Walton, 1997, p. 24). On the one hand, routine deference to experts has political consequences and some scholars even suggest that it undermines rational democratic procedures and communicative action by allowing ideology to substitute for critical discussion (Turner, 2001).[1] On the other hand, volatile controversies over issues with a scientific-technical dimension can result in the suspension of routine deference and increased suspicion towards experts. It is hardly surprising, therefore, that the nature and the proper criteria for identifying expertise have been hotly debated in political and legal contexts. In legal contexts, for instance, the question of the proper criteria for expertise regularly arises in connection with developing the appropriate criteria for certifying expert witnesses.[2]

Expertise is an issue for which philosophical clarification seems appropriate and even essential. The 1993 landmark decision by the U.S. Supreme Court concerning the use of expert witnesses, *Daubert v. Merrell Dow Pharmaceuticals*, appealed to several concepts of philosophical origin, with particular attention given to Karl Popper's notion of "falsification" (Huber and Foster, 1999, pp. 37–68). Philosophers themselves occasionally have been called upon to serve as expert legal witnesses while medical ethicists have advised hospital boards, politicians, and U.S. Presidents concerning such politically sensitive programs as the Human Genome Project and stem cell research (Nussbaum, 2001; Ruse, 1996).[3]

Aside from its social implications, the issue of expertise is also philosophically important for several reasons. One is that it bears on the philosophy of mind. The classical locus of expertise, or "expert knower," is the subject, and the way experts understand and are attuned to the world bears on the nature of subjectivity, intentionality, and rationalist and representational notions of consciousness (Pappas, 1994). The nature of expertise, for instance, is the focal point around which turns the debate over whether intelligence can be successfully disembodied in artificial intelligence (AI) schemes, expert systems, and computer-based distance learning programs (Collins, 1995).

A second reason for the philosophical importance of the question of expertise is that it crystallizes the conflict between two traditions, classical philosophy of science and science and technology studies (henceforth STS). Classical philosophy of science takes expertise for granted and assumes the legitimacy of an expert-lay divide, while STS takes a skeptical approach towards experts for granted, presupposing the need to expose their illegitimacy (Mialet, 1999, pp. 552–553).

Finally, addressing expertise stimulates the interface between phenomenology and the sciences. Though some have argued that phenomenological description is only capable of capturing subjective dimensions of experience, and hence is inappropriate to use when trying to understand science (Latour, 1999, p. 9), scientific practice requires the development, exercise, and coordination of a variety of expert skills that are open to phenomenological clarification.

Nevertheless, philosophers have rarely addressed the subject explicitly, though implicit and unexamined notions of expertise often lurk under rubrics such as "authority," "colonization," "power," and "rational debate" (Turner, 2001). Hubert Dreyfus is one of the few to have overtly addressed the concept, and this essay is devoted to critically appraising his account. This analysis of Dreyfus will proceed in five steps. We shall: (1) place his model of expertise in an embodied context, (2) outline his general conception of skills, (3) summarize his descriptive model of expertise, (4) present his normative theory regarding which expectations about experts are justifiable, and (5) point out certain problems with his account. We argue that Dreyfus, by proposing that fundamental expert characteristics can be specified independently of cultural and historical considerations, demonstrates the importance of phenomenology to the subject by showing persuasively that expertise cannot be examined exhaustively by sociological, historical, and anthropological analyses. But we also identify certain descriptive and normative problems in Dreyfus's account. While Dreyfus shows phenomenologically that experts cannot be reduced to ideologues and artifacts of social networking, he also lacks hermeneutical sensitivity by overstating the independence of the expert and expert decision-making from cultural embeddedness.

1. EXPERTISE AND THE BODY

The significance of Dreyfus's account of expertise can be highlighted by recalling just how the two traditions mentioned above, classical philosophy of science and STS, avoid addressing the issue. Each in effect treats creative expert performance as something extraordinary that 'just happens,' and which poses no special philosophical problems. The goal of traditional philosophy of science, for instance, is the rational reconstruction of the organizational dimensions at the root of science's efficacy and objectivity, with particular emphasis on how its operation is only temporarily disrupted by anomalies. It approaches creativity as a predominantly mental act for which true philosophical discussion can take place only about its products. Where creative ideas come from is not considered a proper epistemological question and is relegated to psychology or history in the framework of the distinction between context of discovery and context of justification (Mialet, 1999, p. 552).

Contemporary STS, on the other hand, treats expertise as "distributed," externalized into particular settings such as laboratory and social networks, and standardized in technologies, criteria of scientificity, protocols for evaluating proof, and the rhetorical means of recruiting allies (Mialet, 1999, p. 552). To do otherwise, according to STS proponents, would risk "naturalizing" expertise, conferring undue authority on experts and leading to the repression of lay knowledge, values, and interests. By focusing on how experts become overly exalted through processes of mediation, STS proponents place a nondistributed sense of expertise into a "black box" (Mialet, 1999, p. 553).[4] While STS theorists occasionally appeal to tacit knowledge in their work, these appeals are not tied to any theory of embodiment.[5]

Although traditional philosophy of science and STS have different motives for wanting to demystify expertise, they produce similar results. For all their sharp disagreements, both traditional philosophy of science and STS refrain from discussing the relation between expertise and the body. They both agree that to demystify expertise and provide an accurate account of science, invariant features of bodily praxis need to be ignored.

The neglect of embodiment broaches the traditional phenomenological theme that practical involvements of living bodies ultimately ground any knowledge that they have about the world, including abstract-scientific knowledge.[6] For phenomenologists, all practical and theoretical activities, no matter how abstract their outcomes, need to be understood on a continuum with basic lifeworld practice. Dreyfus follows in this tradition by rigorously treating expert judgment and behavior as an instance of embodied human performance. Like classical philosophers of science and STS proponents, he seeks to demystify expertise—but he does so by placing expertise on a continuum of lifeworld activities, rather than isolating it from them. Experts merely act the way each of us does when performing mundane tasks: "We are all experts at many tasks and our everyday coping skills function smoothly and transparently so as to free us to be aware of other aspects of our lives where we are not so skillful" (Dreyfus and Dreyfus, 1990, p. 243).

Dreyfus explicitly links his skill model to phenomenological tradition by calling it an explicit development of Merleau-Ponty's notion of the lived body (*le corps vécu*) and the concepts "intentional arc" and "maximal grip."[7] By closely attending to "how our relation to the world is transformed as we acquire a skill," Dreyfus claims that he intends to "lay out more fully than Merleau-Ponty does" how skills are acquired, improved, and used (Dreyfus, 1999a, p. 1). Dreyfus aims to study experts as *embodied, situated subjects*, seeking to note "under what conditions deliberation and choice appear" in order to avoid "making the typical philosophical mistake of reading the structure of deliberation and choice into [his/her] account of everyday coping" (Dreyfus and Dreyfus, 1990, p. 239). But Dreyfus also contends that his model of how experts act is empirically verified by neural network researchers, specifically Walter Freeman in his studies of the brain dynamics underlying perception (1999a; 1999b, pp. 6–10; 1998).[8]

Dreyfus's account offers what might be called a metaphysics of expertise (though he prefers the term 'ontological' to 'metaphysical'). This is in sharp contrast to both traditional philosophy of science and STS. For the former, there is not enough subject matter for a metaphysical account; for the latter, expertise is a culturally dependent phenomenon whose definition changes in relation to the historical transformations that govern how it is perceived.

2. EXPERTISE AND SKILLS

Dreyfus first developed the basis for his descriptive account of expertise with his brother Stuart during the 1960s, when hired by the RAND corporation as a consultant to evaluate their work on artificial intelligence. His research culminated in a paper called "Alchemy and Artificial Intelligence" (Dreyfus, 1967) and the book *What Computers Can't Do* (Dreyfus, 1992).[9] In *Mind Over Machine: The Power of Human Intuition and Expertise in the Era of the Computer*, the two brothers developed a model of expert skill acquisition whose scope was claimed to be universal (Dreyfus and Dreyfus, 1986). They aimed to provide a phenomenological account of how adults acquire skills by instruction in all fields involving skilled performance,[10] whether of the intellectual or motor kind.[11]

This model of expert skill acquisition serves as a touchstone in Dreyfus's career. He uses it, for instance, as the basis to: (1) demythologize the hype associated with artificial intelligence projects, in particular, "expert computer systems" designed to simulate human expertise (Dreyfus and Dreyfus, 1986; Dreyfus, 1992); (2) judge social biases that "endanger" professional experts (such as nurses, doctors, teachers, and scientists) by imposing "rationalization" constraints (Dreyfus and Dreyfus, 1986); (3) explain what is wrong with dominant tendencies in American styles of business management (Dreyfus and Dreyfus, 1986); (4) defend the accuracy of Merleau-Ponty's non-representational account of intentionality and action (Dreyfus, 1998); (5) expose the practical limits of Jürgen Habermas's neo-Kantian conception of ethics (Dreyfus and Dreyfus, 1990); (6) explain the expertise of political action groups (Dreyfus, Spinosa and Flores, 1997); (7) clarify what *techne* and *phronesis* mean for Martin Heidegger and Aristotle, and in doing so, correct his acclaimed, book-length interpretation of Heidegger (Dreyfus, 2000); (8) explain the relevance of Kierkegaard's normative stages of the subject's development to evaluations of the Internet's value as a medium for communication (Dreyfus, 2001); and (9) critique the educational potential of "distance learning" programs (Dreyfus, 2001).

A first key element of Dreyfus's account is his rejection of the common tendency to define experts as sources of information. Expert skills are principally a matter of practical reasoning, of "knowing how" rather than "knowing that" (Ryle, 1984). "Knowing that" is prepositional knowledge of and about things, obtainable through reflection and conscious appreciation. "Knowing how," such as the ability to walk, talk, and drive, involves practical knowledge that is mostly experienced as a "thoughtless mastery of the everyday" and does not require conscious deliberation for successful execution (Dreyfus and Dreyfus, 1990, p. 244). In many instances, knowing how involves the exercise of inarticulable skills of which one cannot fully give an account, though one should not confuse this with Polanyi's concept of "tacit knowledge" (Dreyfus and Dreyfus, 1986, p. 16).[12]

It is possible to suggest hints and maxims that approximate elements of smooth performance, but seeing the point of these hints, and being capable of following these maxims, to a large extent presuppose the skill that the hints and maxims are supposed to account for. Moreover, Dreyfus holds that once skill is acquired, one tends not to follow the maxims used during the initial stages of learning.

Following Heidegger, Dreyfus argues that practical agents tend to reflect only on how their experience is organized during "breakdown" scenarios when pre-thematic styles of environmental coping prove insufficient for accomplishing ordinary goals; people only reflect on how they do what they do when what they do fails to work effectively. The justification for this derivative use of reflection is also practical, based on the agent's experience that to reflect upon what one does and the rules for doing it usually leads to practical problems doing what one wants to do. Going from pre-conscious behavior to a conscious appreciation of the rules followed during particular actions marks the agent's transition from practical to theoretical reasoning, from "knowing how" to "knowing that" (Dreyfus and Dreyfus, 1986, p. 7). For Dreyfus, the most basic tasks requiring skilled performance are best accomplished if one does not consciously focus on what one is doing.

Finally, Dreyfus contends that skills are flexible ranges of response and even physical skills cannot be reduced to a repetitive series of kinesthetic movements. He contends, "A skill, unlike a fixed response or set of responses can be brought to bear in an indefinite number of ways" (Dreyfus, 1992, p. 249). Joseph Rouse elaborates this point in a discussion of Dreyfus. In learning to throw, he writes, what is involved is not a series of repetitive motions, "but a range of responses to throwable things. Learning to throw overhand means that one can also throw sidearm, though the movement is different. Having learned to imitate a fairly limited number of sentences, I can produce an un-limited variety of different ones" (Rouse, 1987, p. 61). Thus to learn a flexible range of response is not to memorize an "actual movement" or "thought pattern" with the intention of repeating them, but to grasp of a field of possibili-ties" (Rouse, 1987, p. 61). Dreyfus, however, no doubt would add that while skills are flexible, there are limits, borderline areas, and marginal spaces; know-ing how to throw a baseball overhand does not enable one to throw a javelin, to throw underhand, or to throw a softball sidearm.

After *Mind Over Machine*, Dreyfus's model of expertise appeared nearly verbatim in numerous articles. In them, he never challenges the core theses about, nor descriptions of, expertise articulated in *Mind Over Machine*, but ex-pands the range of what the model can be applied to, and comments on the epistemological and metaphysical features of expert performance that he pre-viously applied but did not fully articulate.[13]

3. THE DESCRIPTIVE MODEL

Dreyfus's descriptive model of expertise has several key features. A first is that it is supposed to have *phenomenological justification*. Traditionally, the concept of justification relates propositions to a public sphere that can demonstratively verify or refute the content of, or logical or inferential connections in, state-ments. From a phenomenological perspective, propositions are justified when they detail experiential invariants that all subjects are capable of recognizing as according with their own experience. This is signaled in the following invita-tion Dreyfus extends to his readers: "You need not merely accept our word but should check to see if the process by which you yourself acquired various skills reveals a similar pattern" (Dreyfus and Dreyfus, 1986, p. 20). Dreyfus claims to have based his model of expertise on invariant patterns found in descriptions of skill acquisition relayed by first-person testimonies of "airplane pilots, chess players, automobile drivers, and adult learners of a second language" who dis-cussed how they learned to make "unstructured" decisions (Dreyfus and Drey-fus, 1986, p. 20).[14] Thus, his account is not supposed to be vulnerable to accu-sations of mischaracterizing idiosyncratic experts as paradigmatic or begging the question of how one knows what an exemplary expert in a field is. Because of adhering to the phenomenological method and its justificatory mechanisms, his account is also not meant to be susceptible to counterexamples that could disprove its applicability to adults who deliberately seek to acquire skills. It is putatively immune to these types of criticism because his skill model is expected to be meaningful in such a way that all experts can rediscover and verify its essential elements for themselves.

Another key feature is that his model is *developmental*, and envisions skill acquisition as occurring sequentially through five ascending stages: (1) an initial "beginner" phase, (2) an "advanced beginner" phase, (3) a "competent" phase, (4) a "proficient" phase, and (5) finally a culminating "expert" stage. In the first stage, the beginner, who "wants to do a good job," learns a "context free" set of "rules for determining action" and tends to act slowly in remembering how to apply them (Dreyfus and Dreyfus, 1986, p. 21). In the advanced beginner stage, the student who now has more "practical experience in concrete situations" begins to "marginally" improve by recognizing "meaningful additional aspects" of the situation that are not codified by rules (Dreyfus and Dreyfus, 1986, pp. 22–23).

What is particularly interesting about Dreyfus's account is that the learner undergoes not just cognitive and practical transformations but affective ones as well. Beginners and advanced beginners, he claims, typically experience their commitment to a practice as "detached" while a competent performer feels "involved" in the outcome of his or her performance (Dreyfus and Dreyfus, 1986, p. 26). In the competent stage the learner frequently feels "overwhelmed," as if he or she is "on an emotional roller coaster," having to cope with "nerve-wracking and exhausting" aspects of the practice, and feels "overloaded" due to facing too many potentially relevant elements to remember (Dreyfus, 2001, p. 35). Consequently, the competent learner narrows down those elements, devises a "plan" and chooses a "perspective" in order to selectively address "relevant features and aspects" of the situation (Dreyfus and Dreyfus, 1986, pp. 26–27). By making these changes, the competent performer experiences "a kind of elation unknown to the beginner," including "pride" and "fright" (Dreyfus and Dreyfus, 1985, pp. 117–118). Accepting responsibility means "mistakes are felt in the pit of the stomach" (Dreyfus and Dreyfus, 1985, p. 118).

In the proficient stage, the student transcends what Dreyfus calls the "Hamlet model of decision making," the "detached, deliberative, and sometimes agonizing selection of alternatives," which typifies the first three stages of skill acquisition (Dreyfus and Dreyfus, 1986, p. 28). Here the performer's reliance on rules and principles for seeing what goals need to be achieved is largely replaced by "know-how," an "*arational*" grasp of the situation that Dreyfus calls "intuitive behavior"; although the proficient performer must still contemplate and deliberate about what to do to achieve his or her goals (Dreyfus and Dreyfus, 1986, pp. 27–36).[15] "Action becomes easier and less stressful" as the proficient performer "simply sees what needs to be done rather than using a calculative procedure to select one of several possible alternatives" (Dreyfus, 2001, p. 40).

In the final stage, the expert not only sees what needs to be done, but also how to achieve it without deliberation, immediately, yet "unconsciously," recognizing "new situations as similar to whole remembered ones" and intuiting "what to do without recourse to rules" (Dreyfus and Dreyfus, 1986, p. 35)—though, interestingly, the Dreyfus's removed all references to remembered cases in the 1988 paperback edition. Thus, the expert, like masters in the "long Zen tradition" or Luke Skywalker when responding to Obi-Wan Kenobi's advice to "use the force," transcends "trying" or "efforting" and "just responds" (Dreyfus, 1999b, p. 22, n. 13). Dreyfus summarizes the "fluid performance" of expertise as: "*When things are proceeding normally, experts don't solve problems and don't make decisions; they do what normally works*" (Dreyfus and Dreyfus, 1986, pp. 30–31). He even claims that at the expert level the ability to distinguish between subject and object disappears: "The expert driver becomes one with his car, and he experiences himself simply as driving, rather than as driving a car" (Dreyfus and Dreyfus, 1986, p. 30). When the expert experiences the "flow" of peak performance he or she "does not worry about the future and devise plans" (Dreyfus and Dreyfus, 1986, p. 30). By being immersed in the moment, the expert can experience "euphoria," which athletes describe as playing "out of your head" (Dreyfus and Dreyfus, 1986, p. 40).

Yet another key feature of Dreyfus's model has to do with the *demarcations* between beginners and experts. Whereas some STS researchers argue that it is untenable to decisively distinguish the expert from the nonexpert because their apparent differences are social illusions, Dreyfus contends that beginners and experts can be demarcated in three different ways.[16] A first is based on the expert's "immersion into experience and contextual sensitivity." The expert differs from the beginner because he or she no longer relates principally to a practice analytically through context-free features that are recognizable without experience. Instead, through skillful behavior rooted in experience, he or she recognizes important features as contextually sensitive (Dreyfus and Dreyfus, 1986, p. 35). This situational engagement leads to a change in the expert's judgment: "[The] novice and advanced beginner exercise no judgment . . . and those who are proficient or expert make judgments based upon their prior concrete experience in a manner that defies explanation" (Dreyfus and Dreyfus, 1986, p. 36). A second demarcation centers on the temporal connection between action and decision making. While slowly following rules and deliberating characterize the beginner's actions, the expert's actions are immediate and intuitive situational responses. A final way of demarcation concerns affective transformation. In passing through developmental stages the expert's subjectivity and relation to the world are transformed in a manner that qualitatively differs, and can be demarcated, from a beginner's relation to the world. While in early stages the learner is "frustrated" and "overwhelmed," in the last stage the expert, who learned that the outcome of a situation matters by making risky commitments, sheds those affects and enjoys "fluid" and "smooth" performance. In his critique of John Searle's account of the background and its relation to intentionality, Dreyfus describes an expert tennis player as being able to become so absorbed in the "flow" of the game that he or she no longer feels the pressure to win, and only responds to the gestalt tensions on the court (1999b, pp. 4–5).[17] Additionally, the change in affect from beginner to expert corresponds with a change in meaning. While the beginner's attitude is essentially one of interest or curiosity, the expert is fully invested in his or her own being. Dreyfus quotes chess grandmaster Bobby Fischer—chess being the paradigmatic intellectual skill for Dreyfus—to the effect that "chess is life" (Dreyfus and Dreyfus, 1986, p. 33).

Despite these demarcations Dreyfus's account also involves *continuities*. Even the most abstract practice maintains an essential, though sometimes hidden, relation to the lifeworld. This is why he avoids using the contrasting and antonymic terms 'nonexpert' and 'layperson,' and instead uses "beginner," suggesting someone on one end of a spectrum of potentialities.[18]

Dreyfus's model has *foundational implications*. Different fields are organized around different types of essential skills; still, his account is a model of skill acquisition that can be formally specified without reference to any particular field. Of course diverse fields define experts in different ways due to the type of content that typifies the field. For example, a relation to winning defines an expert chess player, whereas a relation to arriving at a destination defines an expert driver. Nevertheless, a fundamental definition of expert as intuitive, committed, rule-transcending subject "whose skill has become so much a part of him that he need be no more aware of it than he is of his own body" applies to both of these (Dreyfus and Dreyfus, 1986, p. 30). At the phenomenological level, the expert chess player and car driver are functionally equivalent *qua* expert (Dreyfus and Dreyfus, 1986, pp. 21–35).

Finally, Dreyfus's account provides a *practical expert point of view*. Based upon the universal scope of his model, he attempts to describe a common cognitive and affective relation to the world that all experts share, which can be evoked as the basis of justifying the expert's commitments. When other theorists have attempted to delineate the "expert point of view," they have mostly done so from a theoretically holistic perspective, pointing to the expert's synoptic command of a field. For example, Scott Brewer argues that while nonexperts can know particular, true things about the methods and facts pertinent to a field, only experts know how the relevant features of a field, such as "enterprise" and "axiological" characteristics relate to provide a shared sense of what, how, and why practitioners in that field can claim to be true (1998, pp. 1568–1593).[19] Brewer notes that his focus on the organization of theoretical knowledge is somewhat Platonic (1998, p. 1591). By contrast, Dreyfus often refers to his skill model as anti-Platonic. Rather than providing a theoretically holistic account of the "expert point of view," he presents a description of the common features of practical understanding. The expert's practical understanding does not come from beliefs or theoretical commitments, but from acquired and embodied skills. Hence, Dreyfus writes, "The moral of the five-stage model is there is more to intelligent behavior than calculative rationality" (Dreyfus and Dreyfus, 1986, p. 36).

4. NORMATIVE IMPLICATIONS

Dreyfus's account describes which expectations about expert services are justifiable, and what can and cannot be legitimately asked of them. Dreyfus's account thus has normative implications, which he discusses empirically in connection with, among others, ballistics examiners, chicken sexers, citizens, judges, nurses, paramedics, physicians, science advisors, and teachers (Dreyfus and Dreyfus, 1986, pp. 196–201; Dreyfus, Spinosa and Flores, 1997, p. 106). A phenomenological understanding of the nature of expert decision making, Dreyfus suggests, needs to be the basis for identifying the means by which we may achieve some of the social and political goals that involve experts, such as consulting experts for personal, institutional, and legal reasons. Although directly deducing normative obligations from a descriptive foundation might suggest that Dreyfus commits the naturalistic fallacy of deducing obligations about what ought to be done from premises that only state what is the case, he contends that the relation between phenomenology and normativity is an issue of "priority."[20]

These normative implications are significant for several reasons. A first is the renowned difficulty experts have communicating with others. Although this is sometimes attributed to the disparity between technical and ordinary language, and sometimes to psychological factors such as arrogance, Dreyfus's account suggests deeper causes, an issue to which we shall return. Another reason the normative implications of his account are significant is the frequency with which experts serving as personal, institutional, and legal advisors are challenged as to their motives and biases, which may arise in connection with: (1) their professional training, which can influence how experts conceptualize and bound the problems they deal with; (2) their employers, which can influence politically the conclusions experts arrive at; (3) economic interests, which are capable of turning experts into hired guns whose recommendations can be purchased for the right price; and (4) the desire for recognition and reputation—another issue to which we shall return. By contrast, Dreyfus suggests that essential normative restrictions surrounding expert performance can be determined without considering the social forces that have sometimes influenced what claims experts make—another issue to which we shall return in our critique.

The normative implications arise from the claim that novices cannot transcend rules through developed intuition in the way that experts can in any field, no matter how dire or pressing the social circumstances might be.[21] It is therefore illegitimate, according to Dreyfus, to expect them to describe their process of decision making in prepositional statements, because their decisions are made on the basis of tacitly operating intuition. Not only does the chess grandmaster act on intuition, so too does the ballistics expert, who cannot propositionally express in a truthful manner how he or she determined whether or not a particular bullet originated in a particular gun (Dreyfus and Dreyfus, 1986, p. 199). Yet in assuming the role of expert legal witness the ballistics expert will be expected to do so, correlating conclusions to rules, and only in so doing will he or she potentially be convincing to a jury. But Dreyfus argues this persuasive power comes at the expense of prioritizing the "form" over the "content" of the explanation (Dreyfus and Dreyfus, 1986, p. 199). Indeed, the expert even "forfeits" expertise when explaining his or her decision-making process based on rules, for that involves abandoning the intuitive experience that guided the decision making process (Dreyfus and Dreyfus, 1986, p. 196). Rational reconstruction of expert decision making, Dreyfus argues, inaccurately represents a process that is in principle unrepresentable. When non-experts demand that experts walk them through their decision making process step by step so that they can follow the expert's chain of deductions and inferences (perhaps hoping to make this chain of deductions and inferences for themselves), they are, according to Dreyfus, no longer allowing the expert to function as expert, but instead, are making the expert produce derivative, and ultimately false, representations of his or her expertise. Hence Dreyfus argues that too much pressure should not be placed upon experts to "rationalize" their "intuitive" process of decision making to nonexperts.[22]

Dreyfus's normative position thus amounts to what Douglas Walton calls "a strong form of the inaccessibility thesis" (henceforth IT): "that expert conclusions cannot be tracked back to some set of premises and inference rules (known facts and rules) that yield the basis of expert judgment" (Walton, 1997, p. 110). According to IT, when experts render a verdict in matters of their expertise, their judgments are inaccessible to nonexperts in the sense that experts cannot propositionally express "a set of laws and initial conditions (principles and facts) that would exhibit an implication of the conclusion [i.e. what the expert concludes] by deductive (or even inductive) steps of logical inference" (Walton 1997, p. 109). Due to the expert's reliance on inexpressible dimensions of intuition, IT suggests that "the most advanced expert in a field, who has achieved outstanding mastery of the skills of her field, may be the least able to communicate her knowledge" (Walton 1997, p. 113). It is not merely that experts will be unable to communicate how their knowledge was achieved to nonexperts; their knowledge will be equally opaque to themselves and other experts.

Dreyfus's chief normative concern, therefore, involves the ways nonexperts can possibly jeopardize experts, not how experts can jeopardize nonexperts. "Experts are an endangered species" (Dreyfus and Dreyfus, 1986, p. 206). He characterizes this as a problem confronting the U.S. in particular: "Demanding that its experts be able to explain how they do their job can seriously penalize a rational culture like ours, in competition with an intuitive culture like Japan's" (Dreyfus and Dreyfus, 1986, p. 196). Furthermore, Dreyfus argues that historical changes have precipitated the expert's current crisis. He contends that recent advances in computer technology and bureaucratic social organization exacerbate this cultural problem: "The desire to rationalize society would have remained but a dream were it not for the invention of the modern digital computer. The increasingly bureaucratic nature of society is heightening the danger that in the future skill and expertise will be lost through overreliance on rationality" (Dreyfus and Dreyfus, 1986, p. 195). In order to solve the problem of the disappearance of expertise Dreyfus seeks to reeducate people on the difference between "knowing how" and "knowing that." The future that Dreyfus speaks of is not distant. He suggests that "within one generation" we may lose "our professional experts" (Dreyfus and Dreyfus, 1986, p. 206). The goal is to get people to appreciate the value of intuition and the limits of rational deliberation.[23]

Dreyfus analyzes expertise from both an *asocial* as well as a *social* perspective. At the *asocial* level, he phenomenologically investigates how all human beings acquire skills regardless of who they are, what field they are apprenticing in, and when and where they learn their skills. At the *social* level, he presents a normative position that dictates how experts should be treated when they are asked to serve as consultants. The arrangement of this particular combination of *social* and *asocial* arguments is predicated upon Dreyfus's foundational assumption that social expectations of what experts *ought* to do should correspond with the embodied limitations that circumscribe what they can *in fact do*. This assumption is the justificatory mechanism behind his views that possessing expert level skills is both a necessary and sufficient condition for being an expert; and it is erroneous to take the ability to rationalize as a sign of expertise.

5. PROBLEMS WITH DREYFUS'S DESCRIPTIVE ACCOUNT

Dreyfus's account, however, admits certain categories of people as experts which do not belong, and omits several which do. Dreyfus's claim, for instance, that adults are "experts" in walking and talking—which as we have seen is essential to his account for it grounds them in the same lifeworld spectrum as conventionally expert behavior—collides with ordinary usage. We do not call people who are merely ambulatory or verbal "expert" walkers or talkers, but reserve the adjective for those who undergo special training, give professional advice, etc. We do not call licensed drivers "experts"—nor even driving enthusiasts or competitive amateurs—even when they have an intuitive relation to operating their vehicles. Rather, we reserve the word for drivers who belong to professional driving organizations, participate in certain kinds of competitions, and so forth.

Meanwhile, Dreyfus's account also excludes certain classes of people from being experts who do belong. One of us, for instance, has drawn a distinction between 'expert x,' which is an adjectival use of 'expert' that stems from the Latin *expertus*, and 'expert in x,' which substantively treats 'expert' as a noun (Selinger, forthcoming). In Dreyfus's terms, 'expert x' corresponds to "knowing how" while 'expert in x' corresponds to "knowing that." Whereas an 'expert x' could be an 'expert farmer,' an 'expert in x' could be an expert "in farming." An expert "in farming" could effectively communicate, coordinate, and synthesize accurate propositional information about farming—could become Secretary of Agriculture—even if terrified of plows and tractors. An 'expert in sports,' who correlates the past behavior of athletes to current situations, could be crippled and lack physical capacity to play the sport; an 'expert in music' could be a terrible musician. Nevertheless, Dreyfus denies that the propositional knowledge definitive of an 'expert in x' is expert knowledge.

> Listening to . . . commentators, who take up at least half the time on erudite talk shows, is like listening to articulate chess kibitzers, who have an opinion on every move, and an array of principles to invoke, but who have not committed themselves to the stress and risks of tournament chess and so have no expertise.
>
> (Dreyfus, Spinosa, and Flores, 1997, p. 87)

Dreyfus associates the knowledge of 'experts in x' ("commentators" or "kibitzers") with the "idle talk" definitive of the "public sphere" that "undermines commitment stemming from practical rationality" (1997, p. 86). The reasoning of an 'expert in x' is "not grounded in local practices" but rather in "abstract solutions" and "anonymous principles" that fail to display "wisdom" (1997, p. 87). Moreover, Dreyfus argues, an 'expert in x' lacks the bodily commitment a genuine 'expert x' possesses; only an 'expert x' affectively cares about the outcome of a situation and experiences the "risk" of performance. But this is an inaccurate characterization. An 'expert in art history' presumably does not spout abstractions and generalities, but helps guide others, and sometimes even artists themselves, to appreciate the meaning in works and the artistic process itself. George Steiner once proposed what amounts to an effective refutation by re-duction of Dreyfus's position here by trying to imagine a society without crit-ics and commentators, in which "all talk *about* the arts, music and literature is prohibited . . . held to be illicit verbiage" (Steiner 1989, pp. 4–5). While Steiner found himself yearning for such a "counter-Platonic republic," like Plato he also recognized the impossibility of his thought-experiment. Ultimately, Steiner re-alized, the result would be to stifle not only the arts and the creative imagina-tion, but the lifeworld itself. Dreyfus could presumably agree, to the extent that his first three stages require verbalization and thus expertise could not be ac-quired in such a world—but the spirit of Steiner's thought-experiment is that even artists, musicians, and writers can only flourish in a sea of high-level dis-course about what they do.

Another category of expertise passed over by Dreyfus's account of skill ac-quisition is the "coach." Dreyfus's account posits the instructor as delivering rules and commands in a standardized way to masses of people at once—and the instructor disappears after the initial stages of learning. Even Dreyfus's slightly revised, seven-stage model that includes copying from a "master," the master is no coach, but simply offers an example of a developed "style" which, in some unspecified way, helps the learner in developing his or her own (2001, pp. 43–46). But even though linguistically mediated, coaching is mischaracter-ized as a dispensing of rules in a standardized way; rather, it involves modeling and demonstration and is addressed to the specific performances of specific embodied learners. Since these performances differ from body to body, the knowledge embodied in coaching cannot be reduced to rules of thumb. Nor does it disappear after the beginning stages; if anything, it grows in impor-tance. Many professional athletes, musicians, actors, and performers of various kinds—including Dreyfus's paradigmatic expert Bobby Fischer—would find it unthinkable to perform without coaches to help evaluate and strengthen their playing. For a coach does more than correct bad performance; a coach also fa-cilitates good performance to become better.

The expert performer and the coach or 'expert in x,' to be sure, are function-ally different. Nevertheless, they are closely related enough to suggest that Drey-fus is relying on a false dichotomy between those who "do" and those who com-ment, kibitz, or at best instruct as a propaedeutic to "doing". (One wonders, perversely, whether the prejudice against teaching evident here isn't an occupa-tional hazard of academics.) A lived, embodied subject lacks an objective pur-chase on its own performing process. Though Dreyfus would clearly disagree, the performing of that subject is always, therefore, in principle open to being better disclosed to outsiders, whose intercessions may then help the subject to perform better. This is not to say, of course, that the outsider's insights are achieved because of an objective stance; rather, it is because the outsider is situated differently. Moreover, coaching involves not only understanding what the performer is doing and what it is to perform well, but also how to intercede best in order to transform the performing of the pupil. Coaching, therefore, is a first person process requiring different types of contextual sensitivity; it has its own techniques, its own intentional arc, and its own quest for maximal grip (which in this context means the ability to optimize the conditions to allow the pupil to go beyond not only his or her present performing ability, but the coach's as well). Coaching, in short, is not just a practice in which one can be expert, but a form of expertise itself—even in Dreyfus's own terms. The expertise is not identical with that of the expert performer, but it is closely related and shares a common end: the achievement of good performance.

The false dichotomy not only exposes a problem with Dreyfus's invocation of the difference between "knowing how" and "knowing that," but a deeper flaw in his account as well. Dreyfus assumes that the body which acquires skill has no relevant biography, gender, race, or age (Young, 1998; Sheets-Johnstone, 2000). He does acknowledge that "cultural styles" affect how skills are learned, noting for example that differences exist between how American and Japanese mothers "handle" their babies (2000, pp. 46–47). However, this notion of "cultural style" is not developed beyond unsubstantiated generalities, and assumes as well the insignificance of any biographic differences existing within individuals sharing a single culture. From Dreyfus's perspective, one develops the affective comportment and intuitive capacity of an expert solely by immersion into a practice; the skill-acquiring body is assumed to be able, in principle at least, to become the locus of intuition without influence by forces external to the practice in which one is apprenticed.

Sheets-Johnstone has sharply critiqued this "adultist" stance. She argues that it is a methodological mistake to take the adult human body as the phenomenological ground zero, for it overlooks "the originating ground of our knowledge, our capacities, our being" (Sheets-Johnstone, 1999, p. 232). The behavior of the adult human body does not simply happen but is itself the outcome of an "apprenticeship" carried out in infancy, an apprenticeship which leaves adult bodies full of meanings and experiences that are essential to all of their future behavior, including even the skills that they go on to learn and their application. "Whatever the particular adult skill-learning situation playing the piano, driving a car, playing chess, making trousers—it is a compound of experiences sedimented with skills and concepts accruing from our history" (Sheets-Johnstone, 2000, p. 359).

To be sure, what Sheets-Johnstone means by apprenticeship (as when she calls an infant apprenticed to its body) is different from what Dreyfus means. For her, apprenticeship involves constant experimenting in the dark, and rules—even if only rules of thumb—flow from rather than precede it. For Dreyfus, on the other hand, apprenticeship is an explicit and deliberate process, and follows the formulation of rules. Nevertheless, she points out something over-looked in Dreyfus's account: that the learning body is always already *embedded*, and does not disembed itself in skill acquisition. In order to learn a skill such as dancing, for instance, I must *already* be able to move my body. Each body that takes up dancing is already embedded in a way of moving, and learning the dance is not a matter of learning to move *tout court*—as learning a new language is much different from learning to speak. Adult human bodies *never* move in a generic way but always in *this* way or *that*, and in a way that reflects how one has been brought up, and has brought oneself up, to move.

Learning dancing, therefore, cannot be said to exemplify the Dreyfusian progression of skill acquisition in which the learner begins by following rules, learns to make exceptions to the rules in the advanced beginner stage, learns to enact spontaneously and fluidly in the higher stages until "called" by the situation, and so forth. Learning dancing as a skill almost always requires coaching by someone who is sensitive to the differences between performances of individual bodies and to ways of adjusting such performances by modeling and demonstration. The learner does not leave this embeddedness behind, departing from what one already has and becoming delivered to a new way of moving, but rather carries it forward and transforms it (Crease, 2002a). Even the affective demeanor and intuitive capacity of a practice are not therefore separable from broader social influences that first do not appear directly connected with the skilled practice. The claims sometimes made in biographies of Bobby Fischer, for instance, that his distinctive aggressive chess playing style was not solely developed by playing many games of American style chess but shaped in part by personal childhood experiences are not implausible and even persuasive; if true, his "expert responses," the choices of specific moves, were not fully and exhaustively forged by contextual sensitivity plus experience, and chess is even more life than Dreyfus suspects.

Oddly, given Dreyfus's philosophical commitments, his account lacks hermeneutical sensitivity. The flaw in his assumption that skilled behavior crystallizes out of contextual sensitivity plus experience without contribution from individual or cultural biography can be traced to a failure to take into account the fact that the embodied subject, *even when behaving expertly*, brings to the situation what has been historically and culturally transmitted to it, and in a way such that the subject can never grasp cognitively all at once. The individual expert performer, as a consequence, does not have a complete purchase on his or her own expert behavior. Therefore, *contra* Dreyfus, it will always be possible in principle for an expert performer to learn about one's own performance from another, contextually sensitive person—though, again, this is not because the other has managed to obtain an objective position, but on the contrary, because the other is differently situated.

6. Problems with Dreyfus's Normative Account

Lack of hermeneutical sensitivity also affects Dreyfus's normative account, for his assumption of the autonomy of expert training suggests a naïveté in his counsel to "trust experts." While a beginner might have entered a training program *culturally* or *situationally embedded* with prejudices, ideologies, or hidden agendas, these would all be left behind by the time one reached the expert stage; an expert's knowledge for Dreyfus, as we saw, crystallizes out of contextual sensitivity plus experience. But if one can never leave the hermeneutic circle, the best that one can do is transform one's embeddedness, rather than extricate oneself from it. The acquisition of expertise is not a transcending of embeddedness and context, but a deepening and extension of one's relationship to it (Crease, 1997). This also means that experts will never be able to free themselves a priori from the suspicion that prejudices, ideologies, or hidden agendas might lurk in the pre-reflective relation that characterizes expertise.[24] Not only is this suspicion to be expected; its absence would be socially dangerous. This gives rise to a *recognition problem* that is such a prominent part of many actual controversies involving a technical dimension where expert advice is required. But there is no room for this recognition problem in Dreyfus's account. He leaves no grounds for understanding how an expert might be legitimately challenged (or instructed, for that matter, as in the case of sensitivity training, nonexpert review panels, etc.). One would never imagine, from Dreyfus's account, that society could possibly be endangered by experts, only how society's expectations and actions could endanger experts. The stories of actual controversies, we will argue, not only shows things do not work the way Dreyfus says they do, but also that it would be less salutary if they did. Such stories, we claim, amount to a counterexample to Dreyfus's normative claims, and point to serious shortcomings in his arguments.

Dreyfus, let us recall, assumes from the start that the people who possess expert level skills are the same people who *should* be socially recognized as experts. It is unproblematic, for him, who "counts as" an expert in a given social situation; he assumes the absence of a recognition problem. This is clear from his examples. On the one hand, he refers to people who are socially recognized as experts, such as airplane pilots, surgeons, and chess masters, to illustrate how embodied expert performance functions (Dreyfus and Dreyfus, 1986, pp. 30–35). These references are descriptively loaded because they use *socially acknowledged* experts as data for *asocial* phenomenological descriptions of expertise. On the other hand, he portrays mundane examples of everyday action, such as driving a car, walking, talking, and carrying on a conversation, as paradigmatic instances of how experts behave, even though these activities would not normally be socially recognized as being performed by experts (Dreyfus and Dreyfus, 1986, p. 30). By arguing that the *same type* of expertise exists in both extraordinary performances of skill that are socially recognized as occurring at the expert level and mundane performances of skill that are not recognized as occurring at the expert level, Dreyfus advances his end of demystifying the seemingly magical quality of expertise and establish continuity between expert and everyday lifeworld activity. In both instances, for him, expertise is a pre-reflexive relation between skill and environment: "An expert's skill has become so much a part of him that he need be no more aware of it than his body" (Dreyfus and Dreyfus, 1986, p. 30).

From Dreyfus's perspective, social problems can arise when social agents are unable to recognize the essential qualities of expertise that are rendered explicit by the *phenomenological* investigator. These social problems arise from a different source than expertise itself; they are brought about by the *interference* of political, social, and cultural involvements. But the hermeneutic circle would turn this around; these involvements *give rise to* recognizing and trusting experts in the first place. Even if one defers to the traditional "experts," that deference, too, has been brought about as a result of one's particular background and lifeworld involvements. For expertise is a two-way relation: the *claim* to expertise itself involves a social demand; it is not merely a neutral identification label but a declaration that others should defer to the expert's judgments. The phenomenon of expertise, therefore, is ultimately and inextricably tied to its social utility; an expert is not only "in" a field but "for" an audience.

An obvious example is the 'selection' problem of experts. In many instances experts who endorse different conclusions within the same field can be pitted against one another as 'counter experts.' A common strategy for 'counter experts' consists in claiming that the judgment proffered by the expert is tainted due to the presence of prejudices, ideologies, or hidden agendas. 'Counter experts' are particularly prevalent in the legal arena where expert witnesses function in accord with the logic of an adversarial system by proffering testimony for both the prosecution and defense.[25]

Stephen Turner's account of expertise addresses the recognition issue by pointing out that, in order for someone to be an expert, he or she needs not only to be skilled, but also to have an audience that socially recognizes his or her type of skills as skilled expertise (Turner, 2001, p. 138).[26] Turner contends that although what Merton calls "cognitive authority" is neither an "object that can be distributed" nor something that can be "simply granted," it nevertheless is "open to resistance and submission" based upon the evaluations of different audiences (Turner, 2001, p. 128).[27] By contrast, Dreyfus, who defines an expert solely on the basis of skill acquisition and use, methodologically excludes the audience from the description of experts and expertise, defining an expert as one with the right affective comportment and intuitive response to the situation.

This methodological move has many advantages over Dreyfus's phenomenology. It explains why: (1) expertise is not a stable property, but can be gained and lost; (2) why discussions of expertise tend not only to focus on epistemological, but also political issues; and (3) why perceptions of expertise can be based on historical transformations. In connecting expertise with these, Turner does not relinquish the philosophical aim of revealing general structures and transcending particularity (i.e. expertise can only be discussed with reference to local features). For example, he provides the following "taxonomy" that accounts for five general kinds of distinguishable experts:

> Experts who are members of groups whose expertise is generally acknowledged, such as physicists; experts whose personal expertise is tested and accepted by individuals, such as the authors of self-help books; members of groups whose expertise is accepted only by particular groups, like theologians whose authority is only accepted by their section; experts whose audience is the public but who derive their support from subsidies from parties interested in the acceptance of their opinions as authoritative; and experts whose audience is bureaucrats with discretionary power, such as experts in public administration whose views are accepted as authoritative by public administrators.
>
> (Turner, 2001, p. 140)

Turner's account thus relates the phenomenon of expertise to changing historical and social perceptions. Dreyfus might well retort that there is no reason why his phenomenological model should be understood as incompatible with Turner's—that Turner simply expands on Dreyfus's model, with Dreyfus describing what expert skills are and Turner describing how different audiences come to recognize these skills as occurring at the expert level. But the critical methodological difference is that Turner does not treat the possession of skill as a necessary *and* sufficient condition of expertise, omitting reference to an audience, while Dreyfus does.[28]

Dreyfus's failure to address the recognition problem is highlighted by the following paradox. On the one hand, he argues that only an expert can recognize another person as a genuine expert, for nonexperts do not know what to look for when evaluating whether someone is skilled.[29] It takes an expert—and only one—to know one. This might be called a *difference* claim (DC): experts are "not like us." DC is fairly innocuous, and even Turner adheres to a version: "[It] is the character of expertise that only other experts may be persuaded by argument of the truth of the claims of experts; the rest of us must accept them as true on different grounds than other experts do" (2001, p. 129). On the other hand, Dreyfus also advances a *similarity* claim (SC), according to which "they"—the experts—are very much "like us." They behave in a similar way the

rest of us do in our everyday activities. At the most basic level of everyday coping, everybody deserves to be characterized as an expert: "Citizens will be
speaking in terms of their expertise, whether they are university professors
who have expertise in foreign cultures doing business with their state or farmers or small-store owners speaking about concrete problems that need legislative solution" (Dreyfus, Spinosa and Flores, 1997, p. 107). Thus we can projec-
tively identify with experts, and understand the kind of knowledge they use in
their judgments. The problem is not only that Dreyfus needs to advance both,
it is that, in the end, he needs SC to trump DC. Otherwise, nonexperts would
lack any basis to recognize, accept, and trust the kind of knowledge that experts
possess.

This point can be exposed most directly by reference to the situation de-
picted in Ibsen's play *Enemy of the People*. The central character, Thomas
Stockmann, is a doctor at a spa on which the livelihood of his town depends.
He thinks an invisible poison is polluting the spa's water, and confirms it by
sending water samples to an expert at the University for analysis. He seeks to
inform members of the community, thinking they will thank him for bringing
the danger to their attention. But a friend warns him not to be so sure how they
will respond: "You're a doctor and a man of science, and to you this business of
the water is something to be considered in isolation. I think you don't perhaps
realize how it's tied up with a lot of other things." Stockmann, citing the expert,
insists that the "shrewd and intelligent" people will be "forced" to accept the
news. The mayor points out that, for citizens, the matter is more economic
than scientific; and indeed, the citizens condemn Stockmann as an "enemy of
the people." The basic conflict thus involves a scientist who accepts expert technical advice as authoritative, versus citizens who do not find that expert advice
authoritative, who find *it* threatening to their world, and who seek guidance
from others.

Ibsen's invented situation is like a model which strips away inessential details to clarify the essential forces of a situation. The situation involves a volatile
controversy with a scientific-technical dimension in which people have lost
confidence in traditional "experts." From the audience position, we see that
two kinds of stories can be told about such a situation from two different perspectives. In the *expert's* perspective, that of Stockmann and the University
chemist, there is no uncertainty or grounds for contestation regarding whom
the expert is, what kind of training that person requires, and what kind of information (the technical issue of the quality of the water) is relevant to maximizing the good of that particular social situation. The experts are like the citizens in that they want to maximize the good of the town; they do so with their
special knowledge. Failure of the citizens to recognize the expert, and defer to
the expert's advice, is due to the peripheral, and even deleterious and corrupt-
ing, influence of economic and political motives and the involvements of the

citizens with politicians, the media, and other nonexpert authorities. In the *citizen's* perspective, on the other hand, things are much more confused. The background and lifeworld involvements of citizens mean that economic and political motives loom much larger than they do for Stockmann and the chemist, and suggest different people to whose advice they should defer in seeking to advance their welfare. In this perspective, the purported "experts"— Stockmann and the chemist—are precisely *not* like the other citizens for they have different agendas (and the chemist is even literally an outsider).

We see the difference between these two situated perspectives clearly from a *third* perspective, sitting in the audience. This third perspective is also not neutral, and its effect is to dramatize the similarity claim; we recognize the common humanity between Stockmann and the citizens, and the relevance of his knowledge to their welfare. One realizes that the welfare of the citizens requires that they defer to the experts, even as one fully appreciates the severity of the conflict and the impossibility of its resolution. But we in the audience have no doubt about who the real experts are, and the difference between essential and peripheral involvements. The person in the audience is not standing anywhere, not situated with respect to this aspect; this third position, in short, is not a hermeneutically sensitive one.

In any real controversy, however, no one occupies the audience position; everyone is as it were "on stage" in a situated position, standing someplace with *particular* involvements which give rise to a *particular* understanding of the situation and, with it, an inclination to accept some people rather than others as authoritative. It is not a priori clear which involvements are essential and which particular; whose actions are in the grip of prejudices, ideologies, and hidden agendas, and whose are in the best interests of society. Meanwhile, experts are also in a particular position, standing someplace and with particular involvements, and the claim to expertise is a charge, a valence, a demand that one should be deferred to. Each real life controversy involving expertise takes the form of a jockeying between those who advance claims of expertise to advance their authority and those seeking the right authorities to whom to defer. This is the hermeneutic predicament: there is no escape from the particular involvements of a given situation. There is and can be no talisman for expertise.

This problem is most visible in connection with volatile public controversies involving a technical dimension—and especially involving public safety— where traditional sources of authority have become distrusted. In these cases, the question of who speaks authoritatively, of who is an expert, is contested. Each citizen, and each person proposed as an expert, has a particular set of involvements, and there is no safe audience position from which to sort out the essential and inessential involvements in an expert's judgment.

Consider, for instance, the case of the closing of the National Tritium Labeling Facility (NTLF)—where, as it happened, Dreyfus's wife Genevieve worked. The facility created unique tagged molecules by putting tritium into specific molecular positions, creating tracers that are used to study mechanisms of biochemical transformation in basic and applied research. But anti-nuclear activists objected to the fact that some tritium was released to the environment in the process. Scientists and local, state, and federal public health experts, after carefully examining the situation, said the emissions were safe, a fraction of the Environmental Protection Agency suggested limit, which in turn was a fraction of the background level. But the anti-nuclear activists sought to discredit those claims, saying that they were either made by those who worked at the facility, or from institutions connected in some way with that facility, or by scientists who knew too much about tritium to be disinterested. The activists were effective at disrupting the normal socially negotiated procedures for who speaks authoritatively about safety, and the facility was closed (Crease, 2002a).

Or again, consider the controversy surrounding the shipment of spent fuel rods from Brookhaven National Laboratory's High Flux Beam Reactor in 1976 (Crease, 1999). The controversy pitted activists, who associated research reactors with power reactors/nuclear weapons/the military, versus scientists for whom such associations made no sense at all. The situation spawned a spectrum of experts of the sort described by Turner's taxonomy. In yet another controversy involving the Brookhaven lab, a program it ran studying the health of Marshall Islanders accidentally exposed to fallout from a nuclear weapons test came under attack, with one complicating factor being that it involved the classic colonialist situation of U.S. scientists working in a third-world country whose language and customs they were not familiar with (Crease, 2002c).

Each of these controversies was complex, and involved high stakes. Each turned on a scientific-technical issue, and thus necessary recourse to expertise. Yet who the "real" experts were deemed to be depended on who one was and where one stood. To understand such controversies involving experts and expertise requires moving beyond the practical expert's point of view.

7. CONCLUSION

Dreyfus's model of expert skill acquisition is philosophically important because it shifts the focus on expertise away from its social and technical externalization in STS, and its relegation to the historical and psychological context of discovery in the classical philosophy of science, to universal structures of embodied cognition and affect. In doing so he explains why experts are not best described as ideologues and why their authority is not exclusively based on social networking. Moreover, by phenomenologically analyzing expertise from a first person perspective, he reveals the limitations of, and sometimes superficial treatment that comes from, investigating expertise from a third person perspective. Thus, he shows that expertise is a prime example of a subject that is essential to science but can only be fully elaborated with the aid of phenomenological tools.

However, both Dreyfus's descriptive model and his normative claims are flawed due to the lack of hermeneutical sensitivity. He assumes, that is, that an expert's knowledge has crystallized out of contextual sensitivity plus experience, and that an expert has shed, during the training process, whatever prejudices, ideologies, or hidden agendas that person might have begun with. This assumption not only flaws in Dreyfus's descriptive account but his normative account as well.

The phenomenological goal is to expose presuppositions that lurk unapprehended in the natural attitude. The phenomenological experience reveals that the most difficult presuppositions to expose and get a grip on are those closest to home. In this spirit, though possibly with a trace of perversity, one might expose this limitation by posing the following question: "Why does Genevieve Dreyfus no longer work at the National Tritium Labeling Facility?"

From Dreyfus's own account, it is because of a breakdown, a corruption of a legitimate and phenomenologically justifiable authority. A group of anti-nuclear crusaders managed to hijack a socially negotiated process and persuade the administrative authorities to overlook robust science and expert advice, and make a purely political decision. But from another, more all-too-common, highly important, and potentially powerful point of view, one could, with hermeneutic sensitivity, tell quite a different story. Individuals do not take in an item of information, even scientific information, nakedly. It matters who conveys that piece of information and in what context. The Berkeley anti-nuclear activists live in a different interpretive world than the scientists, with a different set of supporting behaviors, values, and institutions; their meaning-generating process by which they interpret facts, principles, and their application is very different (Crease, 1999, p. 498). The difference between the scientists and activists thus cannot be regarded as one between knowers and those who corrupt or betray that knowledge, but more like the relation between the members of one culture and another. Surely a description of this hermeneutic predicament, missing from Dreyfus's own, belongs in any account of expertise.

Finally, we believe that when evaluated from an immanent perspective, Dreyfus's account of expertise fails according to his own philosophical standards. In many publications Dreyfus champions Heidegger's hermeneutic approach to phenomenology over Husserl's. He even argues that Heidegger's account of authentic Dasein in Division II of *Being and Time* entails that Dasein is an "expert" (Dreyfus, 2000). However, by *approaching trust as a static scenario* in which a nonexpert solicits advice from an expert, Dreyfus resorts to an implicit Husserlian schema of intersubjectivity. The essence of SC is an analogy: just as I do not expect myself to be able to articulate rules for how to drive or ride a bicycle, I should not expect experts, such as ballistics examiners, nurses, and ecologists, to justify their decisions by referring to rules. Dreyfus implicitly argues that even though trust is established at the social level on the basis of an expert's track record, at the phenomenological level, it is established through what Husserl calls "intersubjective pairing." By making a pre-reflexive analogy from my behavior to the behavior of an expert, I recognize that an expert's behavior is essentially similar to my own. The expert is an expert "alter ego." In the Dreyfusian scheme, I should trust experts because: (1) I trust myself to make decisions in a similarly intuitive manner, and (2) I trust my decisions to be good ones, even though I, like an expert, cannot propositionally justify them according to rules. Even though an expert has more training than I do, our cognitive similarities outweigh our technical differences.

Dreyfus's account of intersubjectivity here, with its lack of hermeneutic sensitivity, recalls Husserl more than Heidegger. Indeed, his portrait of the expert, who masters his or her own relation to expertise by making it through all of the developmental stages, and who feels no need to seek external ratification of his or her own abilities, evokes the caricature of Heidegger according to which authentic Dasein is accorded too much heroic freedom. But Heideggerian sensitivity to the hermeneutic dimensions of worldhood would have to depict an expert as engaged in a much more fragile and vulnerable process, in which expertise does not appear as a destination in which the individual surpasses one's embeddedness in that world. Expertise is always a process of becoming, and, in principle at least, it will always be possible for coaches, commentators, and others whose own expertise overlaps with the expert's to disclose aspects of expert performance which escape the grasp even of that performer.

NOTES

1. Turner discusses the bases of Stanley Fish, Jürgen Habermas and Michel Foucault's concerns about why expertise undermines liberal democracy. He argues that these concerns "depend upon a kind of utopianism about the character of knowledge that social constructionism undermines" (Turner, 2001, p. 147, n. 7).

2. The problem of certifying expert witnesses is frequently discussed in relation to the issue of "junk science" (See Black, Ayala and Saffran-Brinks, 1994; Jasanoff, 1992; Caudill and Redding, 2000; Huber, 1991).

3. When theorists assume an expert social role based upon the prescriptive dimensions of their research then the problem of what Winner calls the "values expert" emerges (Winner, 1995, pp. 65–67). Winner argues that applied theorists not only provide counsel in "interminable" debates, but tend to misconstrue the audiences they address as an overly generalized "we," fail to recognize how social change is instituted, and occasionally help legitimate already made political decisions.

4. "Black box" is an engineering term used in science studies analogously to the Marxist concept of hegemony. "Black boxing," like hegemony, refers to background assumptions that are generally regarded as self-evidently true and not requiring further investigation (Feenberg, 1995, p. 7). To open the "black box" of scientific expertise is to show, through close empirical and conceptual analysis, that what appears to be self-evidently true, culturally sanctioned, and not requiring further investigation about experts is false, hidden from cultural scrutiny, and in dire need of critical analysis. When the "black box" of expertise is opened, STS theorists contend that experts do not emerge as self-sufficient geniuses whose knowledge is infallible, certain, and objective. Scientific experts are revealed rather to be non-extraordinary, biased people whose successes and failures emerge from working within a competitive network of distributed knowledge and prestige. From the STS perspective, it is principally because the network's operation is rarely explicitly described and theoretically examined that nonexperts mistakenly perceive scientific experts as more knowledgeable, authoritative, and trustworthy than is appropriate.

5. STS has occasionally focused on the tacit knowledge involved in the production of scientific results. For example, Harry Collins recently sought to expand Michael Polanyi's classification of tacit knowledge by expanding its range of application beyond skills to the production of scientific results in empirically verifiable cases (2001). Proposed as a general thesis, Polanyi claims that in order to understand how to use a machine one needs to know how its components fulfill their function together (1974). On the basis of field work in the United States and observations of experimental work in Glasgow University, Collins claims that the problem of tacit knowledge accounts for why twenty year old Russian measurements of the quality factors (Q) of sapphire have only just been repeated in the U.S. Collins's focus on tacit knowledge allows him to stress the importance of personal contact and trust between scientists. It also allows him to suggest that information that is not currently contained in experimental reports should be added to increase the likelihood of reproducing scientific findings. Despite some overlap, the examples that Collins and Dreyfus focus on differ in one key respect. Collins focuses on cases in which the tacit knower produces intersubjectively accepted results; his emphasis is on the routine production of these results, and the impediments that prevent them from being easily reproduced. What Collins does not present is a theory of embodiment that explains how knowers are capable of tacitly mastering equipment.

6. Abstract contexts alone do not allow scientists to mathematize, model, and formalize the world. Body oriented skills are used to operate the technological instruments that stabilize phenomena in order for scientists to manipulate and interpret them. Ihde repeatedly argues that as with mundane uses of technology, the technological instruments used in scientific settings extend and transform bodily praxes through "embodiment relations"; they are absorbed and incorporated into bodily experience of the world like Heidegger's hammer or Merleau-Ponty's blind person's cane, and the phenomena that scientists can produce change as the forms of embodiment change (Ihde, 1998, pp. 42–43).

7. Dreyfus defines "intentional arc" and "maximal grip" as follows: "The *intentional arc* names the tight connection between the agent and world, viz. that, as the agent acquires skills, those skills are 'stored,' not as representations in the mind, but as dispositions to respond to the solicitations of the world. *Maximal grip* names the body's tendency to respond to these solicitations in such a way as to bring the current situation closer to the agent's sense

of an optimal gestalt. Neither of these abilities requires mental or brain operations" (1998; 1999a).

8. Although we believe that Dreyfus turns to the theme of "brain topics" simply to establish another perspective that can be made compatible with his phenomenological approach, Sheets-Johnston argues that the reference to neural networks and brain functions contradicts his phenomenological aspirations: "[We] find that any erstwhile sense we might have had of a phenomenological subject has given way to a neurological one, while at the same time we find a phenomenological subject to be ostensibly fully present but transformed in ways utterly foreign to our experience in that it 'presumably senses' its own 'brain dynamics'" (2000, p. 357). Moreover, she argues that, "Not only can they [neural nets] not distinguish between formal and informal learning, but the very vocabulary by which they operate is not simulation-friendly to the latter kind of learning and is thus inappropriate to its description and deflective to its understanding" (Sheets-Johnstone, 2000, pp. 357–358). We believe that what Sheets-Johnston forgets is that some still consider phenomenology to be a subjective form of analysis and that the reason why Dreyfus probably gestures towards neural nets is to demonstrate that phenomenology can reveal objective structures of bodily praxis.

9. During the 1960s, when Dreyfus first formulated his critique of artificial intelligence and its failed hype, the intellectual atmosphere at the Artificial Intelligence Laboratory at the Massachusetts Institute of Technology was overly hostile to recognizing the implications of what he said, and as a result, he almost lost his job. By contrast, Winograd notes that today, "some of the work being done at that laboratory seems to have been affected by . . . Dreyfus" (1995, p. 110).

10. According to Dreyfus, the primary desideratum of phenomenology has always been to adequately describe expertise (even if historical phenomenologists such as Husserl, Heidegger, and Merleau-Ponty did not use the term 'expertise' in their analyses), because at bottom, phenomenology aims at getting behind the prejudices that impede how human experience is understood (Dreyfus and Dreyfus, 1986, pp. 2–5).

11. Sheets-Johnston, though, argues that by analytically separating intellectual from bodily skills Dreyfus assumes a "pernicious" "Cartesian split" between mind and body by forgetting how bodily skills are foundational for intellectual skills (2000, pp. 355–356).

12. Although there are many surface similarities between Dreyfus and Polanyi (Polanyi, 1974) on the issue of tacit knowledge, there also is a notable difference. Dreyfus argues that while Polanyi recognizes that formalisms cannot account for the tacit performance of riding a bicycle, he still believes that such performance is governed by "hidden rules": "The reference to hidden rules shows that Polanyi, like Plato, fails to distinguish between performance and competence, between explanation and understanding, between the rule one is following and the rule which can be used to describe what is happening" (1992, pp. 330–331). Dreyfus also differs from Kuhn's analysis of tacit knowledge since this analysis primarily situates tacit dimensions of knowledge in the general structure of a paradigm (Stengers, 2000, p. 6). For other writings on tacit knowledge see Rawls, 1968; Reber, 1995; Searle, 1983; 1992.

13. The only time Dreyfus alters his five-stage model is when he adds two additional stages—mastery and wisdom—in his recent reflections on the Internet (2001).

14. Dreyfus takes his target group involved in making "unstructured" decisions to be paradigmatic of the "typical learner" and classifies "a common pattern" observable in their be-

havior (Dreyfus and Dreyfus, 1986, p. 20). The adjective "typical" denotes a class of learners who "possesses innate ability" and also have "the opportunity to acquire sufficient experience" (Dreyfus and Dreyfus, 1986, p. 20). "Unstructured" and "structured" are standard terms discussed in theories of information management. They are used to classify organizational differences in the range of decision making, usually with "semi-structured" appearing as a mid-point in this range. When Dreyfus references "structured" decisions, he means the type of decisions that are made when "the goal and what information is relevant are clear, the effects of the decisions are known, and verifiable solutions can be reasoned out" (Dreyfus, 1986, p. 20). In other words, "structured" decisions involve well-understood situations that have a common procedure for handling them. "Structured" decisions can be found in situations that are repetitive, routine, and have the pertinent pieces of evidence remaining stable as time passes. Examples of "structured" decision making can be found in situations where "context-free" deliberation dominates, such as "mathematical manipulations, puzzles, and, in the real world, delivery truck routing and petroleum blending" (Dreyfus and Dreyfus, 1986, p. 20). In contrast with "structured" decisions, Dreyfus characterizes "unstructured" ones as: intuitive, commonsensical, heuristic, and involving trial and error approaches. He states they have a tendency to be ad hoc, are not programmable, and "contain a potentially unlimited number of possibly relevant facts and features, and the ways those elements interrelate and determine other events is unclear" (Dreyfus and Dreyfus, 1986, p. 20). The situations that require "unstructured" decisions typically are elusive ones: where it is not possible to specify in advance most of the decision procedures to follow; where the decision maker must provide judgment, evaluation, and insights into the problem definition whose parameters cannot be precisely identified; where unquantifiable factors are central; and where there is no agreed-upon procedure for making decisions. These types of decision are routinely made by people in "management, nursing, economic forecasting, teaching, and all social interactions" and require "considerable concrete experience with real situations" (Dreyfus and Dreyfus, 1986, p. 20).

15. Dreyfus writes: "Although irrational behavior . . . should generally be avoided, it does not follow that behaving rationally should be regarded as the ultimate goal. A vast area exists between irrational and rational that might be called *arational*. The word rational, deriving from the Latin word *ratio*, meaning to reckon or calculate, has come to be equivalent to calculative thought and so carries with it the connotation of 'combining component parts to obtain a whole'; arational behavior, then, refers to action without conscious analytic decomposition and recombination. *Competent performance is rational; proficiency is transitional; experts act arationally*" (1986, p. 36).

16. As Michael Callon puts this point: "Researchers in the wild participate in the subversion of modern institutional framing by challenging the oppositions that we had come to take for granted, yet that are crucial, such as the distinction between expert and layperson" (Callon, 2001).

17. This emphasis on affect is important for Dreyfus's criticism of so-called "expert" computer systems, which he argues can approximate competent human performance. Skill acquisition for Dreyfus involves not only a cognitive acquisition by the subject but also an affective transformation that computers cannot experience. This point is especially significant in serving to highlight Dreyfus's phenomenological background. Phenomenologists have long argued that to be a subject is to have an intentional relation to the world such that

changes made to the subject correlate to changes made to the subject's world. The subject who goes through the developmental process of expert apprenticeship is not the same subject as the one who began the process, and the world is not meaningful for the subject in the same way; experts and nonexperts are indeed different subjects. They are different types of *people* who deliberate and feel differently, and to whom the world responds differently. Not only can experts do more things than beginners, but their whole affective demeanor changes. The way that experts care about their activities in a practice changes from when they were beginners, progressing from relative detachment to engaged commitment. This is why Dreyfus characterizes the five developmental stages as "qualitatively different perceptions" of what a task is and what mode of decision making is appropriate to handling that task (Dreyfus and Dreyfus, 1986, p. 19).

18. Although not every practice provides every beginner with an opportunity to achieve expert level mastery, many do: "Not all people achieve an expert level in their skills. Some areas of skill have the characteristics that only a very small fraction of beginners can ever master the domain of" (Dreyfus, 1986, p. 21).

19. Brewer, like many other philosophers and legal theorists, appeals to the "point of view" as an analytical device that captures how perspective and justification relate (1998, pp 1568–1570). He turns to the point of view in order to articulate a common theoretical perspective that applies to all scientific experts—and that is arguably generalizable to all experts in all fields of expertise—regardless of which particular scientific field the practitioner is an expert in. He writes: "One invokes a point of view to justify some claim. To serve this justificatory function, the point of view is assumed to be a reliable method for achieving the (explicit or implicit) aims of some rational enterprise" (Brewer, 1998, p. 1575). Brewer's line of reasoning is that one turns to a point of view to rationally justify either a theoretical claim about what ought to be believed or a practical claim about what action ought to occur. What is distinctive about this type of validation is that it relates the justification of a claim to a distinct, yet "reliable" method, which is chosen to achieve a specific cognitive aim. The "reliable method" common to all "rational" points of view is defined in terms of two characteristics: "enterprise" and "axiological" conceptions. An "enterprise" is defined as the choice and use of particular methods of analysis in order to serve specific cognitive goals. He acknowledges that even within the "same generic enterprise" practitioners can disagree about the "proper specific aims of the enterprise" (Brewer, 1998, p. 1571). But such a disagreement will take place within a "holistic" network involving an "axiological" component. When Brewer discusses the "axiological" dimensions of justification, he does so in Larry Laudan's sense of the term. Where Laudan analyzes scientific reasoning, he distinguishes between "factual," "methodological," and "axiological" levels of analysis. Factual analysis focuses on what exists in the world, including theoretical and unobservable entities for scientists. At the methodological level, practitioners in a given field share precise as well as vague rules. For scientists, this can include vague rules, such as avoid ad hoc explanations, to precise rules, such as calibrate instrument 'x' to standard 'y.' The axiological level, which is often explicable in the form of rules, designates cognitive aims. Brewer, like Laudan, argues that the relation between facts, methods, and axiological aims should not be understood as a "simple linear hierarchy," but as "reticular" "constraints [that] are multidirectional within the holistic network of aims, methods, and beliefs" (Brewer, 1998, p. 1575). Facts, methods, and axiological aims relate in a "multidirectional" sense because each can "constrain" the other, with-

out any one of the three being a priori valued more than the others. Facts, methods, and axiological aims relate in a "holistic" sense since the point of view they collectively constitute is comprised of all three features relating together. In other words, a point of view is a complete and systematic perspective, irreducible to isolated observations. Finally, due to the holistic nature of a point of view, Brewer describes its epistemic status as "understanding" and not "knowledge." "The important difference between knowledge and understanding," Miles Burnyeat claims and Brewer repeats, "is that knowledge can be piecemeal, can grasp isolated truths one by one, whereas understanding always involves seeing connections and relations between the items known" (Brewer, 1998, p. 1591).

20. Dreyfus argues that the relation between phenomenology and normativity concerns "priority," in the sense that normative obligations are ascertained by phenomenologists "prioritizing" how agents do in fact respond to concrete situations. This "priority" is also, according to Dreyfus, held by those in the Hegelian tradition of *Sittlichkeit*, such as Bernard Williams, Charles Taylor, and Carol Gilligan. By contrast, Dreyfus argues, a group of theorists in the Kantian tradition of *Moralität*, such as Jürgen Habermas, John Rawls, and Lawrence Kohlberg, "prioritize" detached principles that detail what the right thing to do is over an understanding of the empirical conditions that allow for certain decisions to be made (1990, p. 237).

21. This is an important corrective to Paul Feyerabend, who for example, misses this point. He argues that when the prestige of science does not demand excessive complication, and social circumstances are such that experts cannot be overly esteemed, such as during wartime, medicine is capable of being effectively simplified. He claims that evidence for this can be found in army recruits who have historically proven themselves capable of being instructed as physicians with only half a year of training (Feyerabend, 1987, p. 307). The point that Feyerabend misses is that army recruits may quickly become instructed to be competent at various aspects of medicine, such as triage, but this training does not produce expert physicians or undermine how normal training allows expert physicians to do more than quickly trained ones.

22. It is often desirable that experts defend their recommendations against other experts, or in some way be cross-examined so that those affected can question their presuppositions. If this is taken to mean that the expert must articulate his values, rules, and factual assumptions, examining becomes a futile exercise in rationalization in which expertise is forfeited and time is wasted (Dreyfus and Dreyfus, 1986, p. 196).

23. Dreyfus writes: "Society . . . must encourage its children to cultivate their intuitive capacities in order that they might achieve expertise, not encourage them to become human logic machines. And once expertise has been attained, it must be recognized and valued for what it is. To confuse the common sense, wisdom, and mature judgment of the expert with today's artificial intelligence, or to value them less highly, would be genuine stupidity" (Dreyfus and Dreyfus, 1986, p. 201).

24. At one point, Dreyfus suggests that trust can be obtained legitimately if experts provide a certain type of narrative. When he makes this suggestion, Dreyfus seems to mitigate his IT thesis. He suggests experts can be effective communicators, so long as they do not need to provide deductive accounts of rules they followed when making decisions: "The cross-examination of competing experts in an intuitive culture might take the form of a conflict of interpretation in which each expert is required to produce and defend a

coherent narrative which leads naturally to the acceptance of his point of view" (Dreyfus and Dreyfus, 1986; p. 196). This passage is interesting because it suggests that experts are capable not merely of producing, but "defending" something called "coherent narrative" and that this narrative may be evaluated in an intuitive culture" without endangering expertise. The passage is problematic because Dreyfus fails to explain what a "coherent narrative" is and why it is so efficacious as to "lead naturally" (another undefined phrase) to acceptance of a "point of view." The implication is that somehow experts can avoid the problems of IT, which are connected with what I called a "practical point of view," by presenting another kind of viewpoint in their "coherent narrative." The word "coherent" can be taken to suggest holism, and it is possible that Dreyfus has in mind something like Brewer's "theoretically holistic" expert point of view. But if Dreyfus is evoking theoretical holism, then he fails to explain how he can expect that experts, who according to IT "forfeit" their expertise when producing prepositional content, are capable of providing such a narrative *qua* experts. The guiding presupposition is that whatever this "coherent narrative" is, it is only unproblematic for experts to produce in an "intuitive culture," such as Japan. Not only does this presupposition seem to be predicated upon unsupported cultural essentialism, but it also calls RT into question. Nonexperts may not be able to recognize experts on the basis of deductive procedural steps, but Dreyfus indicates recognition can occur on the basis of a "natural" response to the content of a "coherent narrative." Without explaining what these are, Dreyfus begs the question as to why nonexperts should trust expert intuition.

25. As Shelia Jasanoff points out, the issue of selecting real experts is so difficult in the courtroom that the original 1923 "general acceptance" criteria employed in *Frye v. United States* had to be refined because it "did not provide guidance as to how much agreement was enough, or among who" (Jasanoff, 1995, p. 62). General acceptance, which is an implicit presupposition of RT, did not legally work because it failed to: (1) clarify the degree of consensus required for establishing general acceptance, (2) set guidelines for how the contradictory results produced by variations in boundary work should be resolved, and (3) determine how to weigh results provided by frontier as opposed to established science (Jasanoff, 1995, p. 62). Since general acceptance proved to be a vague and ineffective standard for judging expert consensus in the legal arena, by extension it is problematic for Dreyfus to implicitly connect IT with RT. They are also routinely used in journalism, where sensationalism is generated by showing that a problem is so complicated that experts cannot agree on how to solve it. But for Dreyfus, this is not an issue. Dreyfus overestimates the overall trustworthiness of experts because he analyzes them as a general category and therein fails to recognize that expert consensus is field specific. Since the level of expert consensus is field specific it follows that the trustworthiness of expert intuition is not something that can be addressed in Dreyfus's general terms.

26. This does not mean that an audience of comparable scale recognizes every field of expertise. For example, experts in physics are more widely recognized as possessing expert skills than are theologians, who are only recognized by a particular sect, which Turner calls a "restricted audience" (Turner, 2001, p. 131).

27. For example, when discussing the relation between massage therapists and recognition, Turner notes that some people feel they benefit from massage therapy, whereas others do not find the promise of massage therapy to be fulfilled. The massage therapist is thus only considered to be (or more strongly put, only is) an expert for the former audience: "So mas-

sage therapists have . . . a created audience, a set of followers for whom they are expert be-cause they have proven themselves to this audience by their actions" (Turner, 2001, p. 131).

28. One may nevertheless be concerned that by insisting upon recognition as an essen-tial dimension of expertise Turner undermines the objectivity of expertise. History is replete with examples of people who at one point are acknowledged to be experts and at other mo-ments are denounced as charlatans. Turner acknowledges, "[What] counts as 'expert' is con-ventional, mutable, and shifting, and that people are persuaded of claims of expertise through mutable, shifting conventions" (2001, p. 145). Nevertheless, he claims that the pre-rogative to revise fallible judgments is a crucial part of democratic life and that to insist on a higher standard is "Utopian" (Turner, 2001, p. 146).

29. Again, it is helpful to consider Dreyfus's examples. An example of DC can be found in Dreyfus's description of an experiment in which students, experienced paramedics, and CPR instructors watched videotapes of six exemplars (five students and one experienced paramedic) giving CPR to patients. This target group was then asked which of the exem-plars he or she would choose to save his own life. Dreyfus writes: 'The results were reveal-ing. In the paramedic group, nine out of ten selected an experienced paramedic. The stu-dents chose the paramedic five times out of ten. The instructors, attempting to find a paramedic by looking for the individual closely following the rules they were taught, failed to find the expert because an experienced paramedic has passed beyond the rule-following stage!" (Dreyfus and Dreyfus, 1986, p. 201).

BIBLIOGRAPHY

Barbour, I. (1993). *Ethics in an Age of Technology*. San Francisco: Harper Collins.
Black, B., Ayala, F., and Saffran-Brinks, C. (1994). "Science and the Law in the Wake of Daubert." *Texas Law Review* 72: 715–802.
Brewer, S. (1998). "Scientific Expert Testimony and Intellectual Due Process." *The Yale Law Review* 107 (4): 1535–1681.
Callon, M. (2001). "Researchers in the Wild and the Rise of Technical Democracy." Paper presented at *Knowledge in Plural Contexts*, Science and Technology Studies, Université de Lausanne, Switzerland.
Caudill, D. and Redding, R. (2000). "Junk Philosophy of Science? The Paradox of Expertise and Interdisciplinarity in Federal Courts." *Washington and Lee Law Review* 57 (3): 685–766.
Collins, H. M. (1995). "Humans, Machines, and the Structure of Knowledge." *Stanford Humanities Review* 4 (2): 67–83.
Collins, H. (2001). "Tacit Knowledge, Trust, and the Q of Sapphire." *Social Studies of Science* 31: 71–86.
Crease, R. P. (1997). "Hermeneutics and the Natural Sciences: Introduction." In R. Crease (ed.), *Hermeneutics and the Natural Sciences*. Dordrecht: Kluwer, 1–12.
Crease, R. P. (1999). "Conflicting Interpretations of Risk: The Case of Brookhaven's Spent Fuel Rods." *Technology* 6: 495–500.
Crease, R. P. (2001). "Anxious History: The High Flux Beam Reactor and Brookhaven National Laboratory." *Historical Studies in the Physical and Biological Sciences* 32 (1): 41–56.
Crease, R. P. (2002a). "Compromising Peer Review." *Physics World*, January 2002, 17.

Crease, R. P. (2002b). "The Pleasure of Popular Dance." *Journal of the Philosophy of Sport* 39 (2): 106–120.

Crease, R. P. (2002c). "Fallout: Issues in the Study, Treatment, and Reparations of Exposed Marshall Islanders." In R. Figueroa and S. Harding (eds.), *Exploring Diversity in the Philosophy of Science and Technology*. Routledge (forthcoming).

Dreyfus, H. (1967). "Alchemy and Artificial Intelligence." *Rand*, Paper P3244.

Dreyfus, H. and Dreyfus, S. (1985). "From Socrates to Expert Systems: The Limits of Calculative Rationality." In C. Mitcham and A. Huning (eds.), *Philosophy and Technology 11: Information Technology and Computers in Theory and Practice*. Boston: D. Reidel Publishing Company, 111–130.

Dreyfus, H. and Dreyfus, S. (1986). *Mind Over Machine: The Power of Human Intuition and Expertise in the Era of the Computer*. New York: Free Press.

Dreyfus, H. and Dreyfus, S. (1990). "What is Morality? A Phenomenological Account of the Development of Ethical Expertise." In D. Rasmussen (ed.), *Universalism vs. Communitarianism: Contemporary Debates in Ethics*. Cambridge: MIT Press, 237–264.

Dreyfus, H. (1991). *Being-in-the-World: A Commentary on Heidegger's Being and Time*. Cambridge: MIT Press.

Dreyfus, H. (1992). *What Computers Still Can't Do: A Critique of Artificial Reason*. Cambridge: MIT Press.

Dreyfus, H., Spinosa, C., and Flores, F. (1997). *Disclosing Worlds: Entrepreneurship, Democratic Action, and the Cultivation of Solidarity*. Cambridge, MIT Press.

Dreyfus, H. (1998). "Intelligence Without Representation." *Network for Non-Scholastic Working Paper*, Department of Philosophy, Aarhus University, Denmark.

Dreyfus, H. (1999a). "How Neuroscience Supports Merleau-Ponty's Account of Learning." Paper presented at the *Network for Non-Scholastic Learning* Conference, Sonderborg, Denmark.

Dreyfus, H. (1999b). "The Primacy of Phenomenology over Logical Analysis." *Philosophical Topics* 27 (2): 3–24.

Dreyfus, H. (2000). "Could Anything be More Intelligible than Everyday Intelligibility? Reinterpreting Division I of *Being and Time* in the Light of Division II." In J. Faulconer and M. Wrathall (eds.), *Appropriating Heidegger*. Cambridge: Cambridge University Press, 155–170.

Dreyfus, H. (2001). *On the Internet*. New York: Routledge.

Feyerabend, P. (1987). *Science in a Free Society*. London: Verso.

Feenberg, A. and Hannay, A. (eds.) (1995). *Technology and the Politics of Knowledge*. Bloomington: Indiana University Press.

Huber, P. (1991). *Galileo's Revenge: Junk Science in the Courtroom*. New York: Basic Books.

Huber, P. and Foster, K. (1999). *Judging Science: Scientific Knowledge and the Federal Courts*. Cambridge: MIT Press.

Husserl, E. (1973) *Cartesian Meditations and the Paris Lectures*, ed. S. Strasser. Dordrecht: Kluwer.

Ibsen, H. (1988). *Ibsen: The Complete Major Prose Plays*, trans. R. Fjelde. New York: New American Library.

Ihde, D. (1998). *Expanding Hermeneutics: Visualism in Science*. Evanston: Northwestern University Press.

Jasanoff, S. (1995). *Science at the Bar: Law, Science, and Technology in America*. Cambridge: Harvard University Press.

Latour, B. (1999). *Pandora's Hope: Essays on the Reality of Science Studies*. Cambridge: Harvard University Press.

MacKenzie, D. (1996). *Knowing Machines: Essays on Technological Change*. Cambridge: MIT Press.

Mialet, H. (1999). "Do Angels Have Bodies? Two Stories about Subjectivity in *Science*: The Cases of William X and Mister H." *Social Studies of Science* 29 (4): 551–582.

Pappas, G. (1994). "Experts." *Acta Analytica* 9 (12).

Polanyi, M. (1974). *Personal Knowledge: Towards a Post-Critical Philosophy*. Chicago: University of Chicago Press.

Rawls, J. (1968). "Two Concepts of Rules." In J. Thomson and G. Dworkin (eds.), *Ethics*. New York: Harper & Row, 104–135.

Reber, A. (1995). *Implicit Learning and Tacit Knowledge: An Essay on the Cognitive Unconscious*. New York: Oxford University Press.

Rouse, J. (1987). *Knowledge and Power: Toward a Political Philosophy of Science*. New York: Cornell University Press.

Ryle, G. (1984). *The Concept of Mind*. Chicago: University of Chicago Press.

Schon, D. (1983). *The Reflective Practitioner*. New York: Basic Books.

Searle, J. (1983). *Intentionality: An Essay in the Philosophy of Mind*. New York: Cambridge University Press.

Searle, J. (1992). *The Rediscovery of the Mind*. Cambridge: MIT Press.

Selinger, E. *The Paradox of Expertise*. Ph.D. dissertation, Stony Brook University (forthcoming).

Sheets-Johnstone, M. (1999). *The Primacy of Movement*. Philadelphia: John Benjamins.

Sheets-Johnstone, M. (2000). "Kinetic Tactile-Kinesthetic Bodies: Ontogenetical Foundations of Apprenticeship Learning." *Human Studies* 23: 343–370.

Stengers, I. (2000). *The Invention of Modern Science*, trans. D. Smith. Minneapolis: University of Minnesota Press.

Turner, S. (2001). "What Is the Problem With Experts?" *Social Studies of Science* 31: 123–149.

Young, I. (1998). "Throwing Like a Girl." In D. Welton (ed.), *Body and Flesh: A Philosophical Reader*. Oxford and Maiden, MA: Blackwell Publishers, 259–273.

Walton, D. (1997). *Appeal to Expert Opinion: Arguments from Authority*. University Park: Pennsylvania State University Press.

Williams, R. (1976). *Keywords: A Vocabulary of Culture and Society*. New York: Oxford University Press.

Winner, L. (1995). *Citizen Virtues in a Technological Order*. In A. Feenberg and A. Hannay (Eds.), *Technology and the Politics of Knowledge*. Bloomington: Indiana University Press, pp. 65–84.

Winograd, T. (1995). *Heidegger and the Design of Computer Systems*. In A. Feenberg and A. Hannay (Eds.), *Technology and the Politics of Knowledge*. Bloomington: Indiana University Press, pp. 108–127.

8. DO ANGELS HAVE BODIES?
TWO STORIES ABOUT SUBJECTIVITY IN SCIENCE

THE CASES OF WILLIAM X AND MISTER H

HÉLÈNE MIALET

AS THE TITLE of this paper indicates, I have two stories to tell: that of William X and that of Mister H. The former is William Montel, a world-renowned expert in the field of applied thermodynamics. I have called him 'William X' because no one has ever heard of him, at least among the general reading public, except myself. I spent several months in his research laboratory carrying out the ethnographic study on which the first part of this paper is based.[1] 'Mister H' is Stephen Hawking, probably known throughout the world, almost certainly throughout England and without any doubt in Cambridge. Everyone seems to have seen him there, at least once, except for myself.[2] There are a number of differences between these two scientists, yet they have one point in common: both are recognized as being geniuses. The question is, by whom and why?

LOOKING FOR THE CREATIVE SUBJECT

The questions which originally guided this research can be summarized as follows: How is a new scientific idea conceived?; and Why does one individual rather than another invent? These questions, as they are currently posed, are, to a large extent, informed by a philosophical tradition which has accustomed us to perceiving scientific knowledge as the product of new ideas. In France this trend was inaugurated by Claude Bernard.[3] In the same tradition we find Gaston Bachelard,[4] as well as a certain English-language current represented by Karl Popper,[5] who seems to fit in with this intellectualist tradition in his struggle against logical positivism.[6] Whereas there is common agreement that science feeds on ideas, when we examine these writings to discover how ideas originate, the answer is always the same: the question is beside the point. Thus, although the creative mind is designated as being the basis of the creation of new ideas, it is rejected from philosophical investigation, for what allows for change in science is viewed to be not of the order of science. That is why this question is not of interest to these epistemologists; only psychologists and historians have the right to address it.[7]

On the other hand, those sociologists or historians of science who have indeed studied the question of the production of scientific knowledge—that is to say, of 'science in action'—either in laboratories or by following controversies, have also rejected the questions of interest to us here. By dismantling bit-by-bit all the criteria of scientificity—the reproduction of experiences (Harry Collins), the interpretation of results (Bruno Latour and Steve Woolgar; Martin Rudwick; Simon Schaffer), criteria for evaluating proof (Andy Pickering), to mention but a few—they have challenged the divisions upon which science has traditionally been thought to be based: the context of discovery *versus* that of justification; the cognitive *versus* the social; the subject *versus* the object.[8] By redefining the way in which science is constructed, they have called into question the notion of invention constructed, as we might say, 'negatively' in relation to scientific knowledge. With them, science becomes a practice like any other; similarly, cognitive operations mobilized by scientists are construed as being part of commonplace intellectual processes. That is why sociologists are interested, not in the psychology of scientists, but rather in the objects, techniques and inscriptions that they study in order to produce knowledge.[9] If science does produce different objects, this is related not to the cognitive processes of scientists, but to the objects they manipulate. Creation thus becomes an eminently collective and material process. In this sense, the sources of innovation cannot be localized. Accordingly, Latour and Woolgar, after having deconstructed the process of discovery in their book, *Laboratory Life*, affirm that 'individual thinking', 'the inventive act', the famous 'the day I had the idea of', are all the result of 'a particular form of presentation and simplification of a whole series of material and collective conditions'.[10] Finally, for sociologists and historians, such as Gus Brannigan, inventors are the fruit of processes of arbitrary attribution.[11] What were, for example, the practices of Gregor Mendel, and why did he propose a different perspective? Was it simply by chance or an opportunistic choice? Yet Brannigan does not address this point: he is only interested in how an individual is *constituted* as an inventor; consequently, for him, the question is not how the idea comes to mind, but how the idea comes to society.

Thus, philosophers of science characterize or localize invention, but do not give us the means to study it, since it is mysterious and bears no relation to official science. On the other hand, sociologists of science have given us the means to study invention, but have relegated singular creative thought to the domain of mythical accounts. The singularity of the moment and the act vanishes, while the status of the actor remains problematic. It is almost as if a black box—'a-genius-has-a-new-scientific-idea'—has been opened. What was previously crystallized in an individual—the genius, the intellectual capacities, the ideas, the science—is opened out into the environment, and ends up dissolving the singularity of the individual by emptying her of her innate properties. Pasteur, as described by Latour, is not a body endowed with a mind.

> Or rather, he is far more than a body interacting with other bodies. He is a combination of a large number of varied elements that, through the links between them, produce Pasteur-the-great-researcher; thus he does not exist outside this network which, strictly speaking, constitutes his body and mind.[12]

It is on this question—what is the nature of the actor of invention?—that I will focus.

Indeed, if in the rationalist tradition the driving force of knowledge is inscribed in the subject, this subject is devoid of subjectivity: a subject that is transparent for Descartes, desingularized by Kant, or evacuated with Popper.[13] As for the sociology of science, it reintroduces a flesh-and-blood scientist into the process of knowledge production, but s/he is a subject who is (part of) a 'collective' body. Either the subjectivity is immediately social, in relativist sociology,[14] or, as in the case of actor-network theory, the subject becomes a spokesperson for an association of actors. To cite Latour's example again:

> In the Pasteur-network, we do find a laboratory, docile strains, notebooks, statistics, the Pouilly-le-Fort farm, journalists witnessing spectacular experiments organised by Pasteur, cows dying in infected fields, the French electorate that he strives to convince. . . . A human being is an envelope in so far as s/he is simultaneously distributed in all the elements of which s/he is composed and which at any moment might regain their independence.[15]

Thus, philosophy and sociology, for opposing reasons, deny the necessity of the situated body of the scientist.[16]

THE CASE OF WILLIAM X: AN ETHNOGRAPHIC STUDY

It is by using the tools of sociologists of science—that is to say, the weapons of those who eliminate the questions of interest to me—that I have attempted to characterize invention. I have therefore focused not on the study of a laboratory, but on an individual who works in the scientific and technological research centre of a major French corporation (Elf Aquitaine). His field of study is the characterization of oil fluids, and his main task is to account for the origins and behaviour of oil or gas deposits. Understanding how fluids evolve, and determining their different states (an oil can become a gas, and *vice versa*) has a direct influence on the way in which they are processed. This researcher's field of expertise is thermodynamics; his operative tool is computer modelling.

By choosing this field of investigation, I wanted to satisfy the following 'minimum' conditions:

- observe a scientist in her/his daily work rather than, for example, in an 'artificial' situation of problem-solving,[17] as experimental psychology does;
- avoid observing a scientist who is too well known, so as not to have the impression of 'arriving on the scene too late';
- opt for a scientist doing applied rather than basic research, to avoid problems pertaining to criteria of scientificity and truth which have been extensively debated by sociologists of science, and in which the main issue is discovery (that is, the question of the referent), not invention.

Following these 'minimum conditions' in carrying out my ethnographic study of William X led to a reconsideration of the questions which originally animated my research on the creative subject. Accordingly, in the course of this study, the 'literary' repertoire of the 'lone scientist, who, one day, had an idea' appeared; this time, however, it was not the individual himself who called himself an innovator (as many autobiographies of scientists claim), but rather his surroundings which qualified him as such. Thus, in a strange inversion, I came to study an individual who evaluated the work of others in his capacity as an expert, and instead found myself confronting an individual who was constantly evaluated by others—that is, he was qualified by his co-workers and subordinates to be an innovator—the originator of an important new idea. He was recognized as being both a creator and creative. Indeed, the fact that he constantly had new ideas was precisely what distinguished him from others. But can one really say that an industrial researcher is a creator or even an inventor, especially when the said inventor constantly redistributes the processes of invention to others? This leads us to wonder how a relationship of equivalence is constructed between a mythical repertoire of the lone scientist-genius, and local practices. How, in other words, does an individual become a reference 'in vivo'?

The challenge in my study has been to analyze the oscillation, in a collective context, between an environment—relations among colleagues, instruments and rules of functioning—and the singularization of an individual who gradually distinguishes himself 'in' and 'from' the group (with regard to his creative competence in the development of working tools, organizational structures and human relations). In short, I did not discover an isolated individual totally in control of his destiny in a determined collective, as described in the writings of many scientists and commentators; rather, I sought to understand both how this individual accommodated himself to others, and the collective operations constituting the singularity of his character and his invention.[18] I also wanted to understand how it was possible to invent and innovate in an institutional context with strong technical, economic and organizational constraints. In other words, what skills are needed to be an innovator, and to find oneself in a position to innovate?

From the laboratory to the individual. William X is the head of the Thermodynamics Section of the Oil-Field Department, which is in turn part of the Research and Development Division at Elf Aquitaine's Scientific and Technological Research Centre. His section consists of four engineers and three technicians. William has three functions: development and management of research programmes; in-house services; and research for subsidiaries. Most of his work is carried out in his office. Occasionally he travels to participate in conferences or give lectures, but he never visits oil rigs. Similarly, interaction with colleagues in his own section, with those in other sections (called 'delivery of in-house services'), with subsidiaries (in the Congo, Norway or Russia, with the Boussens or Pau laboratories), and universities (Pau and Toulouse), or the Institut Français du Pétrole (IFP) in Paris, all take place in—or from—his office (by telephone, in writing or through representatives of these different bodies). In most cases, it is therefore others who go to see him (except for members of the IFP, where he lectures), and when he does move about it is to attend local meetings in an advisory capacity. Whenever information is missing or incoherent, or requests are unclear, William is solicited—very much so—to solve or transform problems submitted to him.[19] Sometimes his colleagues come to him for advice. They consider him the expert: 'We go to him because he solves a problem in three hours, whereas we'd take three days' (Arthur).[20] At other times they request his permission, as head of the Section. All files circulate daily or weekly 'through the Section head's office', defining rôles as they pass from office to office, recording the status and activities of each person in the company. Finally, he is also seen as an arbitrator: 'When I'm not sure about what I'm saying in my study, I ask William and he decides' (Patrick).

In his office, William works away with his computer, modelling oil fluids. He comments on curves, calculates discrepancies and reads reports. He skillfully works with the models inscribed in the memories of computers. That is how I first found him, preoccupied, his eyes riveted to a blue graph on the screen. The graph was slightly different from the previous one. 'It's the phase envelope', he says: 'It corresponds to the composition of the fluid, to each pressure and temperature level. A new curve means a new definition of the fluid'. The knowledge seems to be directly related to the application: 'What I calculated was wrong' he says:

> I sold this study. They must have missed a few million tons. . . . Before the development project, there are panels . . . there were disastrous projects. It's not always the quantity you expect. It's all the profits that could go up

in smoke. The field may not be able to produce due to a problem of inter-facial tension that was not estimated correctly. For example, the viscosity may be ten times greater than expected. It's a question of the ratio of oil to gas: one m^3 of fluid at the bottom gives x amount of oil at the surface. If it's a porous volume the field stretches over 40 million m^3. You'll get 0.2 million m^3. A third more. That's the economic margin on 40 billion francs; 15 billion francs will either be in your pocket or will remain in your head.

Although we are focusing entirely on this individual, what is happening in Elf Aquitaine's Thermodynamics Section, in this man-machine 'hand-to-hand' encounter, has a direct impact on the exploitation of oil-fields thousands of kilometres away and, simultaneously, for William, on the judgement of his colleagues only a few metres away. Through good or bad computer simula-tions, ideas may be transformed into oil or revert to the state of a project 'in his mind'. It seems that the laboratory I was looking for when I went into the Sec-tion, into the spatial arrangement of the rooms, in a decor of sterilized work surfaces and tiling, was located elsewhere. It was inscribed in the computer. That is where the transfer of worlds was taking place—thanks to mathemati-cal models, I was told.

To get to the heart of these models, I had to remain attentive to the way in which they were qualified by those who used them. I thus chose to follow the explanations of Arthur, one of William's close collaborators. Arthur introduces me into his world by way of thermodynamics. Its use in oil exploitation is re-cent, he recounts. It is an abstract science. Everyone is put off by it because it seems very unrewarding. Unlike William, who is totally in his element, Arthur doesn't master it. Moreover, he says, it's William who is at the origin of the ap-plication of thermodynamics.

> He was at the PVT lab,[21] he was fed up and offered to apply thermody-namics; for five years he sat alone developing his models, and now he's snowed under with work.[22]

What then is the specificity of this Section? Arthur explains that the Thermo-dynamics Section is called upon whenever hydrocarbons are discovered. It has to describe the initial state of the fluid, and the way in which this fluid will be-have during exploitation. For that purpose it has to use models, which are state equations. Arthur then describes the characteristics of the environment of in-tervention essential for understanding the Section's work. I am thus plunged into technical details:

> The rock reservoir in which the hydrocarbons are trapped may either con-sist of hydrocarbons in a homogeneous phase—that is to say, there is only gas or only oil—or else the pressure and temperature are such that there

are both. Thermodynamics is related to the criticality of fluids—that is to say, volatile oils or condensed gases,[23] for which exchanges between phases are very important.

He added:

> We always reason in terms of phases, that is our criterion. Our problem is to know at which point we have the maximum amount of oil, because what counts for a cubic metre of fluid is what one gets at the surface.

Insufficient knowledge of fluid dynamics may result in a low rate of extraction, and thus in financial loss. Arthur opens a report, shows me a curve; it is the phase envelope. Its shape depends on the composition of the fluid.

> One day William said that the composition of the fluid could be calculated in relation to these measurements. Because an oil is composed of different molecules, there are heavy ones and light ones. . . . The different kinds are separated in terms of their density. That means that in the oil-field there will be heavy oils at the bottom and gas at the top. Originally it was the same fluid, then the different kinds separated. William described the model that has made it possible to calculate that. He's the first person in the world to have identified this phenomenon.

Let us pause for a moment, and reflect on this account. We note that Arthur cannot explain the use of thermodynamic models without referring to William. His name keeps cropping up throughout the discussion; he is no stranger, it seems, to the appearance of models. Apparently he is even at the origin of the application of thermodynamics, and worked alone on this development. He is the first person in the world to have identified a phenomenon and described the corresponding model. Today, individuals reason differently.[24] Reasoning has been affected by the implementation of these models: 'now, we reason in terms of phases'. We thus find the mythical repertoire of invention at the heart of industry: 'One day William said that . . .'; 'he's the first in the world to . . .'; 'he was alone . . .'; 'he identified . . . the phenomenon of segregation of different types and developed the corresponding program, the GER'.[25] For us, this discourse is a narrative, like a children's story, in which the 'typical' phases indicate that we change the world: 'Once upon a time', 'One day'. Thus William is associated with thermodynamics and these models through such linguistic operators, designations and attributions, in a narrative; and, in a similar procedure that we must qualify, what was designated by movements (that is, people go to see William as 'the expert') is now supported by this narrative.

The fact that William had the idea of this model underscores, in his colleagues' view, the greatness of this individual: an idea, 'his' idea, transformed an environment and affected their judgement. Yet if we remain attentive to this operation of designation, we see that it is through the use of the models that the narrative is formulated. It is the requalification of the environment—through the operations of his computer model—which is re-attributed to the novelty of a first idea, and to the intellectual and inventive capacities of its author. This process of attribution redefines the place of the author—and thus, in a sense, simultaneously combines William's competence and his very conditions of existence.

In fact, the model does qualify the oil-field fluid differently. Before and after its application, one does not have the same knowledge of the fluid. It is a new way of making it talk: I had a light oil; all I've got now is tar', says William, manipulating his model while looking at the graph. Or Arthur says: 'The model says there may be oil ten metres from the bottom of the reservoir'.

Arthur shows me a report and a map. In this report are columns and figures: the translation of the constitution of the fluid taken at a given time in a given place. On the map, the fluid is represented by a multitude of scattered points. Drilling indicates its presence in various places on the earth. But is it the same fluid? Points, figures, assertions: that is all we have. It is by using William's mathematical model, resident in each computer, that the properties of the fluid can be calculated and its evolution described:

> We have the pressure of saturation. . . . We use mathematical models, equations which enable us to say, for a given temperature and pressure, whether we have oil or gas. In order to use these models we have to key in the composition of the fluid and its thermodynamic properties. The composition of the fluid is broken down into 15 parts. Why?—it may have many components. C_{12} is a molecule which has 12 carbon atoms. We group together the components which have the same number of carbons. For each component and pseudo-component there is an M (molecular mass), T_c (critical temperature), Omega (acentric factor), V_c (critical volume), and P_c (critical pressure). The model calculates all the properties. For example, at a given temperature and pressure, one has to find the bull pressure.

Arthur sits down at his computer and puts the model into operation. As he works, the tool is objectivized: 'William's model' becomes 'GER'; and GER is capable of 'saying that . . . the oil is there'. A computer tool is capable of transforming the properties of an oil-field fluid thousands of kilometres from an office. This translation is attributed to the capability of an individual.[26] The terror of stormy seas, the violence of oil explosions and the monstrous machines used are all controlled by a model working silently in the heart of the computer. Arthur works with it calmly, sitting in front of his machine. William has been transformed into a fluid:

> Our models are used to restore the properties of fluids *in fine*. . . . Look, I applied *William's model* of segregation by density and I simulated the presumed composition of the fluid in relation to the measurements. They're experimental points resulting from complex operations for sampling fluid indirectly. In relation to the experimental points, there's a curve, and the point of inflection which reflects the oil-gas transition. Other examples where, look here, there's a very clear point. . . . With GER one can see what's happening from the top to the bottom.

We see here a first element which gives rise to this work of attribution. The quality of an individual is redefined by the qualities of a model of which he is said to be the author: the capacity to calculate the properties of a fluid and to predict the place in which a maximum amount of oil can be found. With the model . . . or with William, we can see 'what's happening from the top to the bottom'. As Arthur works with the model, applies it and finds in it new capacities, we see an individual assuming importance because, 'one day', he thought of it.

My first point is to underline the decisive rôle and functioning of the processes of attribution which reveal both the qualities of the object designated as new, and the agent recognized as being its author (in this case William X). It was by successively interviewing the different researchers working in this laboratory and by asking them, not what they thought of this individual, but rather to describe the content of their work, that I gradually discovered how they simultaneously qualified the models they used ('their ability to predict'; 'their ability to produce knowledge'; 'their ability to transform the complex into simple'; 'their operative capacities') and the competence of this individual ('nothing like this was ever done before'; 'it was his idea'; 'he was alone'; 'he's the only one to'; 'his mind's always working'; 'his expertise is a product of his intellectual capacities'; and so forth).

Thus, a proliferation of narratives crystallized around this individual and the object that he was said to have developed. This revealed certain qualities of the subject through the object; the qualities of one were superimposed on the other, and it became possible to see the subject through the object that he had created (a non-neutral object, since it was a so-called 'intelligent' computer program!). Similarly, when the inventor himself qualified the object he had developed, he repeated the very same comments that others had made about him.

Yet in no way does this proliferation of narratives move us away from invention. Saying 'Mr X had the idea of' does not mean that the invention took place at the moment he had the idea; it does, however, take place the moment this narrative is pronounced in a specific context, by making people act in a certain way. To use Austin's concept, this discourse 'performs' the invention by simultaneously presenting the singularity of both object and inventor.[27] Thus, for example, the moment William's colleague manipulated the model, he created his own idea of the person who had developed it. The more he applied the model and changed the invention, the more significant the inventor became. In this colleague's view, the fact that William had had the idea of this model made the greatness of his character stand out. William's idea had transformed an environment and affected his colleague's judgement. Thus, these narratives continually left their imprint on practices, and requalified both inventor and invention. In this sense, one could say that there is no invention without a 'mise en scène'. We have no more need to invoke great causes, for the origin and the meaning are constructed in the accounts.[28] We witness the creation of a plot, and the multiplication of these plots has a fundamental rôle for, to quote Isabelle Stengers,

> ... the beings that science causes to exist are 'invented' in the sense that all their attributes are relative to our stories ... but that is precisely why their existence depends on the multiplication of stories which all relate back to them and designate them as the necessary if not unique condition of their possibility.[29]

Finally, there is a proliferation of non-arbitrary narratives which designate and allow for the existence of individual variations. By an array of interlinking discourses, these operators of individuation cause the person to be isolated and, hence, individualized. This individual incarnates the one who has ideas, who does different things, a figure necessary for the institution. As the Director of the Research Centre says, 'we need Williams'. These narratives tell us that something specific is happening in the body of this individual. In other words, one cannot say either that the invention springs entirely from William (the genius), or that it derives from the social context of which he is part; but rather, in order for this individual to assume the mantle of the inventor, he has to have done—and has to continue to do—something specific.

The body of the scientist: site of distribution and concentration. This assertion enables me to introduce a second point which seems to me fundamental in understanding the 'knowledge process' (*processus de connaissance*): the part played by *the scientist's body*. The actor, in this sense, is not only (as actor-network theory holds) the product of an association of heterogeneous elements constituting his body and mind. In other words, William is *not only* the result of a process of composition, indiscernible from the network of entities on which he acts and which act through him. He can be distinguished from this network precisely because he moves within it in a specific way, mixing with his research object, translating and linking up things which were previously unconnected. He is able to play several rôles at once, his peculiarity as an individual deriving from his capacity to interchange with various elements.[30] Indeed, in the case of William, the presence of the inventor's body is indispensable if we are to understand how certain properties are exchanged, and how an innovative process continues functioning. We see the flesh-and-blood inventor intervening at two specific moments. First, he intervenes every time there is a malfunction in the device he has developed. It is precisely in these moments of problem-solving that we witness a process of subjectification. The fact that he resolves the problem presented to him once again qualifies the inventor. We observe the qualification shifting from a re-qualified environment to a subject. The researcher's skills, enhanced by the computers and the different models they manipulate, make certain operations and simplifications possible. These are re-attributed to the individual's capacity after he has carried out certain operations (other than stabilizing the network, which enables him to carry out these operations).

Second, he has developed a particular skill for resolving these problems: his own specific ability to identify with his research subject. When the researcher is in front of his computer manipulating equations, he has to master his model, analyze curves and know to what their variabilities correspond, have a representation of himself and the fluid he manipulates, and know all the operations to which it is subjected. It is then that he will share with his research object the properties of his own body. In a particular way, he himself becomes, then, an oil fluid. He says that he 'sinks into the oil fluid', that he 'senses inconsistencies', 'endures variations', that he is 'disturbed'. One might say that he is touched in his extended body, just as someone 'becomes his car' when he knows from the sound of the engine where the problem lies.[31]

This ability to solve problems, to make connections that others do not make, is linked to his ability to multiply his fields of intervention.[32] In describing this 'fleshy translation', I would like to show that this is not a metaphorical account, but an attempt to express how William X's body is both concentrated and dispersed at the same time as it is defined by the materials upon which it depends. Indeed, by placing himself at the intersection of different research fields, he is able to translate problems from one field to another, and thus constantly to feed his inventive mechanism and his status. Researchers from other related departments consult him when they encounter thermodynamic problems. As he says:

I don't have to do any specific research. I just have to transcribe in intelligible terms something that can be done elsewhere. It's often quite a tricky job because I realize that there are things that aren't fully understood. But transcribing and trying to transpose them, there is always a question that is more than just a proposition of know-how. That's why I enjoy doing it; it always opens a new door on to a phenomenon that we hadn't taken into account in our problem because it seemed relatively unimportant, but which becomes important in their system.

By consolidating his group from the inside—that is to say, by making it participate in the creation, and by delegating to it his know-how—he extends his recognition. 'When I contract work out', he says, 'I have to motivate the others, to help them to discover what I already know'. By locating his computer programs in different phases of interpretation of oil fluid, he spreads out his action. By placing doctoral students—whom he will eventually bring back to his lab—in well-known research laboratories, he gives himself the possibility of fuelling his theories with new information, and also of spreading his field of influence. In short, it is because he positions himself in several places so as to occupy different spaces, that he is able to feed his know-how,[33] maintain himself at the centre of the network that he has created, and enhance his recognition. We thus see, and this is my third point, contrary to the idea that invention is complete at the moment of inception, that it is constituted through energy expended and maintained over time.[34]

William X: a distributed-centred subject. The fact of re-establishing the importance of the scientist's body in a process of knowledge production (and of showing that the material, human and discursive elements are distributed and concentrated around the scientist's body) enables me to give several elements of an answer to the question that I want to address: What type of subject will be created in a context of knowledge where knowledge is action and no longer contemplation? By rejecting the representation of the creative process as a simple mental operation, I show the redistribution of intelligence in the various practices through which it emerged (William and his inventions cannot function without the entire collective which gives them their power). By pursuing this analysis, I think we can show that this operation of distribution and restoration of practices, far from making us lose the subject, enables us to understand how he is singularized (through processes of attribution), how he singularizes himself, and how he invents. Thus, contrary to what Brannigan or the theory of actor networks seem to claim, these processes of attribution are not arbitrary, they are operative. In other words, far from dissolving the subject by showing that there is not an invention and an inventor, but only 'ways of talking of the invention'—'it's all staged', either 'it's random', or 'it's the objects that make the procedures exist'—we can say that we are observing a collective process of singularization in which the said individual participates, which causes a subject to emerge, and which performs the invention. Thus these accounts enable individual variations to exist. We witness a dual movement of singularization and generalization. It is by singularizing an individual that we generalize his/her ascendancy.

The ultimate aim of this research is the re-introduction of the rôle of a creative actor in the process of invention. But it is an actor endowed with new properties. This subject is closer to the action-without-a-subject of structuralism than it is to classical humanism. Far from being a brilliant scientist revolutionizing the world by the pure force of his reasoning, he is constituted, to adopt Foucault's concept of an author, by our selective operations. At the same time, this subject is closer to the actor network than to the subject of structuralism. He is not only what the structure makes him, for he is capable of playing and melting into the system while simultaneously distinguishing himself by a number of features.[35] And, finally, this subject is closer to the actor of psychology in his capacity to transpose problems, to submerge himself in objects, and to undergo a metamorphosis, than he is to the actor of actor-network theory (a sort of Cubist figure where the actor ends up losing all psychological probability). I have called this subject a 'distributed-centred' subject. I argue, in short, that the more an actor is distributed, socialized, collectivized and multiplied, the more he is singular, an ego, a non-interchangeable body. The greater the number of elements to which the actor is connected, the more innovative s/he will be, or has the potential to be.

INTERMEZZO

From this first case, I have wanted to re-introduce into a problematique of the banalization of scientific production, the emergence of an actor to whom one can attribute the qualities of expertise, creativity—of genius. My study has therefore consisted in putting into relief and describing the ensemble of mediations across which it has been possible to understand how a singular actor is constituted. From the tools and concepts developed in understanding how one constructs this knowing subject in the process of constructing himself—that is, how the collective subject is inscribed in the functioning of a singular actor—I have wanted to shift my focus of analysis back to the question of the pure scientific genius from whom historians and sociologists of science have been trying to escape for over 20 years.

By choosing to study the 'heroic' figure of Stephen Hawking, I have been able to re-establish two variables in this project—the temporal and the public dimensions—at play in the process of attribution, which, as I have indicated, constitutes a crucial part of the inventive trajectory. Indeed, the case study of William X is totally synchronic. I spent several months in his laboratory, but if the study had started two years earlier or later, what kind of figure would I have seen emerging? With Stephen Hawking, it is possible to analyze press reports over a period of about 15 years, and to follow how and why the elements comprising genius changed (perhaps in the mid-1970s the idea of a good scientist was different?). Moreover, William is an acknowledged expert in a world-renowned institution and is internationally recognized in his research field, yet nobody in the general reading public, as I have already noted, has ever heard of him. In the case of Stephen Hawking, everybody knows him, but how do these processes of informal attribution which circulate in the public at large function? Lastly, with him one has—in the negative sense—the corporeal dimension, which is, as I have emphasized, critical to the production of scientific knowledge. Indeed, how can Stephen Hawking, who is totally paralysed, practice his research activity? Drawing on a series of interviews with his assistants and colleagues, I will attempt to answer this question.

With this research framework in place, let us see whether the tools here developed—distribution, attribution, and so on—operate in an analysis of another kind of knowing subject, the renowned Lucasian Professor of Mathematics and best-selling author, Stephen Hawking.

DE-CONSTRUCTION OF A GENIUS: THE FABRICATION OF STEPHEN HAWKING

By way of a prologue, I shall start with a thought experiment suggested by John Locke in his *Essay Concerning Human Understanding*.[36] The experiment is based on the following riddle: What would happen if, instead of eyes, scientists had microscopes in their eye-sockets? The answer: equipped with such prosthetic eyes they would get to the essence of things, for they 'would come nearer the discovery of the texture and motion of the minute parts of corporeal things, and in many of them probably get ideas of their internal constitutions', but they would simultaneously become angels, for then they would be' . . . in a quite different world from other people: nothing would appear the same to . . . [them] and others: the visible ideas of everything would be different'. And Locke adds: 'So that I doubt whether [they] and the rest of men could discourse concerning the objects of sight, or have any communication about colours, their appearances being so wholly different'.[37] Hence, what they would gain in divinity would be lost in humanity, for humans would no longer be able to communicate with them. Thus Locke concludes:

> Since we have some reason . . . to imagine, that spirits can assume to themselves bodies of different bulk, figure, and conformation of parts—whether one great advantage some of them have over us, may not lie in this, that they can so frame and shape to themselves organs of sensation or perception, as

to suit them to their present design, and the circumstances of the object they would consider. . . . The supposition at least, that angels do sometimes assume bodies, needs not startle us; since some of the most ancient and most learned Fathers of the church seemed to believe that they had bodies: and this is certain, that their state and way of existence is unknown to us.[38]

By error or by chance, we think we have discovered an angel. He does not have microscopes for eyes, but he does have a synthesizer for a voice, and instead of his body movements he has a wheelchair and a computer.

Stephen Hawking has undergone a series of trials in his life, starting at the age of 21, when he developed amyotrophic lateral sclerosis (more commonly called Lou Gehrig's disease), characterized by muscular atrophy.[39] In 1985 he contracted pneumonia, underwent a tracheotomy, and consequently lost his voice definitively. In overcoming these ordeals, Stephen Hawking was to become an angel. But whereas we credit him with the ability to attain the essence of things owing to his profound understanding, he was, paradoxically, able to regain contact with humans thanks to technology: his computer. Between Professor Hawking in his wheelchair and the universe there is no mediator—or only one: his mind. As reports in the popular press put it:

Mind over Matter: Stephen Hawking roams the cosmos from the confines of a wheelchair.[40]

Stephen Hawking Probes the Heart of Creation: His scientific genius soars from his severely crippled body—to unfold the deepest mysteries of the universe.[41]

Roaming the Cosmos: Physicist STEPHEN HAWKING is confined to a wheelchair, a virtual prisoner in his own body, but his intellect carries him to the far reaches of the universe.[42]

Reading God's Mind: Confined to his wheelchair, unable even to speak, physicist Stephen Hawking seeks the Grand Unification Theory that will explain the universe.[43]

Thus, Hawking incarnates the mythical figure of the lone genius: a man who wants to—and declares himself capable of—grasping the ultimate laws of the universe with his mind.[44] As with the case of William X, I shall strive, not so much to claim that the genius is socially constructed, as to understand the basis on which he is constituted. My approach will be one of bold constructivism—that is to say, one which includes both the work and the scientist in the collective production of their own grandeur.

The hidden scientist: behind a distributed body. Professor Hawking works in the Department of Applied Mathematics and Theoretical Physics at Cambridge University. He is usually surrounded by four students to whom he assigns different problems, depending on their abilities and qualifications. John is studying virtual black holes (and their predictability), Matthew the creation of black holes, Bob the mathematical formulation of general relativity, and Peter the stability of string vacuums.[45] There is little exchange between these individuals, who interact directly with the scientist. Professor Hawking is the incarnation of the Popperian vision of science.[46] He has, as one of his students notes, 'the intuitive notion of what could be done', 'the ideas' and 'the inventive mind' that his assistants, tirelessly working on proof, try to put into mathematical inscriptions. Alone, the professor cannot manipulate these calculations, for the simple reason, so often overlooked by epistemologists, that in order to do mathematics one needs not only a head but also ten fingers for writing equations, drawing diagrams and using a computer.[47] Totally paralysed in his wheelchair, Stephen Hawking can use only the single finger that his nurse delicately places every morning on the switch linking him to his computer. But a computer cannot calculate on its own; it needs to be operated by a researcher with nimble movements. That is one of the reasons, they say, why Professor Hawking has developed the capacity to visualize mathematics in a geometric form. He also uses a program which transforms words into symbols, for although he cannot produce calculations, he can construct sentences, albeit succinctly. As a result, his students do most of the work themselves.

The computer has, nevertheless, considerably changed his environment. Before he underwent a tracheotomy in 1985 and lost his voice, Stephen Hawking had great difficulty expressing himself. Only a few initiates were still able to understand and interpret what he was trying to say. That was how he dictated his scientific papers. So that he could read, his secretary or one of his students used to hold the book or article in front of him, or he used a music stand and his assistants would turn the pages when he reached the end. He was also able to consult a list of articles on transparencies on the wall. As for the telephone, it remained hopelessly inactive. Today, owing to a whole intelligent system, a communication program called Living Center (given to him by Walt Woltosz, of Words Plus Inc. in Sunnyvale) and a synthesizer (offered by Speech Plus), he can communicate (and now with everyone), write and even read, since many scientific articles are—or will eventually be—available on computer. New articles are sent directly to his electronic archives. Hence he is able to receive feedback, to pinpoint interesting questions and promising methods, and to group these together in the search for new answers. It is possible for him to answer the telephone, which is equipped with a microphone and amplifier so that he need not hold the receiver. Hawking is able to operate this entire system on his own simply by means of a switch (which someone has to place in his hand, curling his fingers around it). The device consists of no more than a highly sensitive on-off switch which operates a cursor. Since the cursor automatically moves up and down, Professor Hawking can select words by pressing on the switch. The synthesizer, and a small personal computer, have been mounted on his wheelchair, so that his voice now follows him wherever he goes. Every year a new student is responsible for the 'mechanical' services involved: taking care of the professor's computer, ensuring that his wheelchair works properly, or even fixing his son's bicycle.

· In this environment, Professor Hawking's secretary also plays a significant rôle. With time, she has developed a peculiar way of interacting with him. She does not wait for him to speak; by merely looking at his eyes she knows his answers, as one does, she explains, with a child before it learns to talk. She recalls, moreover, that this exercise did not seem particularly difficult, for she has four children. Today, she has to show him an invitation received from an eminent professor to attend a conference in Chile. She knows that there is little chance of him going, but 'he's the boss', and 'I need to make him feel that he's the boss'. But what if Professor Hawking were to answer 'yes' (with his eyes)? She knows that this simple word or sign 'yes' would probably give her 18 months of work organizing everything—that is, transforming a single word into Professor Hawking in person in his wheelchair talking through his synthesizer to a crowd of enthusiastic Chilean scientists. She would not be allowed to overlook the slightest practical detail, either about the journey and all the complications that a wheelchair equipped with a computer system entails as regards security, or about the financing of the trip and, above all, its scientific side. She would have to contact all the eminent persons that he would be meeting and know what questions they planned to ask him so that he could prepare the answers in advance. The aim would be to minimize, as far as possible, all unexpected intervention. Professor Hawking would thus, in a sense, be pre-programmed for all the formal meetings he would have. Finally, she takes care of his personal finances and legal problems (by the latter she means conjugal problems, implying his divorce and recent remarriage to his nurse), as well as personal matters such as arranging for his son to go abroad, or simply fetching him from school. She is also responsible for relations with the media, and for storing the dozens of letters received daily from the professor's admirers. Since his book *A Brief History of Time* was published ten years ago, letters which he never sees and is not even aware of would take every waking minute of his every day if he were to read them. Finally, a team of four nurses takes care of nearly all Hawking's physical needs, from brushing his teeth and feeding him, to combing his hair and dressing him.

Thus, in Professor Hawking's case, the motor and cognitive operations unseen by the ethnographer, either because they concern the private person of the scientist (the way in which he brushes his teeth is of little relevance to understanding how he thinks), or because they are normally incorporated in a single body (as, for example, in the body of a healthy physicist—such as William), become visible. They are, in effect, externalized and incorporated in other bodies. The professor is therefore *more* incorporated than any other theoretician, contrary to the image that is given of him. Accordingly, it is possible to 'see' the work of delegation, multiplication and redistribution of skills to people and machines so necessary for understanding the functioning of his brain.[48] As shown above, this mobilization includes that of the computer's competence as well as of the students, spread across very different research fields, who enable him to integrate diverse information and the different facets of a problem; it also includes the competence of his secretary, who sorts and arranges data according to his interests and what his brain can process (she will show him invitations to conferences, but not letters from his admirers). She takes charge of the professional as much as the private side of the person, including the smooth operation of this system based on a privileged relationship with the scientist himself, a kind of court society in which the individual presence of the king determines the activity of the courtiers. All the fundamental aspects of 'pure science'—thought, proof, calculations, the context of discovery and of justification, and the reception of the scientist's work—are thus incorporated and distributed throughout the laboratory.

Yet the Professor Hawking that I have just presented is the fruit of my own construction, based on information gathered during interviews carried out 'around him', for the professor, at least at the time I wrote this paper, was invisible, refusing all written interviews.

Transforming the collective body into a disincorporated brain. Where, then, does he appear, and in what form? The socio-cognitive network or fabric that I have just described will 'magically' disappear in the press.[49] All that will remain is the image of this singular body, the property of a whole media network which exploits and circulates it, making it visible and mobile because it is designated as immobile, invisible and unnecessary. In short, it is because it no longer functions that the scientist's body becomes visible. Left credulously to grapple with this dialectic, we glorify him because he has transcended the conditions imposed on him by his own body, while the prevailing ideology promotes a scientist without a body or self-awareness.[50] For the epistemologist, Stephen Hawking is not disabled: he has become a perfect scientist, a man without a voice, a machine, an angel.

Hence the media, together with the scientist himself—we shall see how he allows or prohibits the media from exploiting his body—transform this collective body into a disincorporated brain. The point of anchor is, therefore, this disability around which all the discourse turns: is it despite or because of it that he has become a genius? The 'owing to' crystallizes a series of negative and positive commentaries: negative, such as that of the physicist Jeremy Dunning-Davies, who attempts to explain the 'Hawking Phenomenon' and sees nothing very special in Professor Hawking 'except for his handicap, blown out of all proportion by the media who, moreover, place him in the tradition of the Newtons and Einsteins when in fact his theories haven't been proved';[51] or that of the journalist Arthur Lubow, who considers that 'Hawking won't revolutionize physics but will leave us the image of his smile'.[52] By contrast, others' comments are more positive—or, rather, as we shall see, they transform the handicap into a positive element. The sick body represents above all an obstacle in the view of everyone except that of the scientist, who sees himself as an exception, and happily agrees with this claim:

> At the age of 35, Stephen Hawking, despite severe physical handicaps, has established himself as one of the world's leading theoretical physicists.[53]

In other accounts, far from Hawking's handicap being treated as an obstacle, it is seen as the source of his performance and creativity. Since he is no longer distracted by daily and worldly occupations shared by the rest of humanity, he can devote himself entirely to thought. He has become a pure cerebral being communicating with the great Universal.[54] People are impressed by his exceptional memory, and compare him to a Mozart composing a symphony in his head; a journalist adds: 'and anyone who saw the lines of complex mathematics covering the blackboard like musical staves at a recent seminar would have appreciated the comparison.'[55]

Stephen Hawking, not simply following in the tradition of Galileo, Newton and Einstein, has in his head alone the cumulative knowledge of some of history's greatest minds; a head supported, as one journalist notes, by an almost non-existent body.[56] Finally, Stephen Hawking is seen through the subject of his research. He himself becomes a black hole:

> There is a phenomenon of relativity known as time dilation, in which time appears to slow down almost to a stop for bodies that approach the speed of light. Hawking alludes, in his book, to what it might be like for an astronaut as he accelerates toward a black hole, as all eternity passes by outside in an instant of his time. *There is a sense in which Hawking himself has experienced a kind of time dilation, an inexplicable slowing of a natural process that has added decades to his expected life span.* The answer he seeks may be almost within his sight. But so, perhaps, is the event horizon.[57]

Thus we witness a process of construction of a genius around the mind/body dichotomy, and a transformation of the person of Stephen Hawking into a pure bodyless cerebral subject, which we end up reading through his research object. As one comes to read William's capacities as an innovator through the capacity of the model he invented, we see the link gradually forming between Hawking's sick body, the disincorporated brain, and the content of the physical theory on which he works.

Yet, while 'the brilliant Hawking' is the fruit of a collective construction,[58] we shall see how he personally intervenes in the process. The newspapers are fed by, and partly organize themselves around, quotations of the scientist— quotations which resemble one another down to the last word because entire passages of his life story are now stored in his computer, as his secretary confirmed. All the answers are ready and waiting to be used. In this sense, talking to Stephen Hawking means speaking to his computer which provides a totally stereotyped version of his life. The scientist's autobiography is henceforth virtually stable. How does he present himself in his writings?

While his mother thinks that he does not feel fundamentally different from others,[59] his ex-wife believes that one of her main occupations was to remind him that he is not God.[60] Between the two extremes we can see how he places himself in the tradition of the greatest. Indeed, Hawking consciously embraces what Stephen Greenblatt has called the process of self-fashioning: 'the power to impose a shape upon oneself is an aspect of the more general power to control identity—that of others at least as often as one's own'.[61] However, as Biagioli notes, in his study of Galileo:

> By emphasizing [this] process of self-fashioning, I don't assume either an already existing 'Galileo' who deploys different tactics in different environments and yet remains always 'true to himself, nor a Galileo who is passively shaped by the context that envelops him. Rather, I want to emphasize how he used the resources he perceived in the surrounding environment to *construct* a new socio-professional identity for himself, to put forward a new natural philosophy and to develop a courtly audience for it.[62]

We see Hawking employing a similar strategy of self-fashioning in inserting himself in the tradition of the greatest scientists. Indeed, as he likes to emphasize, and has included in his autobiography on the Internet, he was born exactly 300 years after the death of Galileo. This statement, first found in a talk given in 1987, appeared in the following form:

> I was born on January 8, 1942, exactly three hundred years to the day after the death of Galileo. However, I estimate that about two hundred thousand other babies were also born that day. I don't know whether any of them were later interested in astronomy.[63]

He was quoted verbatim in an interview granted to *Playboy* magazine in 1990, entitled 'Candid Conversation'. We note in passing the effect produced by the discourse of this genius without a body, framed by the inviting photo of a cur-vaceous blonde. Hence, the quotation was recontextualized, one might say, so that the journalist, by highlighting all the implications of such a declaration, reinforced the filiation and simultaneously the position of the scientist and the impact of his theories:

> J: Galileo was tried and imprisoned for heresy by the Catholic Church for his theories of the universe. Did he have something in common with you?
>
> SH: Yes. However, I estimate that about two hundred thousand other ba-bies were also born on that date. [smiles] And I don't know if any of them were later interested in astronomy.[64]

Bernard Carr, one of his former students, also likes to emphasize this sense of affinity which Hawking has for Galileo. Carr flew with him to Rome when Hawking was to receive the Pius XII medal awarded to a 'young scientist for re-markable work', presented directly by Pope Paul VI. Carr recalls:

> I remember, when we went to the Vatican, he was very keen to go into the archives and see the document which was supposed to be Galileo's recan-tation, when he was put under pressure by the Church to recant on his the-ory that the Earth went around the sun . . . . I think it gave us some plea-sure that the Church finally announced that they'd made a mistake with Galileo, and that in fact Galileo was right. But whether or not the Pope would have approved of what Stephen had discovered, if he actually had understood it, I am not quite sure.[65]

That is also the opinion of Stephen Hawking. He notes that if the hypothesis which he formulated with Jim Hartle of the University of California in 1982–83—on the absence of an edge for calculating the state of the universe, in the framework of the quantum theory of the universe—were correct, there would be no singularities and the laws of science would be applied everywhere, including at the beginning of the Universe. He would thus have succeeded in fulfilling his ambition to discover the origins of the Universe, and would thereby violate the Pope's 1981 prohibition concerning research on the Big Bang itself, 'because it was the moment of creation and thus the work of God'.[66]

Thus Hawking came to write his own *auto-da-fe*: born 300 years after the death of Galileo, the last visible trace of his 'living' body is appended in the form of a signature in the 'golden book' which immortalized him and intro-duced him into the Pantheon of the Great in 1979, when he was installed as Lu-casian Professor of Mathematics, a Chair formerly occupied (as he likes to re-call) by Newton.[67] That was the last time he signed his name:

In 1979 I was elected Lucasian Professor of Mathematics. This is the same chair once held by Isaac Newton. They have a big book which every university teaching officer is supposed to sign. After I had been Lucasian Professor for more than a year, they realized I had never signed. So they brought the book to my office and I signed with some difficulty. That was the last time I signed my name.[68]

The appropriation of his work also passes through the presentation of his sick body which appears in the process of discovery, operating an explicit link between himself and his idea:

> One evening shortly after the birth of my daughter, Lucy, I started to think about black holes as I was getting into bed. *My disability made this rather a slow process, so I had plenty of time.* Suddenly I realized that the area of the event horizon always increases with time. I was so excited with my discovery that I didn't get much sleep that night. The increase in the area of the event horizon suggested that a black hole had a quantity called entropy, which measured the amount of disorder it contained; and if it had an entropy, it must have a temperature. However, if you heat up a poker in a fire, it glows red-hot and emits radiation. But a black hole cannot emit radiation, because nothing can escape from a black hole.[69]

The disease is not mentioned in John Boslough's account of the discovery in his 1985 biography of Hawking.[70] But after publication of *A Brief History of Time* in 1988, Hawking's account was to become the established version. We see how Hawking is going again to be transformed into a subject without a body; thus, for example, in *Stephen Hawking for Beginners*, the first two sentences of the passage quoted above are used, almost word for word:

> One evening in November 1970, shortly after the birth of my daughter, Lucy, I started to think about black holes as I was getting into bed. My disability makes this rather a slow process, so I had plenty of time.

Except this time, the journalists add:

> *He saw in a flash that the surface area of a black hole can never decrease. . . . He didn't need paper and pen, nor a computer—the pictures were in his head.*[71]

Finally, his handicap and the choice of his research subject are explicitly linked. In the introduction to *A Brief History of Time*, he emphasized that, although he had the misfortune of having this motor neuron disease, he had been lucky in every other respect—and notably in his choice of theoretical physics, because everything is in the mind, which is why his infirmity had not been a serious handicap.[72] Thus, when a journalist asked him if his illness affected his choice of work, he answered:

> SH: *Not really, I had decided to work in this field before I knew.* The only thing it may have affected is that I avoid problems with a lot of equations because I cannot easily write them down. I have to look for short cuts.[73]

While, two years later, to essentially the same question:

> J: *Why did you choose theoretical physics for your research ?*

He answered:

> SH: *Because of my disease. I chose my field because I knew I had ALS.* Cosmology, unlike many other disciplines, does not require lecturing. It was a fortunate choice, because it was one of the few areas in which my speech disability was not a serious handicap. I was also fortunate that when I started my research, in 1962, general relativity and cosmology were underdeveloped fields, with little competition, so my disease would not be a serious impediment. There were lots of exciting discoveries to be made, and not many people to make them. Nowadays, there is much more competition. . . . [smiles].[74]

For reasons that were partly practical, the disease was henceforth to become a main factor motivating his choice of a research field. This version has now been established. Hence, the fact that Stephen Hawking was born 300 years after Galileo, his reactions when he learned of his illness (he listened to Wagner and got drunk), the essential rôle played by his wife in his survival, his jokes about his computer which give him an American accent, and his account of his discoveries, are all taken as 'given', and used over and over again. He himself either plays the game and lets the media exploit his writings or, sometimes, rebels and intervenes in the construction of his own myth:

> J: According to newspaper interviews, and a recent 20/20 segment by Hugh Downs on ABC TV, when you got your diagnosis, you simply gave up and went on a drinking binge for a few years.
> SH: It's a good story but it's not true. . . . I took to listening to Wagner, but the reports that I drank heavily are an exaggeration. The trouble is, once one article said it, others copied it because it made a good story. Anything that has appeared in print so many times has to be true.[75]

This statement was, in turn, also taken and re-used countless times before finally becoming part of the myth. Similarly, when he participated in the collective denial of the media construction of his genius around his handicap:

Kevin Berg, a freshman at Seattle Pacific University, asked [him in a conference]: 'How does it feel to be labelled the smartest person in the world?' Hawking rapidly picked out words. He spelled out 'media' and 'hype', which were not included in his computer's library of 3,000 words. 'It's very embarrassing' was his response. 'It's rubbish, just media hype. They just want a hero, and I fill the rôle model of a disabled genius. At least, I am disabled, but I am no genius'.

This declaration was made to an audience of disabled persons, who were thus elevated to the ranks of potential geniuses, while the scientist was simultaneously aggrandized. However, Hawking also embraces this transformation of himself as carried out by the media, for he announced to the very same audience, only moments later that:

Nowadays, muscle power is obsolete. Machines can provide that. What we need is mind power, and disabled people are as good at that as anyone else.[76]

Thus, although Hawking no longer controls his own body, we see how he partly controls his image and the rôle of his body in the construction of his identity. He does so both in his writings—for even if he now refuses all written interviews, his quotations are used—and, above all, in his public speeches to handicapped audiences, in scientific TV programmes, or on *Star Trek* (the American science fiction series).

Where are you, Mister Hawking? We have thus outlined a few processes which contribute to the fabrication of an identity, an ego and a subject, through the relationships woven between the scientist and his peers, the scientist and his work, and the scientist and his research subject. This is partly dependent on the way in which he presents himself—that is to say, on whether or not he plays on the relevance of his handicap; on how he plays on the places where he appears (*Playboy* or *Star Trek*); and on his choice of a public (ally), scientific programmes, or handicapped persons[77]—and partly dependent on moves made in a network that are attributed (in whatever fashion) to Hawking's 'intentions'.

But does this relatively stable form, this identity constructed with this succession of discourses, representations and presentations, really have anything in common with 'the real, unique Mr Hawking', the flesh-and-blood person? It seems that getting closer to him does not enable us to grasp him in his singularity; on the contrary, to approach him is to lose him! For, by mechanizing communication, the computer causes the subject to disappear from the enunciation, and to create a new mediation. Professor Hawking is simultaneously silence and speech. He thus transforms himself into the reflection of all projections—as in all human relationships, but magnified. The machine does not dehumanize the man, but it multiplies his subjectivity. One never knows whether one is dealing with a Hawking who is suffering, or bored, or thinking:

> 'You never know if Stephen is annoyed by your question and thinks you're an idiot, or is it that I think that Stephen thinks I'm an idiot, or is it a joke, or is he really annoyed but wants me to think it's a joke?', Morris reflects. 'It's a hall of mirrors—like your relationship with any other person, but magnified'.

Or:

> In Cambridge, Hawking will sometimes wheel out of a seminar at his Department of Applied Mathematics (DAMPT) in the middle of the speaker's presentation. 'They never know if it's because he's pissed off and thinks it's a boring talk, or if it's just that he needs some physical thing like suction', says Alex Lyons, one of his recent research students.[78]

Thus, the more we work from writings or programmes, the more we will have a stable image and a relatively well-defined ego.[79] The closer we get to the scientist's body, the greater access we will have to the extension of his distributed body, his secretary, computer and students. Finally, when we reach the person himself, we believe we have grasped an individual because we are in the presence of his body, but that is when a multiplicity of Hawkings suddenly appears. In short, the closer we are to the single body, the more we discover a multiplicity of Hawkings; and the more we spread our focus to the multiplicity of constructions and representations of Hawking through the media, the more we have a stable ego. This hypothesis, nevertheless, leaves us with an immense doubt: but where are you then, Mister Hawking?

CONNECTIONS: WILLIAM X AND MISTER H

By way of conclusion, I shall link and compare the two stories narrated here, underscoring the similarities and differences thus revealed. In the case of William X, I studied a collectivity from the inside and saw a singular figure emerge. With Stephen Hawking, by contrast, I started with a singular figure and redistributed it in a collectivity. In both instances, my aim was to show that it was possible to redeploy the mechanisms normally contained in the minds of certain scientists (inventiveness, expertise, genius, creativity, and the like). In this context, Stephen Hawking is more like the inventor working in a firm with a thousand other persons than he is like the solitary scientist he is usually described to be. Moreover, he is even more distributed than William X, in so far as his inability to make his own body work has allowed many operations, normally embodied in a single healthy person, to become visible because externalized and incorporated in other bodies (humans and machines). Thus, both cases, despite their differences, accentuate what I believe to be one of the principal contributions of social and cultural studies of science—namely, the work of redistribution of scientific intelligence and the re-incorporation of competences in machines, tools, practices and social relations.

In this sense, my work is related to the problematique of distributed cognition, in that I have sought to understand the effective modalities of cognitive operations by focusing my study on work in context, and on the rôle played by objects and artefacts in the production of knowledge. Not unlike other similar studies, my objective has been, as Louis Quere has put it, 'to reincorporate the mind, to give it a body, to put it back in an environment, and to reinsert it in a system—of objects, artefacts and cognitive technologies—to which it belongs'.[80] As such, the actor could be said to share certain key aspects of his or her identity with these objects, artefacts and tools. However, if cognition is distributed in the sense that it emerges from an association of related elements, or that it is the result of a kind of collective process, then my project (keeping in mind the importance of the exteriorization of cognitive capacities) aims at going a step further, by re-integrating questions of the specific modalities of the different contributions to cognition as an effect of an emerging collective. In other words, I have tried to show how the movement of (re)-singularization of practices, competences and actors, originated—and operated—in the collective processes of the production of 'new' ideas. In this sense, I have tried to show how certain subjects are at the centre of processes of singularization and distribution that generate prestige and genius.

In the second phase of my inquiry, I explored the processes of operative attribution in the constitution of genius. Both cases revealed the myth of the solitary genius. However, it seemed to me that, rather than telling us that 'the entire invention took place in a single day crystallized in the mind of an individual', the myth had a decisive function in the constitution of the invention. With William, we have followed the movements of the humans and the narratives. In the case of Stephen Hawking, I highlighted other mechanisms of singularization (standardization of autobiographical accounts, the media's representation of the scientist's body, intervention of the scientist in the construction of his own myth, and so on), and showed that far from being a problem identified by the observer *ex post*, genius was a resource deliberately employed by the scientist himself. Finally, in both cases we can see the subject through the object that he developed or 'discovered' (an expert system for William X, a black hole for Stephen Hawking). Thus, for both William and Hawking, distribution and attribution seem to be essential operative concepts in explaining the constitution of knowing subjects.

Of course, the concept of the distributed-centred subject I developed is linked to other concepts created to redefine the paradigm of the subject-object dichotomy (whose foundation has already been put into question by sociologists of science, and by the phenomenon of distributed cognition). Yet there is a crucial difference. As I have tried to show, it is possible to understand the constitution of the knowing subject by using the tools and methods developed by those who would either deny or ignore its relevance. In this sense, my goal has been to try to understand the subject in-the-making; that is, as he or she becomes (productive) through the distribution and re-appropriation of his or her extended body. Indeed, the more a subject is distributed, socialized and collectivized, the more he or she is singular, an ego, a non-interchangeable body. Accordingly, the greater the number of elements to which an actor is connected, the more innovative he or she has the potential to be. Thus I show that the body of the scientist is the crucial site around which tools, techniques, humans and narratives are simultaneously distributed (extended) and concentrated (singularized). And, in this context, a difference does indeed emerge between these two cases. With William X, the singularity of the genius was (re-)attributed to— and situated in—the functioning of the body of the individual (the ability of the scientist to identify with his research object, a capacity to share with his own body the properties of matter). With Stephen Hawking, on the other hand, the singularity of the genius crystallized around a visible body, but a body detached from the very person of the scientist; a body made of paper, of writing and of representation, which became a medium through which his identity is constructed. In the former case, we found a locus for the action: a situated body. In the latter, we no longer know where the 'real Mr Hawking' is. But if we question the very notion of identity, which we started to do in this paper, then we sign the death warrant of an awkward question about the genius: who is, or was, he 'really'?

NOTES

The first part of this paper is derived from my 1994 PhD thesis, submitted at the École des Mines in Paris; I would like to thank my colleagues there, specifically Bruno Latour, Michel Callon, Madeleine Ackrich, Antoine Hennion and Vololona Rabeharisoa. I would especially like to thank Simon Schaffer, with whom I discussed my work as I carried out the research for, and writing of, this paper. I would also like to thank Robin Boast, Yves Cohen, John Forrester, Jean Paul Gaudillière, François Jacq, John Law, Christian Licoppe, Jacques Mirenowitz, Dominique Pestre, Jim Secord and Michael Wintroub. Additionally I would like to thank Michael Lynch, Harry Collins, Yves Cohen, Michael Bravo, Lorraine Daston and Otto Siburn, who gave me the opportunity to present and discuss this work in their seminars. Finally, the comments of William Clark, three anonymous referees for *Social Studies of Science*, and Malcolm Ashmore, were extremely helpful.

1. Hélène Mialet, *Le Sujet de l'Invention, Etude Empirique de la Conception d'une Idée Neuve: Comparaison des Méthodes Philosophiques et Sociologiques* (Paris: unpublished doctoral thesis, Sorbonne Paris I and CSI École Nationale des Mines de Paris, 28 June 1994).

2. After completing this paper, I had the opportunity to meet and interview Stephen Hawking regarding his position as Lucasian Professor of Mathematics at Cambridge University.

3. Claude Bernard, *Cahiers de Notes* (Paris: Gallimard, 1965); Bernard, *Introduction à la Medecine Experimental* (Paris: Flammarion, 1984).

4. Gaston Bachelard, *La Formation de l'Esprit Scientifique* (Paris: Vrin, 1983); Bachelard *Le Nouvel Esprit Scientifique* (Paris: Presses Universitaires de France, 1984); Bachelard, *L'Intuition de l'Instant* (Paris: Stock, 1992).

5. Karl Popper, *The Logic of Scientific Discovery* (London: Hutchinson, 1972); Popper, *Conjectures and Refutations: The Growth of Scientific Knowledge* (London: Routledge & Kegan Paul, 1969); Alain Boyer, 'D'où Viennent les Idées Justes? H. Simon et K. Popper et l'Heuristique', in A. Demailly and J. L. Lemoigne (eds), *Sciences de l'Intelligence Sciences de l'Artificiel* (Lyon: Presse Universitaire de Lyon, 1986), 501–13.

6. Judith Schlanger, *L'Invention Intellectuelle* (Paris: Fayard, 1983).

7. See, for example, Edmond Claparede, *Genese de l'Hypothese: Etude Experimentale* (Genève: Librairie Kunding, 1934); Jacques Hadamard, *Essai sur la Psychologie de l'Invention en Mathématique* (Paris: Gauthiers-Villars, Collection Discours de la Méthode, 1945); Lewis S. Feuer, *Einstein and the Generations of Science* (New York: Basic Books, 1974); Gerald Holton, *The Scientific Imagination: Case Studies* (Cambridge: Cambridge University Press, 1978); Arthur Koestler, *Le Cri d'Archimède* (Paris: Calmann Lévy, 1965); Howard E. Gruber, *Darwin on Man* (Chicago, IL: The University of Chicago Press, 1981); Alexander Kohn, *Fortune or Failure: Missed Opportunities and Chance Discoveries* (Oxford: Blackwell, 1989); Henry Poincare, *Science et Méthode* (Paris: Flammarion, no date); Georges Polya, *Comment Poser et Resoudre un Problème?* (Paris: Jacques Gabay Dunod, 2nd edn, 1989); Yulin Quin and Herbert Simon, 'Laboratory Replication of Scientific Discovery Processes', *Cognitive Science*, Vol. 14 (1990), 281–312; Greg Nowak and Paul Thagard, 'Newton, Descartes and Explanatory Coherence: The Conceptual Structure of the Geological Revolution', in Jeff Shrager and Pat Langley (eds), *Computational Models of Scientific Discovery and Theory Formation* (San Mateo, CA: Morgan Kaufmann Publishers, 1990); William J. J. Gordon, *Synectics: The Development of Creative Capacity* (New York: Collier Books, 1968); Abraham Moles, *Créativité et Méthodes d'Innovation* (Paris: Fayard Mame, 1970); Edward De Bono, *Lateral Thinking: Creativity Step by Step* (New York: Harper & Row, 1973). These historians and psychologists of discovery have elaborated a perspective entirely at odds with that of epistemologists. In the end, however, they do hold one point in common: they consider discovery to be an established fact. Discovery—that is to say, the emergence of a new idea—remains for both a kind of established frontier. It is used by one as a point of departure for reconstructing

rational method 'which allows one to determine when a discovery is a discovery' (Popper), or a kind of end-point from which one can understand how inventors have come to invent: but, in either case, a fundamental question remains unanswered—namely, could it have happened otherwise? The model which emerges from this view is called 'diffusionist'—it is based on the distinction between the elaboration of a new idea, and the time where this idea comes to be generally recognized as being new. Invention thus becomes actual at the moment where the idea is born. To penetrate closer to the heart of these mechanisms is a goal which can only be achieved through a process of banalization. Another approach, that of experimental psychology, aims to outline the protocols of experience by which one can understand the different methods adopted by subjects in the process of resolving a problem. Cognitive scientists, on the other hand, attempt to simulate heuristic methods. In both cases, the problematique of knowledge, of the *specificity* of scientific knowledge, is abandoned. Consequently, invention becomes an act of intelligence which we mobilize each time we are brought to resolve a problem. With Georges Polya, invention becomes the application of heuristic rules; it becomes explicable, decomposable and reproducible. These operations are, therefore, capable of being simulated by a computer (Simon, Thagard), or expressly cultivated (Gordon, Moles, de Bono). The appearance of a new idea thus becomes the fruit of an explicable—indeed, banal—intellectual process. The singularity of the individual who invents is lost insofar as these authors aim to explain creativity in terms of a common intellectual mechanism; it follows from this that the goal behind these studies has been to eliminate the distinction between scientific and non-scientific knowledge.

8. H. M. Collins, 'The Seven Sexes: a Study in the Sociology of a Phenomenon, or the Replication of Experiments in Physics', *Sociology*, Vol. 9 (1975), 205–24; Bruno Latour and Steve Woolgar, *Laboratory Life: The Social Construction of Scientific Facts* (Beverly Hills, CA: Sage Publications, 1979); Latour, *La Science en Action* (Paris: La Découverte, 1989); Martin Rudwick, *The Great Devonian Controversy: The Shaping of Scientific Knowledge Among Gentlemanly Specialists* (Chicago, IL: The University of Chicago Press, 1985); Simon Schaffer, 'Glassworks: Newton's Prisms and the Uses of Experiment', in David Gooding, Trevor Pinch and Schaffer (eds), *The Uses of Experiment* (Cambridge: Cambridge University Press, 1989), 67–104; Andrew Pickering, 'Against Putting the Phenomena First: The Discovery of the Weak Neutral Current', *Studies in History and Philosophy of Science*, Vol. 15 (1984), 87–117.

9. On this point, see also the work of the philosopher and historian of science David Gooding, *Experiment and the Making of Meaning: Human Agency in Scientific Observation and Experimenting* (Dordrecht, Boston, MA & London: Kluwer, 1990).

10. Latour & Woolgar, op. cit. note 8, 171.

11. Augustine Brannigan, *The Social Basis of Scientific Discoveries* (Cambridge: Cambridge University Press, 1981); Bruno Latour, 'Is It Possible to Reconstruct the Research Process?: Sociology of a Brain Peptide', in Karin Knorr, Roger Krohn and Richard Whitley (eds), *The Social Process of Scientific Investigation, Sociology of the Sciences Yearbook*, No. 4 (Dordecht, Boston, MA & London: Reidel, 1981), 53–77; Simon Schaffer, 'Making up Discovery', in Margaret Boden (ed.), *Dimensions of Creativity* (Cambridge, MA: MIT Press, 1994), 13–51.

12. I take this quotation from a draft paper, eventually published as: Michel Callon and John Law, 'After the Individual in Society: Lessons on Collectivity from Science, Technology and Society', *Canadian Journal of Sociology*, Vol. 22, No. 2 (Spring 1995), 165–83, at 169. The published quotation is worded slightly differently; I prefer the draft version.

13. Rene Descartes, *Méditations Métaphysiques* (Paris: Flammarion, 1979); Descartes, *Discours de la Méthode—Pour Bien Conduire sa Raison et Chercher la Vérité dans les Sciences, [plus] la Dioptrique —les Météores et la Geometric qui Sont des Essais de cette Méthode* (Paris: Fayard, 1987); Emmanuel Kant, *Critique de la Raison Pure* (Paris: Presses Universitaires de France, 1986); Karl Popper, *Objective Knowledge: An Evolutionary Approach* (Oxford: Clarendon Press, 1972); also see the analysis of Evelyn Fox Keller, 'The Paradox of Scientific Subjectivity', in Allan Megill (ed.), *Rethinking Objectivity* (Durham, NC & London: Duke University Press, 1994), 313–31; and Henry Michel, *Généalogie de la Psychanalyse* (Paris: Presses Universitaires de France, 1985), at 61: 'Une subjectivité privée de sa dimension d'intériorité radicale, réduite à un voir, à une condition de l'objectivité et de la représentation'. Also see Allan Megill, 'Introduction: Four Senses of Objectivity', in Megill (ed.), *Rethinking Objectivity*, op. cit. 1–20, at 10: ' . . . Insofar as one stresses the universality of the categories—their sharedness by all rational beings—one will see Kant as a theorist of absolute objectivity, an objectivity stripped of everything personal and idiosyncratic'.

14. See, for example, H. M. Collins, *Changing Order: Replication and Induction in Scientific Practice* (Chicago, IL: The University of Chicago Press, 2nd edn, 1992), and Barry Barnes, *Scientific Knowledge and Sociological Theory* (London: Routledge & Kegan Paul, 1974).

15. Published as Callon & Law, op. cit. note 12, 169. Again, my quotation is taken from the draft; the published version is slightly different.

16. Some readers may be surprised to hear that sociologists of science 'deny the necessity of the situated body of the scientist' when one of the principal contributions of the social and cultural studies of science (which I will emphasize and show again in this paper) has been the work of 're-incorporating' scientific intelligence into its environment. For a recent example, see Christopher Lawrence and Steven Shapin (eds), *Science Incarnate: Historical Embodiments of Natural Knowledge* (Chicago, IL: The University of Chicago Press, 1998). Nevertheless, I would like to underline that insofar as scientific intelligence has been re-incorporated into the social world (as an antidote to rationalist/individualist conceptions of science, for example), the 'situated and singular' character of a body endowed with its own idiosyncratic competences tends either to be dissolved into a 'collectivity', or it is 'black-boxed' as, say, 'tacit knowledge'. See Helene Mialet, 'Une Nouvelle Figure du Sujet', *Les Cahiers de Philosophie* (forthcoming, 1999). Concerning the critique of tacit knowledge as a 'black box', see the work of the historian of science, Otto Sibum, and how he deals with the problem of reintroducing the power of the productive body, in his 'Les Gestes de la Mesure: Joule, les Pratiques de la Brasserie et la Science', *Annales Histoire, Sciences Sociales*, Nos 4–5 (Juillet–Octobre 1998), 745–74.

17. I also created an experimental protocol, but not an artificial situation in which I would have defined the constraints needed to achieve the results I wanted (for example, by starting with the principle that invention is a psychological phenomenon which is observed for the purpose of solving a problem, the individual will fulfill the function chosen for her/him). In an ethnographic study, it is the individual caught up within her/his own institutional, personal and technical constraints, who will define the main features of the problem.

18. In 'collective' I include the individual himself, who is also party to this operation of singularization. Since both my subjects are male, I will from now on use the words 'he', 'him', 'his' and 'himself.'

19. During my study, I was struck by William's availability for his colleagues, and by the number of informal and formal requests from other services which he (unlike his colleagues) centralized. Sometimes he dispatched them elsewhere.

20. I quote from interviews with William and his colleagues: Arthur and Patrick. These taped statements and interviews were conducted during my study of William's research group at Elf Aquitaine, as occasion arose.

21. 'PVT' means 'Pressure Volume Temperature'. This acronym encompasses a series of operations and analyses defining thermodynamic values characteristic of an oil-field fluid. The PVT laboratory at Boussens (France) is responsible for these measurements.

22. Arthur was to mention this again at the end of my stay: 'In any case, we all agree, you've got close to a great scientist, a great person. You know, during his leisure time, he reads books on the philosophy of science. He's really an all-rounder. A few years ago, there were staff cutbacks here, and William came to me saying "I'm useless". He had to become known. He had ideas, he developed tools and you can see how he's in demand now. He's the first in the world to have developed the GER [model] . . . it's a pleasure working with a guy like that.'

23. 'Condensed gas' is a fluid that is potentially very rich in oil.

24. This is not dissimilar from the kind of 'paradigm shift' described at length and in detail by Thomas S. Kuhn in his *The Structure of Scientific Revolutions* (Chicago, IL: The University of Chicago Press, 2nd edn, 1970).

25. 'GER' (Gravitation Effect Reservoir) is the name of William's model.

26. On this point, see Donald MacKenzie's analysis of the charisma of Seymour Cray. The engineer's charisma, he explains, must embody itself in the machine, its success (or glorious failure), and the people who worked with him and contributed to its success. 'They [his co-workers], and the wider world, may be happy to attribute authorship of the machine and of its success to the charismatic engineer, but without them the machine and its success would not exist': D. MacKenzie, *Knowing Machines: Essays on Technical Change* (Cambridge, MA: MIT Press, 1996), 131–59, at 156.

27. John L. Austin, *How to Do Things with Words* (New York: Oxford University Press, 1965).

28. Brannigan, op. cit. note 11, 167–71.

29. Isabelle Stengers, *L'Invention des Sciences Modernes* (Paris: La Decouverte, 1993), 113.

30. Erving Goffman, *The Presentation of the Self in Everyday Life* (Garden City, NY: Doubleday, 1959).

31. See Michael Polanyi, *The Tacit Dimension* (Gloucester, MA: Peter Smith, 1983), 15–16; Polanyi, *Personal Knowledge: Towards a Post-Critical Philosophy* (Chicago, IL: The University of Chicago Press, 1974), 55–63, 174–76; Polanyi, *The Study of Man: The Lindsay Memorial Lectures* (Chicago, IL: The University of Chicago Press), 25, 30–31.

32. According to Howard E. Gruber's fascinating analysis of Darwin's creativity, '. . . the reorganization of thought in partially independent systems developing unevenly and each following somewhat different laws provides the setting for the chancy interactions that give each creative process its unique flavor': see H. E. Gruber, *Darwin on Man* (Chicago, IL: The University of Chicago Press, 1981), 251; see also 257: The individual's work '. . . is organised in a number of enterprises, forming an ensemble that expresses his unique purposes. Some of these enterprises may be shared or at least readily shareable with some of his contempo-

raries, others not. But the uniqueness of the single enterprise is not, in any case, the important point. The individual is always the unique host of a living network of enterprises'.

33. Gruber makes a similar point: see ibid., 113.

34. As opposed to a diffusionist model: see B. Latour, *Les Microbes: Guerre et Paix suivi de Irréductions* (Paris: Métaillé, Coll. Pandore, 1984), 281.

35. As Foucault says: 'I would like to know whether one cannot discover the system of regularity of constraint, which makes science possible, somewhere else, even outside the human mind, in social forms, in the relations of productions, in the class of struggles etc.': in Arnold Davidson (ed.), *Foucault and His Interlocutors* (Chicago, IL: The University of Chicago Press, 1997), 123. And also: '... In the history of science or in the history of thought, we place more emphasis on individual creation and we had kept aside and left in the shadows the communal, general rules, which obscurely manifest themselves through every scientific discovery, every scientific innovation, or even every philosophical innovation' (ibid., 119).

36. John Locke, *An Essay Concerning Human Understanding* (Oxford: Oxford University Press, 1988), Chapter 23, 160.

37. Ibid., 162.

38. Ibid., 163–64.

39. The doctor gave him two years to live after the first symptoms appeared; he has now been living with the disease for over 30 years.

40. Jeremy Hornsby and Ian Ridpath, 'Mind over Matter', *Sunday Telegraph Magazine* (London, 28 October 1979), 44–50, quote at 44.

41. John Boslough, 'Stephen Hawking Probes the Heart of Creation', *Reader's Digest*, Vol. 124 (February 1984), 39–45, quote at 39.

42. Leon Jaroff, 'Roaming the Cosmos', *Time* (8 February 1988), 34–36, quote at 34.

43. Jerry Adler, Gerald C. Lubenow and Maggie Malone, 'Reading God's Mind', *Newsweek* (13 June 1988), 36–39, quote at 36.

44. Simon Schaffer shows how the term 'genius' was applied to natural philosophers in the late 18th century. The Romantic genius, as natural philosopher, he explains, possessed extraordinary, even mystical, power which enabled him to divine nature and discover its secrets. See Simon Schaffer, 'Genius in Romantic Natural Philosophy', in Andrew Cunningham and Nicholas Jardine (eds), *Romanticism and the Sciences* (Cambridge: Cambridge University Press, 1990), 82–98. For an interesting essay on the theme of solitude, see Steven Shapin, 'The Mind Is Its Own Place: Science and Solitude in Seventeenth-Century England', *Science in Context*, Vol. 4 (1991), 191–218.

45. The research for this paper was conducted in 1996. For reasons of privacy, I have changed the names of Hawking's students.

46. As, for instance, in the famous distinction between the context of discovery and the context of justification, borrowed from Reichenbach and applied by Popper in both his *Logic of Scientific Discovery* and his *Conjectures and Refutations*, opera cit., note 5.

47. See, for example, Jack Goody, *The Domestication of the Savage Mind* (Cambridge: Cambridge University Press, 1977).

48. For an interesting ethnography of the distributed body, see Stefan Hirschauer, 'The Manufacture of Bodies in Surgery', *Social Studies of Science*, Vol. 21, No. 2 (May 1991), 279–319. For another conception of distributed cognition, see Edwin Hutchins, 'How a

Cockpit Remembers its Speeds', *Cognitive Science*, Vol. 19, No. 3 (July–September 1985), 265–88; Hutchins, *Cognition in the Wild* (Cambridge, MA: MIT Press, 1990); Lucy A. Suchman, *Plans and Situated Actions: The Problem of Human-Machine Communication* (Cambridge: Cambridge University Press, 1987); also see *Reseaux* (CNET), No. 85 (Septembre–Octobre 1997), which discusses cooperation in the context of work.

49. With the exception of certain details concerning his private life and the rôle of his nurses. The computer is generally perceived more as a means of communication than as a tool for working.

50. See Schaffer, op. cit. note 44; William Clark, 'On the Ironic Specimen of the Doctor of Philosophy', *Science in Context*, Vol. 5, No. 1 (Spring 1992), 97–137.

51. Jeremy Dunning-Davies, 'Popular Status and Scientific Influence: Another Angle on "The Hawking Phenomenon"', *Public Understanding of Science*, Vol. 2, No. 1 (January 1993), 85–86.

52. Arthur Lubow, 'Heart and Mind', *Vanity Fair* (June 1992), 44–53, paraphrasing from 53.

53. Ian Ridpath, 'Black Hole Explorer', *New Scientist* (4 May 1978), 307–9, at 307.

54. For an analysis of the repertoires 'for speaking about the bodily circumstances that either assisted or handicapped the processes by which genuine knowledge was to be attained', see Lawrence & Shapin (eds), op. cit. note 16, 1, and particularly Steven Shapin's contribution, 'The Philosopher and the Chicken: On the Dietetics of Disembodied Knowledge', 21–51, and Janet Browne's, 'I Could Have Retched All Night: Charles Darwin and His Body', 240–88.

55. Ridpath, op. cit. note 53, 308. But the fact that it is science remains essential. As another scientific journalist told me: 'I don't expect that you would be doing the same project if Hawking had been a composer writing music. It would be interesting if Hawking had been writing music in his head and dictating to somebody or a computer, that would not be the same thing, people think science is different from other aspects of culture' (interview, September 1996).

56. Bryan Appleyard, 'A Master of the Universe', *Sunday Times Magazine* (London, 19 June 1988), 26–30, at 26.

57. Adler, Lubenow & Malone, op. cit. note 43, 39 (my emphasis).

58. Regarding the collective construction of the genius, see in particular Geoffrey Cantor, 'The Scientist as Hero: Public Images of Michael Faraday', in Richard Yeo and Michael Shortland (eds), *Telling Lives in Science: Essays on Scientific Biography* (Cambridge: Cambridge University Press, 1996), 171–95; R. Yeo, 'Genius, Method, and Morality: Images of Newton in Britain, 1760–1860', *Science in Context*, Vol. 2, No. 2 (Autumn 1988), 257–84; Nathalie Heinich, *La gloire de Van Gogh: Essai d'anthropologie de l'admiration* (Paris: Editions de Minuit, 1991).

59. In Stephen Hawking (ed.), *Qui êtes vous Mr Hawking?* (Paris: Odile Jacob, 1994), 204. This book was billed as *A Reader's Companion to A Brief History of Time*; it was edited by Hawking and prepared by Gene Stone. I take my quotations from the English-language edition, published in the USA by Bantam Books (New York, 1992).

60. Lubow, op. cit. note 52, 53.

61. Stephen Greenblatt, *Renaissance Self-fashioning: From More to Shakespeare* (Chicago, IL: The University of Chicago Press, 1980), 1.

62. Mario Biagioli, *Galileo, Courtier: The Practice of Science in the Culture of Absolutism* (Chicago, IL: The University of Chicago Press, 1993), 5.

63. Talk given to the International Motor Neurone Disease Society in Zurich (1987).

64. 'Playboy Interview: Stephen Hawking—Candid Conversation', *Playboy*, Vol. 37, No. 4 (April 1990), 63–74, at 64.

65. *Reader's Companion*, op. cit. note 59, 119.

66. Ibid., paraphrasing from 120–21.

67. As Stephen Greenblatt says: 'Self-fashioning for such figures involves submission to an absolute power or authority situated at least partially outside the self, God, a sacred book, an institution such as church, court, colonial or military administration': Greenblatt, op. cit. note 61, 9.

68. *Reader's Companion*, op. cit. note 59, 151–52.

69. Ibid., 92 (my emphasis). The first two sentences of this quotation effectively repeat sentences from Stephen Hawking, *A Brief History of Time* (London & New York: Bantam Books, 1988; 2nd edn 1996), 113: 'However, one evening in November that year, shortly after the birth of my daughter, Lucy, I started to think about black holes as I was getting into bed. My disability makes this rather a slow process, so I had plenty of time.'

70. John Boslough, *Beyond the Black Hole: Stephen Hawking's Universe* (London: Collins, 1985), 63–64.

71. Joseph P. McEvoy and Oscar Zarate, *Stephen Hawking for Beginners* (New York: Totem Books, 1995), 124 (my emphasis).

72. In *A Brief History of Time*, op. cit. note 69, 49, Hawking also tells of how, just after he had been told of his disease, when he was a young doctoral student looking for a subject for his thesis, he discovered the works of Penrose on the gravitational collapse of bodies.

73. Appleyard, op. cit. note 56, 29 (my emphasis).

74. Quoted from Hawking's *Playboy* interview, op. cit. note 64, 68; see also Stephen Hawking, *Black Holes and Baby Universes, and Other Essays* (New York & London: Bantam Books, 1994), 23.

75. *Playboy*, op. cit. note 64, 66–68; see also Hawking, op. cit. note 74, 23.

76. These two quotations are taken from: Lisa Kremer, 'The Smartest Person in the World Refuses to be Trapped by Fate', *Morning News Tribune* (2 July 1993), electronically published on the Web. In 1996, when I visited the site, its address was: http://www.astro.nwu.edu/lentz/astro/hawking-2.html—but that address is no longer operating: it is now http://weber.u.washington.edu/d27/doit/Press/hawking3.html.

77. See Ingunn Moser and John Law, 'Good Passages, Bad Passages', and Law, 'Making Voices: Disability, Technology and Articulation' (unpublished manuscripts), which Professor Law (Department of Sociology, Lancaster University) has kindly shared with me.

78. Both quotations from Lubow, op. cit. note 52, 47.

79. Errol Morris, the producer of the film adaptation of *A Brief History of Time*, faced with the multiplicity of Hawkings that he will have to present, says that it will be the creative and courageous scientist who will be remembered: How much Hawking will there be in the movie? 'The documentary provides, in Morris's words, "very little biography, but a biographical sketch" that suggests Hawking's persistence, discipline and creativity, following the onset of the progressive illness . . .': Lubow, op. cit. note 52, 46.

80. Louis Quéré, 'La Situation Toujours Négligée?', *Reseaux*, Vol. 85 (1997), 163–92, at 177.

9. MORAL KNOWLEDGE AS PRACTICAL KNOWLEDGE
JULIA ANNAS

I. DIFFERENT PERSPECTIVES

In the area of moral epistemology, there is an interesting problem facing the person in my area, ancient philosophy, who hopes to write a historical paper which will engage with our current philosophical concerns. Not only are ancient ethical theories very different in structure and concerns from modern ones (though with the rapid growth of virtue ethics this is becoming less true), but the concerns and emphases of ancient epistemology are very different from those of modern theories of knowledge. Some may think that they are so different that they are useful to our own discussions only by way of contrast. I am more sanguine, but I am quite aware that this essay's contribution to modern debates does not fall within the established modern traditions of discussing moral epistemology.

Because ancient moral epistemology is rather different from modern kinds, two kinds of danger arise when we try to compare them. On the one hand, we may produce a historical account which fails to engage with modern concerns. On the other hand, we can pose a philosophical question in terms of modern assumptions about knowledge, and then find that the ancients' answer to it appears naive or off the point.

While both can be unhelpful, the second is likely to be more so than the first. In what follows, I shall begin with a passage from a modern author which displays a dramatic misunderstanding of a famous ancient position. I shall then try to isolate the assumptions which prevent a better understanding. However, I will also, and mainly, be concerned to bring out one aspect of the ancient position which I think marks not just a significant difference between ancient and modern approaches to moral epistemology, but also a weakness in the modern approaches, and a point where we might actually learn from the ancients.

II. EXAMINING ASSUMPTIONS

In his influential book *Ethics: Inventing Right and Wrong*,[1] John L. Mackie argues that our intuitive confidence that there are what he calls "objective values" is misplaced; a very simple argument shows that there can be no such things.[2] For, if there were any objective values, they would be extremely "queer" things:

If there were objective values, then they would be entities or qualities or relations of a very strange sort, utterly different from anything else in the
universe. Correspondingly, if we were aware of them, it would have to be
by some special faculty of moral perception or intuition, utterly different
from our ordinary ways of knowing everything else.[3]

Since, it is assumed, this is unacceptable, we conclude that there are no objective values—that is, there is nothing which is both a value, directing our actions, and objective.

At once we can see that most moral philosophers of the past have been in
error on this point, with an especially spectacular form of the error to be found
in Plato.

The main tradition of European moral philosophy from Plato onwards
has combined the view that moral values are objective with the recognition that moral judgements are partly prescriptive or directive or action-
guiding. Values themselves have been seen as at once prescriptive and objective. In Plato's theory the Forms, and in particular the Form of the
Good, are external, extra-mental, realities. They are a very central struc-
tural element in the fabric of the world. But it is held also that just knowing them or 'seeing' them will not merely tell men [sic] what to do but will
ensure that they do it, overruling any contrary inclinations. . . . Being acquainted with the Forms of the Good and Justice and Beauty and the rest
[the rulers of the *Republic*] will, by this knowledge alone, without any further motivation, be impelled to pursue and promote these ideals.[4]

And later on:

Plato's Forms give a dramatic picture of what objective values would have
to be. The Form of the Good is such that knowledge of it provides the
knower with both a direction and an overriding motive; something's being
good both tells the person who knows this to pursue it and makes him pursue it.[5]

Mackie allows that,

It may be thought that the argument from queerness is given an unfair
start if we thus relate it to what are admittedly among the wilder products
of philosophical fancy—Platonic Forms, non-natural qualities, self-
evident relations of fitness, faculties of intuition, and the like.[6]

This hostile portrayal of the Forms of moral qualities as bizarre entities picked out by an equally peculiar faculty of intuition is clearly a coarse and imperceptive interpretation of Plato.[7] What is of interest here, however, is not just for specialists to complain that Mackie's interpretation of Plato is terrible, but to try to isolate the assumptions that lead to it, or at least help to produce it. For what Mackie is missing is very important—and not just Mackie, of course; I have focused on him as being representative of (and also, to a certain extent, responsible for) a widespread view of the available options in moral epistemology. I think that this view is impoverished, and that one significant indication of this impoverishment is precisely its peculiar view of what is going on in the case of Plato and the moral Forms.

A number of the assumptions that Mackie makes are rather obvious. (In what follows, I shall concentrate on the epistemological rather than the metaphysical side of the queerness argument.) A crucial assumption is that our epistemological access to values is "utterly different from our ordinary ways of knowing everything else." On this view, values are not part of the world that we have access to either empirically or by reasoning.[8] First, the brisk exclusion of values from the area to which we have empirical access, in turn, presupposes that our access to that area is epistemologically unproblematic and that there are no serious problems in demarcating what that area contains. Second, the exclusion of values from the area to which we have access through reasoning also presupposes a highly determinate philosophical account of what reason and reasoning are, one in which reason itself is assumed to have no motivating force. On such an account, reasoning can lead to conclusions, for example, but a completely different kind of factor is required to get those conclusions to lead to action. A third assumption is that values motivate by "prescribing"; ethics is envisaged as telling others (and presumably also yourself) what to do. Mackie thus assumes that ethics is primarily about action, and about getting others and yourself to act in certain ways.

Mackie cites A. J. Ayer and C. L. Stevenson as influences on his thinking, although he rejects logical positivism as a characterization of his position, preferring to call it empiricist.[9] It can also, more broadly but defensibly, be called scientistic, for it assumes that we have a clear notion of observation and explanation which excludes values without the need for any argument. Such a notion relies on the idea that the sciences (envisaged as a unified "science" in a way that Mackie seems not to regard as controversial) will demarcate such a notion for us, and that this demarcation will be deferentially accepted by people outside the sciences.[10]

In any case, whatever we call the position, Mackie's assumptions are widely shared in contemporary philosophy (though in the broader culture they cannot be taken for granted, and the idea in particular is widely rejected that we defer to science in defining value-excluding notions of experience and observation). Someone who shared these assumptions but was a more sympathetic interpreter than Mackie might come out with a less absurd interpretation of Plato's moral Forms; nonetheless, it would be hard for such a person to do justice to what I shall now try to present as a leading point about moral epistemology in Plato's arguments for Forms. This will take up Sections III through V, and I will return to the empirical assumptions in Section VI.

III. Searching for Moral Knowledge

I will start by giving a straightforward (and probably embarrassingly simple) account of the argument, which appears more than once in Plato's Socratic dialogues, to get us to recognize what Plato (sometimes) calls Forms.[11] Then I will bring out the important feature of this and similar arguments; this is the feature on which I want to focus. I should emphasize at the start that this account is not idiosyncratic or presented as in any way original; it could reasonably be regarded as a standard account among philosophers in the area of ancient philosophy.

In the *Laches*, two generals, who are both unquestionably brave, have gotten into a discussion about bravery. Will a fashionable new course of training help to make young men brave or not? Disconcertingly, the brave generals disagree, and Socrates, in his usual annoying way, gets them to examine the question of what bravery is and try to come up with a satisfactory account of it. (We know, of course, since this is a Socratic dialogue, that they will not succeed; part of the point of the dialogue is to get the reader engaged in that task.) It is not surprising that the value term that is under discussion is a virtue term, since virtue is a central notion in all of ancient ethics.

Laches, one of the generals, tries to give an account of bravery by pointing out that people are brave who stand up to the enemy in battle and do not break the ranks by running away (190e). Socrates points out that while this is true, courage can also be also shown by people who do what looks like the opposite—namely, strategic retreat (191b–c). He then sets out his conditions for an adequate understanding of courage:

I wanted to learn from you not only what it is to be brave as an infantry-man [i.e., standing firm], but also as a cavalryman [i.e., strategic retreat] and in general as anyone fighting in an army; and to learn about the brave not just in a war but in dangers at sea and when facing illness or poverty, and in public life; and, further, not just the brave in face of pains or fears, but those who are intelligent at battling with desires and pleasures, whether by standing firm or by strategic retreat. . . . All these people are brave, but some possess courage in pleasures, others in pains, some in desires and others in fears, while others again possess cowardice in these circumstances. . . . What each of these is—that's what I wanted to find out.

(191c–e)

It is clear enough that we do not understand what courage is, and that we cannot give an adequate account of what courage is just by having some moral beliefs, that is, some beliefs about courage and courageous people. It is worth pointing out that Socrates does not suggest that these beliefs are *false*. Soldiers who stand firm in the ranks and fight are indeed brave. However, if we rely on these beliefs, we are let down, and this is because, while they are true, they are unsatisfactory in being ungrounded and disunified. These two defects are connected. If we look at the various ways in which Socrates indicates that people are brave, we find a ragbag. What does standing firm in the ranks have to do with firmly facing financial ruin? Nothing obvious. Why, then, do we think that in both these situations someone can *brave*? The reason is, again, not obvious. We would reasonably think that unless the concept is confused, there would be something common to brave ways of fighting in battle and brave ways of facing poverty. What this is, however, cannot be read off from the contexts of bravery themselves. Beliefs about bravery are formed and picked up from our experience, but our experience itself does not tell us how to unify them, nor, hence, how to understand what their basis is. Until we understand what it is that we understand bravery to be, we will not understand why we recognize the examples of bravery that we do. We will also be unable to improve on our present understanding of bravery by extending our use of the term, for how could we be sure that a new example was an example of bravery or not?

Hence, to understand what bravery is, we must use our minds rather than relying on our experience. (Note that here and in what follows, "experience" is used in the everyday sense, not in a restricted philosophical sense restricting it to what empiricism or some other theory decrees—the input of the sense organs, for example.) We pick up, just from looking or from having others tell us, that certain types of action are brave.[12] But, although we are not wrong, we lack understanding of bravery until we have used our minds to think through what bravery is in a way which goes beyond what we can learn from experience. Hence what bravery is, the nature of bravery—or, if you prefer, the Form of bravery—is something that can be grasped only by the mind. But, whether you call this a Form or not, the performance involved is quite ordinary. We have found nothing so far to justify talk of bizarre intuitions of mysterious objects. We simply have the rather straightforward thought that to find what fighting in battle bravely and facing cancer bravely have in common, we need to use our minds to think about the matter rather than just relying on our experiences or reports from others about theirs. The beliefs we acquire from these experiences need to be thought through for us to be able to unify and hence understand them.

The difference between moral belief and moral knowledge, then, is that a moral belief is an isolated grasp of a particular fact or set of facts, while moral knowledge is understanding of what underlies and unifies our bringing different types of fact together as examples of bravery or whatever. Knowledge is not thought of here, as it is in some modern theories, as being an improved state with respect to the very same particular fact about which the person previously had a belief. Such a conception of knowledge is familiar to ancient philosophers, and even occurs elsewhere in Plato. What is operative here, however, is something different; knowledge as understanding. Rather than focusing on individual beliefs, concern with understanding focuses on what unifies a number of possibly disparate-looking beliefs so that they can be grasped as examples of the subject in question. Understanding is achieved over an area or subject matter when you can grasp what relates the different beliefs to make them understandable as a unified subject matter. This is the more dominant notion of knowledge in Plato, and indeed in ancient epistemology more generally.[13]

In the *Laches*, then, we see an ordinary brave but unreflective person coming to see that he has true beliefs about bravery but lacks knowledge—that is, lacks understanding of what unifies his beliefs as beliefs about bravery. A person who comes to have this understanding (Socrates does not, for all the attempts in the dialogue to find this knowledge fail) would have moral knowledge. As I have stressed, this is a mundane enough accomplishment in principle, even though the dialogue's failure to come up with any moral knowledge indicates that the task is difficult.[14]

Some may complain that what we have just seen is too mundane to be what Plato means by coming to grasp a Form. It is true that in some other dialogues Plato says much less mundane things about Forms; they are sometimes characterized as the objects of intellectual aspiration and said (in the *Phaedo* and the *Republic* for example) to be eternal and unchanging.[15] Furthermore, Plato in some dialogues raises the level of understanding that is required for someone to have a grasp of Forms; I shall return to this point later. Even where Forms are thought of in more metaphysically elevated ways, however, there is still nothing to justify talk of strange intuitions. To have grasp of a Form always requires you to use your mind by thinking: in some dialogues, thinking at great length and with great complexity and formality. Nothing could be further from the idea of effortlessness and passivity suggested by the idea of intuition.[16] The notion of a peculiar intuition is a construct of the empiricist assumptions we saw at work in Mackie's interpretation, and does not correspond to anything in Plato.

IV. Expertise as a Model

Why does Plato think it so obvious that what we should do, faced by piecemeal moral beliefs, is to try to understand them by thinking through their unifying basis? Why does our search for moral knowledge take this form?

We cannot miss the way that Socrates, in the shorter Socratic dialogues, constantly appeals, in his search for moral knowledge, to examples of various types of skill. Flute-players, shoemakers, launderers, doctors, and navigators regularly crop up as relevant comparisons in the search to understand the moral terms we use. Indeed, the interlocutor in one dialogue complains about this, on the elitist ground that these occupations are socially beneath the kind of thing he wants to talk about.[17]

"Skill" here is *technê*, which is also sometimes translated as "art" or, more recently, as "expertise"; someone who is skilled is an expert in some area. The appeals to *technê* knowledge are appeals to free knowledge of the expert—more precisely, to the knowledge of the expert in a practical field. (In other dialogues Plato appeals to the knowledge of experts in fields like mathematics, but I am concerned here only with his appeal to expert knowledge as a kind of practical knowledge.)[18] When Socrates is trying to understand what courage or piety is, the kind of knowledge—that is, understanding—that he is looking for is illuminated by the knowledge of the practical expert, because practical expertise is, at least in these dialogues, Plato's model for knowledge.[19]

What is so appealing about practical expertise as a model for knowledge? It brings together a number of features which further specify the kind of unified understanding of piecemeal beliefs already mentioned. First, skill or expertise is teachable; there is conveyable intellectual content which a learner can learn from a teacher.[20] A skill is intellectually complex and requires thought to acquire; it is not just something which can be picked up casually from experience. Hence, it is contrasted with a "knack" (*empeiria*, literally "experience"), which you can pick up just by copying other people without thinking much about it.

Second, a skill or expertise demands a complete understanding of the relevant field. For example, to be an expert in French requires that you have a grasp of everything relevant to the understanding of French, such as grammar, syntax, vocabulary, and so on. The field is defined by what is relevant; understanding modern French does not require understanding medieval French, for example, or modern Italian (though these might be useful). Hence, this requirement can be seen to be an expansion of the demand, already seen, that our understanding of a subject matter should unify the various beliefs that we have about that subject matter. Learning French involves acquiring numerous beliefs about vocabulary, word order, and so on; understanding French is achieved only when all of these beliefs are brought together in a way which gives the learner control of French, the whole subject matter, in a unified way.[21] (This example also brings out a further point, namely, that when this point is achieved, you are no longer dependent on the books, etc. from which you originally learned; the understanding is now yours in a way which is no longer dependent on its sources.)

Finally, skill or expertise requires that the expert, unlike the mere muddler or the person with the unintellectual knack, be able to "give an account" (*logon didonai*) of what it is that she is expert in.[22] The expert, but not the dabbler, can explain why she is doing what she is doing; instead of being stuck with inarticulacy, or being reduced to saying that "it feels right this way," she can explain why this is, here and now, the appropriate thing to do in these circumstances. What such an explanation will look like will of course vary with the skill in question: "You have to use the subjunctive here because . . . ," "You can't have the electricity line going there because . . . ," and "You need to steer the boat to the right here because . . ." all appeal to different kinds of considerations, but what is being appealed to is the understanding which the expert has of French, electricity, or navigation.

This demand for articulate explanation is what would be expected given the point about teachability; the teacher gets the expertise across to the apprentice first by example and then by explanation of what has been done. And what the expert teaches and explains is the understanding of what unifies the relevant subject matter. The different requirements are themselves comprehensibly unified. Moreover, they give, together, a reasonable picture of what it is to have a practical expertise. There is nothing particularly time-bound about this idea, although of course Plato's examples belong to the ancient world.[23]

When Socrates seeks moral knowledge, then, it is only to be expected that this will be seen on the model of practical expertise, since this is the model for knowledge in general. We should note that it is particularly appropriate for the specific kind of moral knowledge that is being sought, knowledge of various virtues, since the virtues themselves have the intellectual structure of skills. Plato in fact thinks of virtue as a special kind of skill, an idea which had a continuing appeal in ancient ethics. Aristotle, for example, holds that virtues are *like* skills, while the Stoics revert to the more straightforward thesis that virtue is itself a skill. Coming to understand what courage is, is thus the same land of performance as coming to understand any subject matter which is an area of expertise.[24] But, although the attraction of the idea that virtue is like skill may explain the enthusiasm of Plato and other philosophers for the idea, it does not itself account for the use of practical expertise as a model for knowledge. The model is sufficiently attractive in itself as a model for the kind of knowledge expressed in living well for it not to be dependent on virtue in particular as its application in the moral case.

There is one important disanalogy between particular skills and moral knowledge conceived of as a skill. Particular skills have an aim—mastery of French, fixing the car—which is local; it is an aim which you might well in other circumstances not have, and from which you can readily become motivationally detached, since the aim is one you have conditionally on your having certain broader interests and needs. When moral knowledge is thought of as a skill, however, its object is global—namely, your life as a whole. The Stoics make this point by calling virtue the "skill in living." When you start reflecting morally, you already have a life; you have a particular family, country, job, and the results of some moral education. If you are not content to drift along and wish to live your own life, you come to think of these circumstances as raw material which can be formed by understanding into a result which is the product of reflection, which is your life and what you have made of it, rather than just being what circumstances have left you with. The object of the "skill of living" is thus not optional, as are the objects of local skills. The options are between drifting along and trying to live your life on the basis of overall understanding of it, since you are going to live your life whether you think about it or not; this is not something from which you can become motivationally detached by a shift in interests.[25]

Moral knowledge, then, is thought of as explicated on the model of expertise, because both are cases of practical knowledge, and expertise appears as illustrative of the kind of understanding that the moral person has. The moral person who lives well and does the right thing is thought of as being relevantly like the skilled worker who does a good job and knows the appropriate thing to do. There are two relevant points here. One is the thought of skill as the specific model for moral understanding. There are many factors tending to make this seem alien and even peculiar to us, some deriving from the different roles of skill in the ancient and modern worlds, some from modern misapprehensions about skills.[26] In the present context I shall pass these over, since they are not relevant to my particular point in this essay, which concerns moral knowledge rather than the particular application of the idea of skill. The point I shall pursue here is the more general one, that practical expertise serves as a useful model for moral knowledge because both are examples of *practical knowledge*, knowledge expressed in action.

One reason, then—perhaps the main one—why Mackie's interpretation of Plato is so grotesque is that the ancients do not worry about moral knowledge within assumptions that make it problematic how there can be any such thing. For it is a kind of practical knowledge—like expertise, in fact, a kind of practical knowledge that we are already familiar with and good at identifying.[27] Different philosophers diverge on just how similar moral knowledge is to expertise. But they do not doubt that it is a kind of practical knowledge, and that our familiarity with expertise helps us to understand it.

V. Practical Knowledge

When it comes to practical expertise, we are in an advantageous position: we have some! In fact, we have quite a lot. You or I may not individually have much, but nobody can sensibly deny that there is some. When I mess up the computer, find a leak under the sink, or find that the car will not start, I take my problems to the relevant experts. They are practical experts; I want not a theoretical computer whiz, but someone to fix the software, not a Ph.D. in engineering, but someone who can stop the leak, fix the oil gauge, and so on. Serious skepticism about the existence of practical experts of this kind does not get off the ground. (This is a major reason why discussions of moral knowledge in ancient ethics are not structured by skeptical concerns.) I may wonder whether my plumber is really an expert, but there is something deeply wrong about the idea that I might hesitate to call *anyone* listed under "Plumbers" in the Yellow Pages to fix my leaking pipe, on the grounds of doubt as to whether there was such a thing as *knowledge* of plumbing.

Nowadays, few philosophers think that this point about the obvious existence of practical knowledge as manifested in plumbers and car mechanics has much bearing on issues of moral epistemology. Coming from study of a moral tradition where practical expertise is standardly found very relevant to discussions of moral knowledge, I find this noteworthy, and suspect that it is explained more by the persistence of assumptions that come from outside the subject than by serious reflection on the nature of moral knowledge itself. Moral knowledge is, after all, practical knowledge, whatever account we give of this; it is knowledge of what to do, and results, often, in doing something. An obvious starting-point for understanding what moral knowledge is would appear to be some practical knowledge where there actually is agreement that such a thing exists, that we have it, and that we can say some useful things about its structure, how it is acquired, and how it demands that we reflect on our experience. The kind of practical knowledge displayed in plumbing may be mundane, but it offers us just this kind of advantage. For we can then go on to ask what makes moral knowledge similar to, and different from, the easier case—a task developed by ancient moral theorists from Plato onwards.

Perhaps this advantage is not apparent because homely performances like fixing the computer are not thought of as amounting to *knowledge*. Here we can only ask why. Prephilosophically, we have no problems in saying that the mechanic knows how to mend the car—*if*, that is, he is an expert; that is how we distinguish experts at car repair from mechanics who are not experts.

It is possible, of course, to introduce a special philosophical category of "knowing how" and oppose it to "knowing that," where it is the "knowing that" category that brings in familiar epistemological issues, and "knowing how" is supposed to be different in kind. Labeling the issue, however, does not much further us. Either "knowing how" involves "knowing that" or it does not. If it does not, then what we think of as practical knowledge is being construed as a kind of inarticulate practical knack, an ability to manipulate the world which is not at a sufficiently rational level to be judged epistemically. This, however, would amount to saying that there is no such thing as practical expertise, only knacks—that there is no significant difference between the inarticulate practitioner and the expert in the field. This is ridiculous. However, if "knowing how" does involve "knowing that" in some way, then we have not evaded any of the issues that arise when we speak of practical knowledge.

We do (some of us) have practical knowledge—namely, expertise. The significance of this for moral epistemology has been insufficiently appreciated in modern discussions.

VI. PRACTICAL KNOWLEDGE AND VALUES

This brings me back to Mackie and the assumptions which drive his interpretation of Plato's moral Forms as peculiar entities which are supposed to exert a peculiar force making people act in accordance with their intuitions of them. Mackie failed to see that there is nothing in the least peculiar about coming to understand a moral Form, any more than there is about coming to understand French or computer languages. They are all examples of practical knowledge. They all involve using your mind to get understanding of a subject matter about which (especially in learning which) we have beliefs which on their own are piecemeal; coming to understand them, we unify our understanding of the subject matter and come to be able not only to act accordingly but also to explain why our decisions and actions are as they are. Coming to understand a moral Form is harder than the other cases, because morality is harder than French or electronics; but this in itself is no reason to deny that both are examples of practical knowledge.

I suggested earlier that Mackie's caricature Forms were constructs of his assumptions, particularly the assumption that values are, unproblematically, not accessible to experience and not accessible to reasoning. These in turn rested on very strong and narrow assumptions about experience and reasoning. It is worth asking what happens if we approach practical expertise in a way structured by these assumptions. For the argument is not presented as being directed at specific features of values that make them *moral* values, but just at *values*.[28] It is "objective values" which are supposed to be so mysterious, since we need them to explain how the "authoritative prescriptivity" of the pattern of reasoning is to be made available to the agent.

Take the expert plumber. Faced by a leak, he exercises his expertise by working out what the problem is which he has to solve, and then figures out the best solution to it in the circumstances. Working through this problem he comes to the banal conclusion, say, that he should turn off the water near the leak. This is a response to his expert appraisal of the problem, and it appeals to him with the "authoritative prescriptivity" of being the best immediate step in solving the problem; he turns off the water there (and not somewhere else). On Mackie's view, there is a deep mystery in the view that the plumber is motivated to do something (i.e., turn off the water here rather than there) because it is (part of) the best way of mending the leak. There is, of course, a "natural fact" accessible through the normal workings of the senses and low-level reasoning, consisting of the fact that turning off the water here bears such and such causal relations to the leak and further actions involving it. The plumber, however, is motivated not just by this natural fact, but by the further fact of its being what he, here and now, ought to do (given that it is the best immediate step in solving the problem of fixing the leak). What is the connection between this natural fact and the "practical fact" that he ought to do it? It must somehow be the case that he ought to do it *because* it bears such and such causal relations to the leak. This is the conclusion of his practical deliberations. "But just what *in the world* is signified by this 'because'?"[29]

In his eagerness to show that there are no objective values "in the world," Mackie has produced an argument which (if it is regarded as successful) not only debunks the objects of moral knowledge, as intended, but subjects ordinary practical expertise to the same objections. This is not surprising if moral knowledge and practical expertise are both kinds of practical knowledge, as we prephilosophically assume and as appears obvious to philosophers who, like the ancients, are not working within the framework of Mackie's assumptions.

This point may have passed unremarked because of the apparent implausibility, to philosophers, of the point that moral reasoning could be relevantly like anything as ordinary as practical expertise. (I shall return to this in Section VIII.) It may also be at least in part due to another assumption, Mackie's second, which is very common in much modern discussion of practical reasoning. This assumption is that there is a prima facie problem as to any kind of reasoning's motivating us to act. Since we do act, we must be motivated by something different, and this is assumed to be desire. It is accordingly thought simpler to interpret any presumed case of reason's being practical as a case where what provides the practical force is really a desire, and reason is limited to the means-end task of figuring out how to fulfill the desire.[30] If we have good reason to make this assumption, then practical reasoning like the plumber's will be no exception.

This approach, however, fails, because it simply misrepresents the reasoning of the practical expert. The expert's deliberations are typically exercised in problem-solving, figuring out what to do in providing a solution. Faced by the leak, the expert plumber, regarding it as a problem to be solved, deliberates as to the best way to do it; having worked this out, he then acts accordingly, since this is what he ought to do. Nothing in this story requires us to bring in the plumber's *desires*. The plumber shuts off the water where he does because this is what he ought to do as part of the expert way of fixing the leak; he does not shut off the water as a means to satisfying a desire. The suggestion is grotesque—for one thing, desires usually signal their presence by pain and frustration until they are fulfilled, something we have no reason to ascribe to the plumber. Experts deliberate about the objects of their expertise, not about how to fulfill their desires (of course they might do the latter, but not in a way relevant to the exercise of their expertise).

The standard response to this is to appeal to the common orthodoxy that there *must* be a desire, or the plumber would not have shut off the water; for the action must have been brought about by a combination of belief and desire, and here the items we need are a desire to fix the leak and a belief that shutting off the water here is a means to fixing the leak and hence to satisfying that desire.

However, this case makes three points about the response uncomfortably clear. First, it obviously trivializes the notion of desire, since a desire is invoked mechanically to account for action, with no independent grounds whatever.[31] Second, it is not adequate to say (perhaps by way of response to the first point) that the notion of desire in question is a theoretical one. If we take desire just to be what the theory invokes to bring about action in every case, it is completely unclear how it is to fulfill this function, if not by way of having at least some of the characteristics of a real desire—for example, pressuring us to fulfill it by way of getting rid of the pain and frustration it produces until fulfilled. Standard invocation of a desire in order to provide the needed motivation to bring about action often blur together two ideas: of a real desire, something that can reasonably be seen as actually getting the person to do something, and of a theoretical item posited *a priori* to meet the demands of a theory regardless of whether the conditions for there being a real desire are met or not.

Finally, a third problem with the response is that the case of the practical expert is one in which it is particularly unconvincing that what the person does is brought about by desires—either real desires or the fictitious desires postulated by a theory that demands a desire for every action. It just is not plausible that either kind of desire is had by the expert translator who uses the Italian subjunctive to translate an English expression containing no subjunctive, the expert papyrologist who conjectures the presence of Greek letters from marks on a papyrus, or the forensic expert who detects blood on the murder weapon. They have all solved problems; it is not plausible that they have devoted their energies to fulfilling their desires.[32]

The idea that practical expertise is to be accounted for in terms of purely means-end reasoning requires us to misdescribe how experts actually think, and to posit fictitious and irrelevant desires. Of course the plumber has some desires: to take on the job in the first place, for example. But no desires that can convincingly be ascribed to the plumber figure in his expert practical reasoning as to how to fix the leak.

A defender of the orthodoxy will object that it is, in fact, the plumber's desire to take on the job which sets the end that his deliberations achieve. The reasoning he employs results in the achievement of the end set by that original desire (even if we allow that we do not need *further* desires once that is granted, such as the desire to turn off the water). This, however, misconceives the expert's end. His end as an expert is established by his response to the problem. His desire to take on the job, if he has it, gives him an end he has as somebody with bills to pay. The translator, papyrologist, and forensic expert all no doubt have desires to make money, get a reputation, or the like. As experts their ends are set by their responses to the problems to be solved: the English to be made comprehensible to Italians, the holes in the papyrus text to be filled in, and so on. This is in part what distinguishes the expert's reasoning, for the desire to make money can be fulfilled perfectly well by amateur, nonexpert deliberation. If what is done is done expertly, it is done to achieve the expert's end of responding to the problem by solving it; desires do not form a part of this and do not set it going.

"Objective values" are in play in ordinary practical expertise as well as in moral and aesthetic reasoning. We cannot avoid this conclusion by regarding the former as not grand enough to be brought into comparison with the latter. Nor can we avoid it by claiming that practical expertise is merely a matter of purely means-end reasoning. Any gains aimed at in making this move are vastly outweighed by the costs of inventing implausible, fictitious desires for the purely means-end reasoning to be the means to fulfill. If we honestly recognize ordinary practical expertise for what it is, we will see that the kind of argument so familiar from Mackie and others is an attack on the whole idea of practical knowledge, not just moral knowledge in particular: the problem is that of how objective values can motivate, and it is the practicality of practical knowledge that is in question, not the specific demands of morality.

VII. Skepticism and Practical Knowledge

I have already mentioned that Plato and other ancient philosophers are not motivated by skepticism when they discuss moral knowledge, even though they are alert to it elsewhere. It is already clear why this is the case. We are not normally skeptical that there is such a thing as the expertise that gets a battery replaced, restores a crashed hard drive, or builds a house. It is clear by now that widespread modern assumptions about values imply that we ought to be. I am suggesting, obviously, that this might well make us more critical of those assumptions.

There are plenty of places where skepticism about moral knowledge in particular might reasonably arise. We might not be sure that we adequately understand the moral beliefs that we have; they might contain unresolved problems and conflicts.[33] And there are ways, stressed by Aristotle, in which moral knowledge looks different from expertise. Perhaps the special features of morality limit the usefulness of using expertise to illuminate it. (This particular issue was explored in depth in ancient debates.) These problems get going, however, on the basis of a generally unskeptical acceptance of the existence of expertise. This surely reflects something important about our everyday reasonings about morality. Our everyday moral discourse takes for granted that we have a lot of true moral beliefs. The problem comes when we try to give an articulate account of what we have these beliefs about. We have the model of practical expertise; the problems come with morality, because coming to understand morality is harder than it is with electronics or engineering. But this is just what we would expect if moral knowledge is a kind of practical knowledge; other forms of practical knowledge, then, are still a useful beginning for understanding the moral kind.

A point of contrast between ancient and modern approaches to moral knowledge can be seen if we consider the role often played in modern discussions by the figure of the amoralist. This is the person who, in all other respects normal, is fully aware of and appreciates moral considerations, but is motivationally indifferent to them. The amoralist has often been used to undermine different forms of internalism,[34] but whatever the use to which the amoralist is put in the argument, the assumption is that there is nothing conceptually incoherent or confused about the idea of a person who is fully aware of all the relevant moral considerations, perfectly conscious of which is the most salient, and ready to accept that as a result of this she ought to make a moral response (acting, responding, intervening, and so on), and who yet fails to make this response, not as a result of ignorance, stupidity, misinformation, or failures of reasoning, but simply because all these conditions leave open the possibility of the failure of the relevant motivation.

Could there be such a person as the "apracticalist"? This would be the expert who responds to the problem of the crashed computer, fully realizes that the best way to solve the problem of fixing the computer involves reinstalling the software, is fully cognizant of all the factors pointing to this, yet remains aloof and motivationally detached from actually reinstalling the software, but not as a result of ignorance, failure of reasoning, and so on. How do we make sense of this? In the case of the expert, failure to actually reinstall the software implies either imperfect understanding of what was the right way to proceed or a rethinking of what the best way to fix the computer is. This is just a corollary of the point that expertise is expressed in the problem-solving kind of deliberation.

Faced by an actual expert who knows what to do but fails actually to do it, we are not generally puzzled, because we take it that the aims of local skills and kinds of expertise are likewise local. You can become detached motivationally from one or more of them if your interests shift (i.e., you are no longer concerned to achieve the aim) or your other interests become more pressing (i.e., you want to complete the job but are distracted by other matters). Failure to act on a local kind of practical knowledge shows only that the expertise in question is local.

But this is, of course, the respect in which local skills differ from moral knowledge where that is thought of as a skill. Its aim is not local, nor is it conditional on your other interests and concerns; hence, it is not similarly unproblematic for you to become motivationally detached from that aim. Living your life well is not an aim from which you can become detached by boredom or the prospect of better pay elsewhere So if the apracticalist is not, while the amoralist is, a coherent and threatening idea, this must rest on some *other* difference between moral knowledge and other forms of practical knowledge. Such a difference could of course be established, but to avoid begging the question it should not rest on the postulation of the kind of assumption that we saw at the beginning.

VIII. Conclusion

I am quite aware that the conclusions of this essay will not recommend themselves to philosophers who share Mackie's framework of assumptions or related ones. It might also be argued that to think of moral knowledge as primarily practical knowledge may commend itself to the ancients, but does not fit modern ways of thinking of morality. Indeed, I stressed at the beginning of this essay that ancient and modern ways of thinking of epistemology are rather different. It would be a mistake, however, to think that the only results of this inquiry are archaeological. To think of moral knowledge as a kind of practical knowledge is not just intuitive to the ancients; it is intuitive for us also, and the right contrast is between this prephilosophically appealing view and a view of moral knowledge which depends on very specific philosophical assumptions about things other than morality.

Moral knowledge is knowledge which is, among other things, about how to act; it is also knowledge that is put into practice. This is a simple enough thought, and it is significant that so many modern positions in moral epistemology not only refuse to start here, as we prephilosophically do, but make the practicality of moral knowledge mysterious, or even regard it as forcing us to accept debunking conclusions about value.

One assumption of Mackie's which has not yet much figured in this essay is the third one, that morality is basically a matter of telling others (and yourself) what to do. Mackie acknowledges the influence of Stevenson's emotive theory of ethical terms, and it is familiar that theories which conclude that there is something deeply faulty about our everyday assumptions that people and actions really are good or bad also tend to conclude that part or all of the function of ethical terms is nondescriptive: they express attitudes, or "prescribe" what to do, and so on. There is a model here of moral discourse as fundamentally focused on pressuring people (others and oneself) to act in certain ways. The more emphasis is put on the supposed nondescriptive element in moral terms, the more moral discourse is presented as a matter of influencing people to act in certain ways, ways that ought have a contingent and possibly remote connection to whatever descriptive content moral terms are allowed. This way of thinking, a descendant of logical positivism, is eerily postmodern in the way it represents moral discourse as really being a power struggle.

By contrast, to think of moral knowledge as being, like expertise, a kind of practical knowledge is to think of it as a humbler, more cooperative kind of activity. When Aristotle says that learning to be just is like learning to be a builder, this may strike us not only as insufficiently grand, but as missing the combative element so familiar from modern analyses. The apprentice builder is learning something from teachers or role models who are better at it than she is; she is learning to think for herself about what she has learned from others. She does not start off pressuring others to think differently; if she does think that she and others should think in a way different from the one she has learned, this comes later, when she has learned something to disagree with. Similarly, the moral learner is coming to think for himself about what he has learned from others. When he has done this, of course, he will be in a position to tell others what to do, and to try to get them to do it if they disagree; but on the expertise model, conflict, or pressuring people to act, comes later, and presupposes that the critic has already learned something from others.

Perhaps this humbler model has little appeal to philosophers because it is simply assumed that moral discourse is fundamentally a matter of conflict, what matters being winning, getting others to accept your view. Perhaps, on the other hand, the expertise model has advantages in reminding us that we do all, in fact, learn to be moral before we get to the point of criticizing what we have learned and telling others what to do. Perhaps there is something to be said for respecting this idea of becoming moral as a process of learning in which we all start as pupils or apprentices, and where it is up to us to become experts.[35]

If we start with something like Mackie's assumptions, then moral discourse will indeed seem to be a set of pressures brought to bear by some people on other people, and the moral learning we have done, before we started to reflect about the nature of moral knowledge, will appear to be nothing but the acquisition of error, something to be wiped out rather than respected. Perhaps it should even be by-passed for future enlightened generations, if this could be done, as in Neurath's chilling proposal that moral education should be done in terms that do not even allow the learner to formulate "incorrect" concepts.[36]

I have argued in this essay that if we go back to ancient views of moral epistemology we will see the attractions of an alternative view. We do not, in principle, need to go back to the ancients to discover this, but it is helpful to enable us to see both that the hold of a common modern set of assumptions is not inescapable, and that an alternative model is not only possible but was actual for quite some time. This is one way in which, I hope, paying attention to past philosophers, rather than patronizing them, can help us to understand where we are coming from, to ask whether we might find a starting-point other than the ones familiarized by tradition, and to enrich our view of the options that are open to us.

NOTES

1. J. L. Mackie, *Ethics: Inventing Right and Wrong* (Harmondsworth, NY; Penguin, 1977).

2. I am aware that discussion of moral realism has moved on since Mackie, but there has been surprisingly little change on the point I wish to examine. It is often assumed that the realist options are either a kind of minimalist realism, making only the internal commitments of moral discourse (for an example of this kind of realism, see Ronald Dworkin, "Objectivity and Truth; You'd Better Believe It," *Philosophy and Public Affairs* 25, no.2 [1996]: 87–139), or a more substantial realism, making metaphysical commitments external to moral discourse, and this option is readily labelled "Platonism."

3. Mackie, *Ethics*, 38. Mackie frequently makes disparaging references to "intuition," which he gives no account of. He assumes that we know what intuition is from the "intuitionists," but this is not a great deal of help, since theories calling themselves intuitionist have diverged widely over what they take intuition to be and what kind of items (e.g., principles, kinds of value, individual features of situations) they take to be the objects of intuition.

4. Ibid., 23–24.

5. Ibid., 40.

6. Ibid, 41.

7. This harsh verdict is reasonable in that Mackie did, after all, read Plato in the original and was not dependent on possibly misleading translations. Mackie is also insensitive to Plato's uses of argument, reading the dialogues crudely as ways of putting forward positive ideas of Plato's own. Mackie regards the arguments of the *Protagoras*, for example, as Plato's own attempts to put forward a positive position, and deals with the obvious resulting problems by cheerfully ascribing to Plato "a lot of sophistry" (Mackie, *Ethics*, 187).

8. More expansively, we cannot access them by "sensory perception or introspection or the framing and confirming of explanatory hypotheses or inference or logical construction or conceptual analysis, or any combination of these" (ibid., 39).

9. In his notes to chapter 1 of *Ethics*, Mackie says that his views were first written down in 1941, and cites Ayer's *Language, Truth, and Logic* and Stevenson's *Ethics and Language* as helping to determine the main outlines of his position (ibid., 241). He rejects characterization as a logical positivist on the grounds that he rejects both the verifiability principle as a criterion for descriptive meaning and the idea that moral judgments lack descriptive meaning (ibid., 39–40). Mackie's characterization of his position as empiricist can be faulted on the ground that many positions describing themselves as empiricist start from much less restrictive assumptions. Also, although in *Ethics* he holds that morality has to be "made" and that we "have to decide" what moral views to "adopt" (ibid., 106), he writes a whole book on ethics, thus insulating his practice from his theoretical view. I have made some remarks on this issue in my "Doing Without Objective Values: Ancient and Modem Strategies," in Stephen Everson, ed,, *Ethics*, vol. 4 of *Companions to Ancient Thought* (Cambridge: Cambridge University Press, 1998), 193–220, esp. 216.

10. For a valuable discussion of these claims, see Tom Sowell, *Scientism* (London: Routledge, 1991). Scientism takes many forms; here I am concerned only with assumptions made by philosophers who give science primacy in defining our epistemological concepts. These assumptions are not shared in the wider culture, even where that culture contains other kinds of deference toward science.

11. Actually the terms *eidos* and *idea* do not occur very often; Plato is not so much arguing for a new notion to which he will give a new technical name, as drawing out the implications of ways of thinking which have appeal anyway, and rendering these more precisely.

12. Note that it is types of action which are in play from the start. The argument has nothing to do with supposed deficiencies of particular actions.

13. Knowledge as understanding and knowledge as improved true belief are both prominent in ancient epistemology. As noted, Plato is mostly concerned with the former, though the latter is in question in the *Meno* and at the end of the *Theaetetus*. For a good introduction to these different strands in ancient epistemology, see Stephen Everson, ed., *Epistemology*, vol. 1 of *Companions to Ancient Thought* (Cambridge: Cambridge University Press, 1990), particularly Everson's introduction.

14. The *Laches* is particularly interesting in that Socrates' actual demand seems controversial; do we really show courage in resisting pleasures and temptations, as well as in resisting pains and discouraging factors? The claim that we do comes from thinking of courage as involving the idea of endurance more centrally than the idea of resisting unpleasant or dangerous circumstance. This is something on which people may reasonably disagree.

15. We should remember, however, that in the first part of the *Parmenides* Plato has Socrates put forward an account of Forms much as they are found in the *Phadeo* and the *Republic*, and then has the account subjected to the same kind of relentless criticism that Socrates elsewhere directs at the positions of others. Furthermore, as in these other cases, the objections are nowhere met. Plato was obviously aware of intellectual difficulties in his account of Forms, and he continued to work toward more satisfactory characterizations.

What we rather misleadingly call "the theory" of Forms is a cluster of ideas which Plato engaged with in different ways at different times.

16. When Plato uses language that suggests insight or grasp, this is not as an alternative to hard and effortful thinking, but suggestive of what it is to come to understand something by thinking hard about it. Perception is a misleading metaphor for moral thinking if it is used (as it sometimes has been) to suggest that no effort is required, or that getting it right is easy and obvious. Such ideas could scarcely be further from Plato's insistence that moral thinking is hard and requires effort which many people are too lazy, or too focused on material success, to make.

17. Callicles, at Plato, *Gorgias*, 490e–491a.

18. The appeal to mathematics in particular goes with a greatly raised standard for what is to count as having understanding of the kind which is sought.

19. This point has been recognized in the secondary literature for some time now. See Paul Woodruff, "Plato's Early Theory of Knowledge," in Everson, ed., *Epistemology*, 60–84, for a very useful discussion of the many issues involved.

20. See Plato, *Meno*, 89e ff. and Plato, *Protagoras*, 319e ff. The obvious and agreed teachability of kinds of expertise comes in when these discussions center on the teachability or otherwise of virtue.

21. In the case of the productive skills, exercise of the skill produces a unified object whose organization reflects the expert's unified grasp of her skill and its requirements; see Plato, *Gorgias*, 503d–504b.

22. On this point, see ibid., 465a, 501a.

23. A common objection is that we are in fact prepared (even if the ancients were not) to call someone an expert if she can reliably come out with a certain performance, whether or not she displays articulate understanding. Thus we are (allegedly) prepared to call someone an expert gardener even if they are stubbornly unwilling to say anything about the basis of their success, and to call someone an expert who knows lots of facts about, say, ranches in the Southwest between 1880 and 1885, despite never relating this to any wider context. There is more than one response to this alleged problem. First, we might be missing the term "expert" in using it of these people, since they lack articulate understanding of what they are doing. It might also be the case that we are assuming that the people in question do have a unified understanding of what they do, but for some reason cannot, or are unwilling to, articulate it. If we thought the gardener really unable to relate frosts to early plantings, or the history buff completely ignorant of other contemporary facts, I doubt that we would call them experts.

24. My "Virtue as a Skill," *International Journal of Philosophical Studies* 3, no. 2 (1995): 227–43, traces some of the implications of this idea for virtue as a concept in a theory of morality.

25. Hence, it is ineffective to object that moral knowledge cannot be thought of as a skill because the aims of local skills are obviously optional. Nor does the analogy import the claim that the object of moral thinking is already fixed, as with local skills. Clearly, living your life in a good way is not an aim which can be clearly defined in advance, unlike, say, the aims of plumbing or car repair.

26. For discussion of these various factors, see Annas, "Virtue as a Skill."

27. When we first begin to acquire moral knowledge, how can we be sure that we have identified the right moral experts? We can't; the person whom we take as an expert and follow initially might turn out to be a fraud, or flawed. We can identify the expert confidently only to the extent that we ourselves are progressing toward the relevant understanding. This does not make the account circular; though it does indicate that it requires a "bootstrapping" progress where by you become surer that you have got the aim correctly specified as you progress in the skill of acquiring the aim. This is true of some local skills as well. If this is found objectionable with respect to moral knowledge, this must depend on an assumption such that moral considerations must be available equally to all, however much (or little) effort they make to understand them.

28. The inconclusive "argument from relativity" is directed at "moral codes," while the "argument from queerness" is directed at objective values.

29. Mackie, *Ethics*, 41.

30. This idea is often associated with "Humean" theories of reasoning, but I shall not go into the question of whether any of this relates to Hume, or into the larger matter of exactly which modern theories are held to be Humean.

31. In this part of the discussion, I have tried to use the notion of a desire as that figures in the recent literature. I cannot resist, however, pointing out that in ancient discussions, desire is understood in a more restricted way (one which in fact answers to our modern folk psychology better than the modern philosophical concept). In ancient ethics, a desire is not just any goal-directed, forward-looking, or "mind to the world" motivation, but a subset of these which also have a significant backward-looking element, since *a desire signals a perceived lack or need.* (It is because of this fact, indeed, that desires typically have the phenomenology they do, namely, that of producing frustration until satisfied.) For example, when the world produces the irritating state of me that consists in hunger, I will (normally) have a desire to eat, indicated by certain feelings of pain and frustration, and the satisfaction of this desire will remove those feelings. As sources of motivation, desires will typically be reactive and short-term. Given this picture, there is little temptation to represent all forward-looking motivation as brought about by desire. I think, though, that it is of interest to show that even the broad concept of desire, incorporating no reference to need and with no explanation of desire's phenomenology, is not well-suited to account for the plumber's practical motivation.

32. I am not under the illusion of having made a contribution to the vast modern secondary literature devolved to the discussion of Humean versus non-Humean theories of motivation. I have simply tried to indicate what the major issues are for the kind of position I am sketching, and to make some comments from outside the modern orthodoxy.

33. Cf. note 14 above, concerning the understanding of courage.

34. "Internalism" has been used for a variety of different positions; here I take it to be the position that fully to understand a moral requirement is to be (to some degree) motivated by it. See David O. Brink, *Moral Realism and the Foundations of Ethics* (Cambridge: Cambridge University Press, 1989), 45ff.

35. For discussion of respect for the idea of moral learning, see the excellent comments in Rosalind Hursthouse, *On Virtue Ethics* (Oxford: Oxford University Press, 1999), 12–16.

36. "Must everyone in turn go through metaphysics as through a childhood disease? . . . No. . . . *Every child can in principle learn to apply the language of physicalism correctly from*

the outset. . . . A new generation educated according to unified science will not understand the difference between the 'mental' and the 'physical' sciences, or between 'philosophy of nature' and of 'culture.'" Otto Neurath, "Unified Science and Psychology," in Brian McGuinness, ed., *Unified Science: The Vienna Circle Monograph Series* (Dordrecht, The Netherlands: Reidel, 1987), 1–23. On page 8 Neurath says, without apparent irony, that he himself has long employed "an *index verborum prohibitorum*" such as "norm," "categorical imperative," and "intuition." Such is the world without Mackie's "errors."

10. ON INTERACTIONAL EXPERTISE

PRAGMATIC AND ONTOLOGICAL CONSIDERATIONS

EVAN SELINGER AND JOHN MIX

INTRODUCTION[1]

Clearly, experts matter to us. Many routine decisions, technical and non-technical in nature, are executed only after consultation with, and perhaps deference to, experts. By contrast, not every theory of expertise is significant, and some are riddled with insurmountable difficulties.[2] The topic of review here is the conceptual and pragmatic import of Harry Collins's latest investigation into a third form of knowledge, "interactional expertise."

On one level, Collins is beyond critique. Contrary to what some scholars in the science and technology studies community might espouse, his concept of the interactional expert designates a real group of knowledgeable and skillful people.[3] Moreover, as a corrective to Hubert Dreyfus's untenable portrayal of commentators, Collins makes a genuine contribution to the phenomenological discussion of expertise.[4] And yet, there is a sense in which these compliments conceal the limitations of Collins's account. Based on his critical evaluation of current sociological practice as limited because it is not normative, it appears that Collins does not wish to advance a purely conceptual argument. Rather, he invites us to approach the subject matter from a pragmatic vantage point.

If it is legitimate to take a pragmatic stance, then we are bound to inquire into whether the concept of interactional expertise can be applied in practice to produce useful consequences. For the pragmatist, it is ineffectual to only appeal to a category in the abstract. As is well-known, John Dewey advocated that social scientists should not limit themselves to a descriptive enterprise; they should, instead, implement their findings so as to develop new ways to cope successfully with the world in which we live. A pragmatist of this kind would thus want to know if the outcome of the debate between formalists and informalists, as structured by Collins, is directly relevant to worldly concerns, as, perhaps, a strategic counterpoint to some aspect of institutional dogma. In order to discern if the concept of interactional expertise has useful cognitive potential, a pragmatist would ask if acknowledging the existence of a thoroughly demarcated notion of interactional expertise could make a noteworthy

Response to H. M. Collins

difference in our decision-making processes, e.g., when we try to make the best judgment concerning whom to turn to for advice and in what situation. With the pragmatist's concerns in mind, we proceed as follows. First, we inquire into whether Collins's views enrich the phenomenological literature on expertise. Second, we critically assess how Collins differentiates interactional from contributory expertise. Third, we evaluate Collins's analysis of the embodied dimensions of interactional expertise. We conclude by appraising the epistemological status of Collins's account of interactional expertise in light of the pragmatic and ontological objections that are considered here.

LINGUISTIC SOCIALIZATION: EXPANDING THE SCOPE OF A SHARED PRACTICE

Collins wants to close what he calls "the practice gap." This entails differentiating between three ways of trying to learn to pass a Turing test where the medium is ordinary conversation: (a) being fully physically immersed in a form-of-life, (b) being linguistically socialized, without sharing physical activity, and (c) acquiring discrete propositional knowledge. For Collins, the import of differentiating between these three approaches to the acquisition of conversational ability is that it enables him to distinguish between two types of expertise: interactional expertise and contributory expertise. A contributory expert is a practitioner in a field who learns to make contributions to that field via (a). By contrast, an interactional expert is someone who can talk competently about aspects of a field (e.g. pass on information, assume a devil's advocate position, and make judgments on a peer review committee), but only learns about the field from talking with people who have acquired contributory expertise. In other words, while the interactional expert has quite a bit of tacit (non-propositional) knowledge, he is not a direct practitioner in a field; interactional expertise only requires (b), but not (a). What this means for Collins is that someone who lacks full physical immersion in a field can, through linguistic socialization, become so conversant about that field that, under the conditions of a Turing test, it would be hard for authorities in that field to differentiate between an interactional expert and a contributory expert.

Because Collins places his analysis of interactional expertise within the context of a perennial disagreement between formalists and informalists, we need to ask first if, in fact, he is advancing the debate at all. More specifically, we need to understand how Collins's account of language and cognition differs in either substance or tone from alternative, informal accounts. A brief comparison with Dreyfus will serve this purpose.

Collins claims that "it is possible to learn to say everything that can be said about bicycle riding, car-driving, or the use of a stick by a blind man, without ever having ridden a bike, driven a car, or been blind and used a stick" (p. 125). In other words, for Collins, it is possible to immerse oneself in the linguistic culture pertaining to a practical domain without immersing oneself in the corresponding physical activities. Dreyfus, however, articulates the opposite position:

> There is surely a way that two expert surgeons can use language to point out important aspects of a situation to each other during a delicate operation. Such authentic language would presuppose a shared background understanding and only make sense to experts currently involved in a shared situation.
>
> (Dreyfus 2000a, p. 308)

Based on Collins's analysis of interactional expertise, we can see that in this passage, as elsewhere, Dreyfus confuses process and person. He conflates the process of acquiring surgical language with the substantive issue of determining just what kind of person can acquire this language. On the process side, Dreyfus rightly insists: "An expert surgeon does not have names for the thousands of different ways she knows how to cut, and even if she did, she could not pass on in the public language a description of each specific situation in which each way is appropriate" (Dreyfus 2000a, p. 307). Dreyfus's analysis of process is correct because the apprenticeship in which surgical language is acquired is a course of action in which the subject acquires tacit knowledge and not discrete medical propositions. But Dreyfus errs in insinuating that only a surgeon can have the appropriate background understanding to experience the "authentic" language of surgery as meaningful.

Even though intersubjective experience and personal commitment are required to obtain surgical language, Dreyfus fails to establish that direct surgical experience is the only type of experience that can provide an individual practitioner with the prerequisites that would enable him to come to experience surgical language as meaningful. We can agree with Dreyfus that surgical language could not be acquired if someone's only experience of it came from reading medical books. But Collins's analysis of interactional expertise establishes the possibility that, for example, a sociologist who does not have any direct experience practicing surgery, but who, nevertheless, studies surgery from a third-person perspective (and has been linguistically socialized into the surgical community) would be quite capable of acquiring "authentic" surgical language; such a person could know how to give a proper response when a surgeon linguistically refers to a significant dimension of a subtle operation. What we learn from Collins, therefore, is that Dreyfus underestimates how much one can learn from linguistic socialization. But in granting this epistemological point to Collins, we nevertheless remain concerned about other dimensions of his discussion of interactional expertise.

Making Interactional Expertise More Interactive

In a reflective moment, Collins admits that the need to bolster the expertise of his discipline, sociology, was one motivation that prompted his reflections into interactional expertise: "The rhetorical trick was to persuade sociologists to reflect upon their expertises" (p. 125). Unfortunately, in our estimation, Collins remains too focused upon sociology, and more specifically, too preoccupied with the view that the sociologist ought to follow a principle of "meta-alternation."[5] When he classifies sociologists, journalists, science administrators, and activists as interactional experts, he overemphasizes their similarities, underplays their differences, and creates too restrictive a conceptual category. In fact, Collins comes close to recognizing this point himself when, after mentioning the domain of literary criticism, he asks: "In such domains is there a distinction between contributory and interactional expertise and what would it mean?" (p. 125). To clarify this point further, it will be useful to examine the manner in which he defines interactional expertise by negatively contrasting it with contributory expertise.

At various moments of his analysis, Collins suggests that the best way to understand what an interactional expert is, is to conceive of the interactional expert as occupying a middle-range on a continuum of knowledge and skill, somewhere between having no expertise and being a contributory expert: he defines interactional expertise as being "opposed"[6] to contributory expertise; he claims that contributory expertise is more difficult to attain than interactional expertise; he proposes a Turing test for identifying interactional experts in which they are judged by how well they can converse with contributory experts; and he suggests that while the interactional expert can take a devil's advocate position that makes "the conversational partner think hard about the science," it is only a contributory expert who can advance the knowledge in the field being discussed. While this description might hold true for the sociologist, it becomes problematic when we consider other groups that he classifies as being constituted by interactional experts, such as activists. Activists often develop interactional expertise to affect how a practice should be understood and conducted. They aim to make contributions in a field by means of the persuasive powers that their apprenticeship into linguistic socialization sometimes affords.

Rather than considering activists in the abstract, we will draw from Steven Epstein's work on AIDS activism. Epstein notes that during the early stages of the pandemic, before non-experts had any credibility within the relevant scientific communities, the activists "reconstituted themselves as a new species of expert—as laypeople who could speak credibly about science in dialogue with the scientific research community" (Epstein 2000, p. 20). The AIDS activists thus deserve to be considered interactional experts. In the absence of becoming full blown practitioners, they mastered the language of medical research and could pass Collins's Turing test: "Through a wide variety of methods— including attending scientific conferences, scrutinizing research protocols, and learning from sympathetic professionals both inside and outside the movement—the core treatment activists gained a working knowledge of the medical vocabulary" (Epstein 2000, p. 20). And yet, if we classified the activists as interactional in Collins's sense of the term, we would run the risk of overlooking exactly why it is that Epstein uses the phrase "activist-experts" to describe them (Epstein 2000, p. 16).

Epstein notes that the activists were able to bring about "changes in the epistemological practice of science" (Epstein 2000, p. 16). Consider the following claim:

> . . . activists yoked together methodological (or epistemological) arguments and moral (or political) arguments. . . . For example, activists insisted that the inclusion of women and people of color in clinical trials was not only morally necessary (to ensure equal access to potentially promising therapies) but was also scientifically advisable (to produce more fully generalizable data about drug safety and efficiency in different populations).
>
> (Epstein 2000, p. 21)

While Collins would have to classify the AIDS activists as interactional and not contributory experts, Epstein shows that they helped make scientific contributions by bringing their linguistic abilities to bear upon how research was conducted; thus, they are not, in a limited area, the epistemic inferiors of the contributory experts: "As Dr. John Phair (1994), a former chair of the Executive Committee of the AIDS Clinical Trials Group, commented in 1994: 'I would put them up against—in this limited area—many, many physicians, including physicians working in AIDS [care]'" (Epstein 2000, p. 20). Furthermore, if we appraised the expertise of the activists by measuring how well they could converse via a Turing test with contributory experts who had already made up their minds as to what counts as medical knowledge, we would miss the fact that their value as interactional experts lies precisely in their ability to 'interact' with contributory experts in a way that provides the latter with a new understanding of how to best contribute to the advancement of medical science. In this context, 'contributory' and 'interactional' expertise are not concepts that can be understood when arranged in a continuum from less to more knowledge: instead, the two terms are relational, the contributory achieves its identity from interacting with the interactional and the interactional achieves its identity from interacting with the contributory. As interactional experts, the AIDS activists were not—as Collins suggests interactional experts are prone to doing—trying to get scientists interested in them by conveying the thoughts of other scientists. They were invested in presenting their own ideas, ideas that became expressible because of their linguistic socialization, ideas that had significance due to epistemological as well as normative commitments.[7]

What this analysis suggests is that in at least some cases it is a mistake to place interactional and contributory expertise on a continuum, as if, in most interesting cases, an interactional expert is someone who is best understood as someone who 'almost' has contributory expertise. What Collins needs to do, in subsequent work, is articulate why he believes it is best to judge interactional expertise on the basis of the standard of contributory expertise when, in fact, the value of interactional expertise can lie precisely in its ability to provide reasons for the contributory expert to reassess his own practice. In some instances, the proof of interactional expertise will not be decided upon by how well someone can fool a contributory expert, but rather, it will emerge from a dialogue in which both interactional and contributory expert leave the conversation changed, acknowledging that some presupposition needs to be reexamined. Indeed, if disciplinary boundaries are fluid, then there are fluid ways of becoming expert in a field, and proof of expertise, of any stripe, will come from continuous and ever changing standards of negotiation. Thus, Collins errs when he allows himself to be guided by the question: "How much knowledge do you need to have in order to do the sociology of a scientific domain?" Because such a question achieves its intelligibility by asking for a quantified answer, it misrepresents the nature of academic discourse. To believe that recognition of being knowledgeable can be quantified in advance of a conversation is to provide an arbitrary standard, one that has the potential to impede the kinds of conversations that are possible.

In this context, it also becomes clear that Collins needs to be more precise about why it is that 'contribution' functions as the definitive marker that separates contributory from interactional experts. This precision can, in part, be gained by tying his account to the following kinds of questions. Can a nonexpert accidentally contribute to science? Does a patient who donates his body to science contribute to the advancement of science? Does a science fiction writer whose work inspires a contributory expert contribute to science? Does a lab manager who secures funding and supervises operations contribute to science? What questions such as these suggest is that the notion of 'contribution' is too vague to be of use to his account until it becomes reformed by further discussion of the difference between 'direct' and 'indirect' contributions.

Finally, it is worth noting that even if we did not raise all of the former objections, we would still find Collins's appeal to a Turing test problematic on at least two grounds. First of all, Collins presupposes that contributory experts are necessarily good conversationalists concerning the fields that they contribute to. In fact, despite characterizing being able to speak a language as a 'social skill,' Collins makes the rather strange claim:

> Thus, though Collins can do a reasonably good job of talking gravitational wave physics, even though he cannot do gravitational wave physics, he probably cannot talk it as well as most of those who do it. The scientists just practice talking it much more because they do it most of the time! Likewise, though Madeleine could acquire a pretty good ability to talk tennis talk even though she will never walk the tennis walk, she probably won't be as good a tennis talker as Martina Navratilova because Navratilova does a lot more tennis talking in the normal course of things.
>
> (p. 125)

This claim is problematic because it assumes that Martina Navratilova is as good a talker as she is a tennis player. But what evidence licenses this inference that "talking it" is a necessary consequence of "walking it"? Perhaps when it comes to expression and explanation of her abilities, she is no better than Madeleine. Indeed, communicative ability is no less a talent than playing tennis or doing gravitational wave physics; in some cases, it is even an area of expertise. While someone who never played a sport can be a good commentator (as per Collins's views on interactional expertise), every sports fan knows that many of the top players are rather inarticulate about the game that they excel in *qua* player. Dreyfus's philosophy of expertise suggests one reason why this might be the case. Whether experts really act intuitively and without following rules, or, whether Dreyfus's critics are right, and they act by following rules unconsciously, is moot here. The relevant phenomenon that Dreyfus captures is the gap between performance of a task and ability to speak about the task: "You probably know how to ride a bicycle. Does that mean you can formulate specific rules that would successfully teach someone else how to do it? How would you explain the difference between the feeling of falling slightly off balance when turning?" (Dreyfus and Dreyfus 1986, p. 16). In this passage, Dreyfus not only suggests that experts do not follow rules when they act *qua* experts, but, moreover, he indicates that they make immediate perceptual discriminations that are difficult to express linguistically. If Collins does not agree with Dreyfus on this matter, he owes us a more explicit rebuttal.

The second problem with the Turing test concerns Collins's presuppositions about what the test measures and how it is to be administered. By conceiving of the test as an ideal that allows us to abstract from the prejudices that attend to everyday processes of social recognition, Collins black-boxes the question of establishing the mindset of the interrogator, and correlatively, of thinking through the relation between the interrogator's prejudices and the validity of the test itself. This issue becomes more pronounced when we consider the fact that Collins seems to have two different ideals concerning the aim of the test, one adversarial and the other friendly: sometimes he describes the test as simply measuring the ability to hold a conversation with practitioners,[8] and at other times, he suggests that the test is to be conducted such that the interrogator is charged with the task of discovering who among the participants is not a genuine practitioner.[9] And once these points are granted, then we have to consider the possibility that a properly prepared novice might know all of the answers that the interrogator is looking for, and that the genuine practitioner might be mistaken for a novice if his theories contradict the theoretical commitments maintained by the interrogator.

Embodiment and Phenomenology

Collins first expresses interest in the topics of embodiment and the method of phenomenological analysis in the context of a debate with Dreyfus about artificial intelligence. Whereas Dreyfus argued that computers could not become intelligent and expert (in the same manner that humans are intelligent and expert) because they do not possess human bodies, Collins countered that the problem of socialization and not embodiment is more decisive for determining what a computer can and cannot do than Dreyfus acknowledges. During this exchange, an interpretative dispute over Madeleine, one of Oliver Sacks's patients, proved crucial. As will be evident shortly, the issue of how best to characterize Madeleine's relation to the human form of life remains highly relevant.[10]

Collins characterizes Madeleine in different terms in different essays. Sacks informs us that Madeleine is a "congenitally blind woman with cerebral palsy," who, for most of her life, experienced her hands as "useless godforsaken lumps of dough" (Sacks 1998, p. 59). In his previous work, Collins transformed this description to suit his analytical purposes by identifying Madeleine with his "minimal embodiment thesis": Madeleine is described as a 'brain' that is endowed with "some sensory inputs" (Collins 2000, p. 188) and as almost completely disembodied.[11] In his latest essay, Collins writes about Madeleine in a more cautious tone; he searches for the features that can best satisfy the conditions of minimal embodiment, rather than presupposing what they are. But we should not be misled by the appearance of congeniality; Collins remains committed to "embrainment" over "embodiment." Evidence for this commitment can be found in his claim that empathy and imagination do not require much of a body: "but the notion of interactional expertise suggests that empathy and imagination do not require a body, at least not much of one" (p. 125).

By highlighting the problems with this claim, we will be in a good position to show why Collins's account of Madeleine is problematic.

An account of empathy that runs contrary to Collins can be found in Evan Thompson's essay "Empathy and Consciousness." There, he writes:

> Affect has numerous dimensions that bind together virtually every aspect of the organism—the psychosomatic networks of the nervous system, immune system, and endocrine system; physiological changes in the autonomic nervous system, the limbic system, and the superior cortex; facial-motor changes and global differential motor readiness for withdrawal; subjective experience along a pleasure-displeasure valence axis; social signaling and coupling; and conscious evaluation and assessment. . . . Thus the affective mind isn't in the head, but in the whole body; and affective states are emergent in this reciprocal, co-determination sense: they arise from neural and somatic activity that itself is conditioned by the ongoing embodied awareness and action of the whole animal or person.
>
> (Thompson 2001, p. 4)

In order to provide a more concrete sense of what Thompson means when he claims that affect is experienced in the whole body, and not just in the head, all we need to do is consider how much bodily activity a human being, even someone as disabled as Madeleine, must initially experience in order to subsequently empathize with someone who is feeling scared.

Without a body that is capable of experiencing and coordinating complex physiological functions, the brain would be incapable of experiencing fear. Depending on the intensity of the experience, fear can affect the human body in many different ways: alertness increases, the pupils widen in order to let in more light, the adrenal glands begin to pump more adrenaline and other hormones into the bloodstream, the heart races, the muscles tense, the blood pressure rises, digestion slows, the liver converts starches to sugar to generate more energy, and sweat production increases, sometimes leaving the hair on our bodies standing on end. This range of bodily activity, including Madeleine's, is ultimately rooted in our shared evolutionary history, a history in which the survival value of the human experience of fear has been embedded in our ability to coordinate unconscious physiologic reactions with external demands. For example, if one experiences fear in connection with the perception of a predator in a particular location, the memory of this experience may unconsciously trigger a rapid heartbeat (as well as other signaling physiological responses) when one enters a similar environment, even though one may not immediately recognize the need to be vigilant in that environment. Evolution has not only demanded that the human organism be capable of recognizing real threats and responding appropriately to them, but also that human organism has the capacity to be aware of potential threats. What these considerations of the human experience of fear show, is that to empathize with someone who is fearful, a person must experience more than mere brain activity and even more than brain activity augmented by "minimal" bodily assistance. A phenomenologist would therefore find it acceptable to acknowledge that the brain serves an essential mediating role in conscious experience. However, these remarks on fear suggest that it is of crucial epistemological and ontological importance to avoid the reductive tendency of envisioning processes that are mediated by the brain as processes that are reducible to primarily brain activity.

Another problem with Collins's account is that he consistently conflates 'embodiment' with 'bodily features.' He acts as if it is possible to adequately discuss the relation between embodiment and language acquisition by locating the "minimal set of necessary features that enable language of any kind to be learned" (p. 125). By appealing to Madeleine, the Nunez story, and the prelingually deaf, Collins acts as if he has established that movable limbs, eye sight, and hearing are not necessary features that a human needs to possess to learn a language. And yet, while Collins's analysis of these examples may inform us about the relation between *specific bodily features* and the capacity for language acquisition, it does *not* at all address the matter of *embodiment* and *language acquisition*. The reason for this occlusion is that embodiment is ultimately a matter of the holistic interrelation (what Walter Freeman calls "circular causality") between an organism and its environment, and not a matter of agency being actualized from isolatable and discrete body parts—not even a brain. When Collins describes "the minimal bodily requirements that Dreyfus sets out—having an inside and outside, having a front and back, experience of moving, and so forth," he obscures what a phenomenologist such as Dreyfus means when he discusses the importance of human beings relating to the world in a manner in which the orientation of front and back is significant. Collins's statement creates the misleading impression that what Dreyfus is claiming is that without 'having' a front and a back, a being could not become linguistically socialized into our world. But, of course, computers 'have' fronts and backs (as well as insides and outsides, etc.) and, at least so far, they have not been capable of becoming linguistically socialized. If Dreyfus is right, this difficulty arises partially because of the fact that although computers literally 'have' these features, they do not live them the way that humans do. Indeed, even if, as per some science fiction scenario, some awareness of these features were programmed in, computers still wouldn't know and rely upon these features in the way that humans do. This point can be clarified by briefly touching upon the work of Samuel Todes, a theorist from whom Dreyfus derives much of his discussion of embodiment.

In *Body and World* Todes appeals to the familiar phenomenological distinction between the "body-as-object" and the "body-as-subject" to advance the claim that while humans can experience inert moments in which front and back and left and right "appear to be merely two sides of our body," such an experience is not typical of waking life, during which we do not merely notice, but actively "produce" the holistically unified spatial and temporal fields that allow objects to appear to us as significant. Dreyfus summarizes the import of Todes's argument as follows:

> To make his case that, in structuring the spatiotemporal field the body plays a fundamental role in unifying *all* human experience, Todes goes beyond Merleau-Ponty's account of the body as a pure 'I can' that responds to the world's solicitations. . . . Since the body moves forward more effectively than backwards, it opens a *horizontal field* that organizes experience into what can be coped with directly, what can be reached with effort, and what is over the perceptual horizon. Furthermore, the front/back asymmetry of the active body—the fact that it can cope well only with what is in front of it—organizes the temporal field. In everyday coping, what has yet to be faced is experienced as the future, what is currently being faced and dealt with makes up the pragmatic present, and what already has been faced and is behind us is experienced as both spatially and temporally past.
>
> (Todes 2001, pp. xix–xx)

Because Dreyfus characterizes Todes's position as applicable to 'all' human beings, its scope ought to extend to Madeleine; although disabled, she is, after all, human. But how can this be true? Isn't Todes's position based on movement, a capacity that Madeleine lacks? "Without our moving in it," Todes writes, "there would be no apparent spatiotemporal field in which objects might appear" (p. 49).

To justify locating Madeleine's sense of embodiment as one that is situated within our shared human spatiotemporal field, it is useful to consider Todes's analysis of someone who is imprisoned:

> Someone can be aware of his circumstances without being aware of as much movement as possible, e.g., if he is imprisoned. But then he is always privately aware of the restriction of his movement, and this restriction diminishes his sense of time. He drags out his days; time tends to be unreal; he tends to lose track of objective time because he tends to lose the practical subjective sense of time that gives import to the objective measurement of it. But so long as he retains some sense of his ability to move, if only within the confines of his cell, he retains some sense of time. Thus if any quadrant of this spatiotemporal field exists, it all exists. And if no quadrant exists, nothing can appear to us nor can we feel alive.
>
> (Todes 2001, p. 50)

It would be imprudent to deny that Madeleine has "some sense" of her ability to move, and this is because Collins errs in suggesting that Madeleine is "minimally embodied." Even if, for the sake of argument, we only take into account the standard five senses (sight, smell, hearing, taste and touch), we must immediately concede that Madeleine still has 4/5 of them. But how can we grant Madeleine a sense of touch when she could not initially move her arms and hands? The critical point to emphasize is that Madeleine has Cerebral Palsy; she is not a quadriplegic. What Madeleine suffers from is a condition that affects bodily movement and muscle coordination. Her sensory receptors are still intact, although this would not be the case if she suffered from a more severe form of paralysis in which the sensory network of touch is completed severed. Granted, the fingertips have a high concentration of nerve endings and are therefore an important source of sensation in able-bodied people. However, the sense of touch is actually distributed throughout the entire body, wherever nerve endings in the skin can transmit their sensations to the brain. Indeed, even the hairs on the skin play a relevant role here insofar as they can magnify sensitivity. In this context, it is important to note that each *square inch* of skin contains approximately:

78 nerves,
650 sweat glands,
19–20 blood vessels,
78 sensory apparatuses to detect heat,
13 sensory apparatuses to detect cold,
1,300 nerve endings to record pain,
19,500 sensory cells at the ends of nerve fibers,
160–165 pressure apparatuses for the sense of touch,
95–100 sebaceous glands,
65 hairs and hair muscles and
19,500,000 cells

(Texas Education Agency, Nd April 2004).

In the light of these considerations, Madeleine would most certainly be aware of any physical movement, either in her vicinity or to her person; she would be aware of self-ambulatory movement or movement imposed by an outside force.

To further clarify this point, it will be useful to consider what Madeleine would sense as her day begins. In doing so, it will become clear that Collins's ontological atomism fails to capture the holistic nature of embodied experience because his analysis of minimal embodiment is guided by the erroneous assumption that human experience is founded upon eliminable (or additive) body parts that lack essential unity.

Madeleine would wake up lying in bed, waiting for the routine to begin. She would certainly hear her caretaker walking down the hall, most likely already knowing who it was by the sound and speed of her footsteps. She would know when the caretaker had entered the room. She could feel the blankets being thrown off of her. She could hear the blinds being drawn open, and could possibly feel the warmth of the sun hitting her face, or hear rain, and from these experiences, know what the weather was outside. She would sense two arms coming underneath her and lifting her up and into a chair, then knowing that she was now sitting in a chair and no longer lying in bed. She could sense herself being pushed out of her bedroom through the passageways of the house, from one room to another, not only by the wind on the front of her face, but also by the change in sounds. She would be able to hear if others were there in the house. She could feel the warm water as she was washed, and smell the soap; she might enjoy the feeling of the towel as she was dried off, or she might feel discomfort and demand that they start using fabric softener. She would smell the coffee brewing and hear the sizzle of the frying pan, knowing that breakfast was almost ready and that the hunger pains she feels now would soon stop. She would certainly have a sense of her front from her back by the way in which she is carried, placed in a chair, moved from room to room, fed her meals and spoken to, given a hug or even kissed. All of these embodied experiences occur, and we are only one hour into the day! Thus, Madeleine may not have the benefit of experiencing all of her senses, but she is far from "minimally embodied" and certainly has a sense of her spatiotemporal field. According to Todes's analysis, once we accept that Madeleine has some sense of her ability to move as a human, we necessarily must accept that the entire quadrant of the spatiotemporal field exists for her.

Conclusion: Exorcising Collins's Demons

It would seem that in recent years, Laplace's famous demon has spawned children, and that these offspring have now arrived here, to baffle and confound Collins and his coterie of interactional experts. Their names are not of demonic origin, but are instead far more prosaic: "properly conducted Turing Test," "minimally embodied," "contribution," and "domains." These demons need to be vanquished, which is to say, more rigorously circumscribed, before we can proceed and ever hope to apply the idea of the interactional expert in any practical sense.

"Laplace's genius in solving problems," John Collier suggests, "was matched by his brilliance in selecting problems that are tractable to his methodology" (Collier 2002). Indeed, much of his success "comes from looking only where the light of the methodology shines brightly" (Collier 2002). As with Laplace's theory of causal determinism, so too with Collins's current theory of interactional expertise—both accounts seem to work only when one grants unsubstantiated assumptions that are tethered to idealized scenarios. Once these conditions are interrogated, rather than presumed, the account begins to lose coherence. If and when the day arrives that we can all agree upon distinct parameters for these demons, they will then, but only then, be properly exorcised. The question is: can that day ever arrive? For now, it seems that the best way to think about the interactional expert is to note that he does most certainly exist, but in a state much like a quantum particle. Heisenberg's uncertainty principle suggests that the more accurately one tries to measure the position of a particle, the less accurately its speed can be measured. So too with the interactional expert—the more accurately we try to measure his position on a continuum of knowledge, the less accurately we can measure the relevant testing variables.

The most tenacious of all of the aforementioned demon siblings is the one called "minimally embodied." Collins suggests that a girl born and raised in the weightless conditions of outer space, "Weightless Wanda," would be able to say everything about weight that could be said. That may very well be true, if we assume that a crew of people who knew what the experience of gravity is like raised Wanda (as per Collins's "collective embodiment thesis"). These ex-earthlings would provide for Wanda something like a gravity society. They could explain and teach Wanda the concepts of 'gravity' and 'weight' by filling her imagination with stories, similes and metaphors. But for every parable they told to Wanda, it would only have as much meaning as the vocabulary that they used, which in this case, would frequently (if not always) refer to some dimension of a shared physical embodiment. The experiences or awareness of some of the senses would have to be drawn upon to "fill in the blanks" of Wanda's lack of gravitational understanding. The most obvious of these senses would be sight. Because of the specifics that define our own physical immersion on this planet (i.e., our scale in it, our evolved upright posture, etc.), we find ourselves oriented towards a horizon line, perceiving everything as up or down in relation to a perceptual base. Even in space, where the ideas of 'up' and 'down' become completely relational, the ex-earth crew would still have many of the habits of orienting themselves to what was considered up and what was down, as for example, when the space craft stood still, waiting for launch on *terra firma*. Wanda would soon find herself adopting these habits and expressions, even if they were of no real meaning to her. If, on the return trip, her family decided it best to prepare Wanda for what to expect after splashdown, all of the explanations of gravity would be in relation to an artificial or viewable horizon line.

But now, let's consider a different example, imagining that it was no ordinary child born on that spaceship, but instead, that it was Madeleine herself that grew to the age of eight in the great expanse. Madeleine, or better yet,

Space Traveler Madeleine, would be blind and unable to use her hands, but further, she would have never felt the pressure of her own weight. Indeed, Space Traveler Madeleine will have never felt tethered to a two-dimensional surface. Without her sight, she would be unable to see a horizon line, artificial or otherwise, to orient herself to. It thus seems unlikely to expect that she would be able to understand the crew's stories and explanation of gravity.[12] When Space Traveler Madeleine "dropped" something, it would not fall down, but instead, it would fall away, if it even moved at all. Everything would come to her or leave from her in a 360-degree sphere that is not bisected by a horizon. In this context, Space Traveler Madeleine begins to better approximate a genuine sense of minimally embodied; she is not only without sight, but furthermore, a large portion of her "touch" sensory matrix has been taken away. Even with just this slight difference, we can begin to see the many difficulties that would arise when trying to teach Madeleine to engage in gravity discourse. We believe that if Space Traveler Madeleine were given a Turing test before re-entry that focused on gravity, even after the crew tried to linguistically socialize her, she would most certainly fail. Of course, we may be wrong. But if we are, the burden of proof falls upon Collins to explain how, considering Space Traveler Madeleine's minimally embodied relation to her environment, she might succeed.

In light of how difficult we anticipate it being for Collins to respond to all of our objections, including the Space Traveler Madeleine scenario, we contend that if he intends for his analysis to appeal to a pragmatist, then he needs to more rigorously articulate the nature and scope of interactional expertise. As it currently stands, Collins's account is too vague; inadequately circumscribed borders bound the epistemological region that he focuses upon. In the process of describing his investigation as a work in progress, he plants too many escape routes. As our reference to Laplace suggests, Collins fails to clearly inform us what evidence could contradict his claims, therein potentially presenting us with a concept that can always be validated against counterargument through *ad hoc* stipulation. Additionally, by reifying the concept of interactional expertise so as to create epistemological equivalence between activists, sociologists, critics, journalists, and some science administrators, he potentially undermines the value that a more rigorously construed concept might have. He mistakes a paradigm case that might fall within the parameters of a carefully defined concept of interactional expertise—the knowledge that the sociologist of science typically possesses—for a justifying mechanism that can be appealed to, to set the parameters for defining interactional expertise. Consequently, he miscalculates the value of his proposed Turing-test criteria for identifying an interactional expert. While it may make sense to deem the sociologist of science an interactional expert only if he can pass a Turing-test with contributory experts in the field of science that he is studying, such a standard is less valuable, and in some cases inapplicable, when we consider critically some of the other groups that Collins identifies with interactional expertise, particularly activists. In short, although Collins has good reasons for advocating a Turing test—at the most basic level, it appears to enable us to transcend the prejudices that attend to "real" processes of social recognition— his choice, nevertheless, belies an alternative agenda: defending sociological practice as a practice of "meta-alternation."

Collins could, of course, maintain that we are missing the foundational nature of his project. He could insist that his goal is modest: initially present the general outlines of the concept of interactional expertise, subsequently refine its parameters, and then, finally discern how it can be usefully applied. This rationale, however, would also prove to be problematic. As we have already discussed, when Collins defines interactional expertise, he not only investigates what an interactional expert knows and how an interactional expert acquires knowledge, but he also provides an argument that one need only be minimally embodied to become an interactional expert. As our phenomenological analysis demonstrates, Collins relies upon a fundamentally flawed understanding of what it means to be embodied: (1) he mistakes discrete and isolatable body parts for the genuine experience of embodiment, an experience that is best understood as an ecological relation between an organism and its environment; (2) he overestimates the amount of embodiment that is lost when a particular capacity for sensory experience is inoperative; and (3) he overvalues the cognitive abilities of the brain, by mistaking processes that are mediated by the brain for processes that are reducible to primarily brain activity. The analogy that he wishes to draw between interactional expertise and the bodily comportment of a severely disabled woman named Madeleine is simply untenable. Moreover, by couching his explanation of interactional expertise in minimally embodied terms he renders the possibility of generating the requisite explanatory power (i.e. explaining how interactional expertise is acquired and who can acquire it) more difficult than it needs to be.

Ultimately, Collins's misunderstanding of interactional expertise can be attributed to three mistakes:

- He misunderstands the nature of embodiment.
- He unduly limits the contributory potential of interactional expertise.
- He misunderstands the relation between embodiment and interactional expertise.

Regarding the last point, Collins has good reasons for trying to understand how embodied an interactional expert must be. He may be concerned that some STS theorists, such as Michel Callon and Bruno Latour, have gone too far in postulating symmetry between humans and other kinds of "actants." In this respect, it is understandable that Collins might want to establish that scallops, speed bumps, and bacteria cannot be interactional experts. However, even if Collins did not have a flawed, ontologically atomistic understanding of embodiment, and even if his analysis did not depend upon fictional contrivances that are wholly unrelated to embodied experience, his account of interactional expertise would still remain problematic.

NOTES

1. Good philosophical partners matter as well. This paper could not have been written without helpful pointers from: Casper Bruun Jensen, Robert Crease, Timothy Engström, Shaun Gallagher, Lisa Hermsen, Don Ihde, John Sanders, Brian Schroeder, Robert Scharff, Noreen Selinger, and Evan's Philosophy of Science class (Winter 2003–04). Throughout the process, Collins proved to be the ideal interlocutor: insightful, witty, and gracious.

2. For critiques of Paul Feyerabend and Steve Fuller's views on expertise, see Selinger (2003a) and (2003b).

3. In this context, Collins makes a genuine contribution to the field of Science and Technology Studies (henceforth, STS). Collins characterizes his project of constructing a new conceptual category, interactional expertise, as a scholarly ambition that was prompted by reflection upon the epistemological and normative limits of "second wave" STS. Since 'expertise' is a term that is not only associated with knowledge, skill, and authority, but also with hierarchy—elitism, paternalism, and power—the deconstructive project that typifies "wave two" initially served a valuable end. What "wave two" established, through empirical means, is that: techno-science *qua* practice should be understood primarily as a social activity; and it is not possible for theorists to ascertain clear and unchanging demarcation criterion that separates constitutive (internal) and contextual (external) techno-scientific values. The problem with the "wave two" movement, however, is that while some of its studies provided compelling evidence that the basis of technical decision-making should be widened beyond the core of certified experts, the historical and sociological orientation of STS practitioners did not facilitate adequate reflection upon conceptual and social-political matters. Specifically, "wave two" theorists never provided a satisfactory answer to the questions: how much stakeholder participation is justified, and in which dimensions of techno-scientific culture? Furthermore, by pursuing a deconstructive strategy, "wave two" theorists ran the risk of overstating their case. They frequently failed to treat expertise as *real*, as a relation to skill that is marred epistemologically when exhaustively reduced to shifting historical perspectives and competing social attributions. As Collins and his collaborator Robert Evans astutely observe: "By emphasizing the ways in which scientific knowledge is like other forms of knowledge, sociologists have become uncertain about how to speak about what makes it different; in the same way they have become unable to distinguish between experts and non-experts" (Collins and Evans 2002, p. 238). In light of these considerations, Collins is entirely justified in calling for STS theorists to enter into a "third wave" of research. He deserves praise for (1) attempting to establish a criterion that distinguishes experts from non-experts and (2) attempting to discern if any type of expertise exists that occupies the middle range of the expert/non-expert continuum.

4. Although Collins does not directly criticize Dreyfus's evaluation of commentators, the limits of Dreyfus's phenomenological approach are immediately recognizable when his assessment of the matter is juxtaposed with Collins's views on interactional expertise. Dreyfus's judgment of commentators is succinctly encapsulated in the following passage, a passage in which he suggests that full immersion into a form of life is a necessary condition for the acquisition of any type of expertise pertaining to a field: "Listening to . . . commentators, who take up at least half the time on erudite talk shows, is like listening to articulate chess kibitzers, who have an opinion on every move, and an array of principles to invoke, but who have not committed themselves to the stress and risks of tournament chess and so have no expertise" (Dreyfus, Spinosa and Flores 1997, p. 87). Dreyfus associates the knowledge of commentators or, more pejoratively put, "kibitzers," with the "idle talk" definitive of the "public sphere" that "undermines commitment stemming from practical rationality" (Dreyfus, Spinosa and Flores 1997, p. 86). Dreyfus is obligated to maintain this position because of the binary opposition that runs throughout his work, an opposition between tacit and prepositional knowledge. Associating tacit knowledge with contextually sensitive, intuitive perceptions, Dreyfus views the linguistic claims that commentators proffer as propositional bits of information; they are "abstract solutions" and "anonymous principles" that fail to display "wisdom" because they fail to originate in the context of an engaged commitment to some local practice (Dreyfus, Spinosa and Flores 1997, p. 87). Hence, for Dreyfus, those who have "not committed themselves to the stress and risks of tournament chess" have "no expertise," not even, it would seem, the expertise that others would find befitting of a chess commentator, critic, or analyst. For more on this problem, see Selinger and Crease 2002.

5. In "Epistemological Chicken," Collins and Steven Yearly articulate a view of "meta-alternation" in which the sociologist allows himself to reduce natural explanations to social explanations for methodological reasons, although, in the end, allowing scientists to continue practicing as naïve realists.

6. Collins writes: "What you get from immersing yourself in the linguistic culture pertaining to a practical domain rather than the practice itself is what I call 'interactional expertise.' This is opposed to 'contributory expertise'—which is what you have if you immerse yourself in the culture in a full-blown way" (p. 125).

7. Collins writes: "Eventually the scientists will become interested in what you know, not as a scientist in your own right, but as a person who is able to convey the scientific thoughts and activities of others" (p. 125).

8. Collins writes: "We tend to believe that only those who, as we might be inclined to put it, 'share the form of life' of the relevant activity would be able to understand it fully. We then argue that if we were not accomplished practitioners our lack of understanding could be revealed if it was subject to the right kind of test. If, say, we were placed in a Turing-Test-like situation, where our ability to hold a discussion in some domain in which we were a novice were tested against the ability of a genuine practitioner in that domain, we would be correctly identified as a novice (if the judge, who must also be an accomplished practitioner in the domain, were to ask the right questions)" (p. 125).

9. Collins writes: "Collins (1990) argued that a spy pretending to be a native of a city which he or she had never visited would always be caught out by a native of the city if the native asked the right questions. Collins did not take into account the possibility that the spy would learn all there was to say about the city just by talking long enough to other natives

of the city. Collins and Kusch's (1998, p. 34) account of love-letter writing is better because a distinction is drawn between knowing the emotion of love and being able to discourse successfully in the language of love as a result of immersion in a society in which the idea of love is prevalent" (p. 125).

10. For more on the topic of Madeleine, see Selinger 2003c.

11. Collins writes: "Many mistakes are made through failing to distinguish between individual competence and embedding within society. For example, those who are unable to use their bodies may still come to understand concepts that are related to the style of embodiment of the majority of the members of the embedding society" (Collins and Kusch 1998, p. 94, Note 4). Since this passage refers back to Collins's review of *What Computers Still Can't Do*, it is clear that he is discussing Madeleine. Without using qualifying words, such as 'fully' or 'easily,' Collins's use of the phrase "unable to use" suggests that Madeleine is completely unable to use her body. This, of course, is not true. Madeleine can use her body, albeit in a limited and disabled way.

12. Clearly, our scenario, like the ones before it, makes certain narrative assumptions that could be revised—although such revision would only point towards the centrality of our claims about embodiment. For example, if Space Traveler Madeleine were tethered to a bed, then she could likely come to understand and talk about gravity because she would have an embodied orientation towards a two-dimensional plane. (Although, it remains questionable as to whether this orientation is comparably analogous to experiencing gravity, a scenario in which one's entire body is oriented towards an invariant horizon. After all, if one is incompletely tethered to a rope that straps one in, one's limbs could still float up and away, etc.) Indeed, it seems that any attempt to further refine the scenario would entail additional mimicry or analogues of embodied experience. What this example shows, therefore, is that Madeleine as she currently is, is far from minimally embodied, and that it takes more imagination than Collins acknowledges to even begin to approach something like "minimal embodiment."

REFERENCES

Collins, H. M. 1990. *Artificial Experts: Social Knowledge and Intelligent Machines.* Cambridge: MIT Press.

Collins, H. M. 1992. Dreyfus, forms of life, and a simple test for machine intelligence. *Social Studies of Science* 22: 726–739.

Collins, H. M. 1996. Embedded or embodied? A review of Hubert Dreyfus' *What Computers Still Can't Do. Artificial Intelligence* 80: 99–117.

Collins, H. M. 2000. Four kinds of knowledge, two (or maybe three) kinds of embodiment, and the question of artificial intelligence. In: M. Wrathall and J. Malpas (eds), *Heidegger, Coping and Cognitive Science: Essays in Honor of Hubert L. Dreyfus*, vol. 2, pp. 179–195. Cambridge: MIT Press.

Collins, H. M. Interactional expertise as a third kind of knowledge. *Phenomenology and Cognitive Science* 3: 125–143.

Collins, H. M. and Evans, R. 2002. The third wave of science studies: Studies of expertise and experience. *Social Studies of Science* 32: 235–296.

Collins, H. M. and Kusch, M. 1998. *The Shape of Actions: What Humans and Machines Can Do.* Massachusetts: MIT Press.

Collins, H. M. and Yearley, S. 1992. Epistemological chicken. In: A. Pickering (ed), *Science as Practice and Culture*, Chicago: University of Chicago Press.

Collier, J. 2002. Holism and emergence: dynamical complexity beats laplace's demon. http://www.nu.ac.za/undphil/collier/papers/Vienna-Laplace.PDF.

Dreyfus, H. 1992a. *What Computers Still Can't Do*. Massachusetts: MIT Press.

Dreyfus, H. 1992b. Response to Collins, *Artificial Experts. Social Studies of Science* 22: 717–726.

Dreyfus, H. 2000a. Response to Carmen Taylor. In: M. Wrathall and J. Malpas (eds), *Heidegger, Authenticity, and Modernity: Essays in Honor of Hubert L. Dreyfus*, vol. 1, pp. 306–312. Cambridge: MIT Press.

Dreyfus, H. 2000b. Response to Collins. In: M. Wrathall and J. Malpas (eds), *Heidegger, Coping and Cognitive Science: Essays in Honor of Hubert L. Dreyfus*, vol. 2, pp. 345–346. Cambridge: MIT Press.

Dreyfus, H. 2001. *On the Internet*. New York: Routledge.

Dreyfus, H. and Dreyfus, S. 1986. *Mind Over Machine: The Power of Human Intuition and Expertise in the Era of the Computer*. New York: Free Press.

Dreyfus, H., Spinosa, C. and Flores, F. 1997. *Disclosing New Worlds: Entrepreneurship, Democratic Action, and the Cultivation of Solidarity*. Massachusetts: MIT Press.

Epstein, S. 2000. Democracy, expertise, and AIDS treatment activism. In: D. Kleinman (ed), *Science, Technology, and Democracy*, pp. 15–32. New York: SUNY Press.

Sacks, O. 1998. *The Man Who Mistook His Wife for a Hat and Other Clinical Tales*. New York Simon and Schuster.

Sanders, J. 1997. An ontology of affordances. *Ecological Psychology* 9(1): 97–112.

Selinger, E. 2003a. Feyerabend's democratic argument against experts. *Critical Review* 15(3–4): 359–373.

Selinger, E. 2003b. Expertise and public ignorance. *Critical Review* 15(3–4): 375–385.

Selinger, E. 2003c. The necessity of embodiment: The Dreyfus-Collins debate. *Philosophy Today* 57(3): 266–279.

Selinger, E. and Crease, R. 2002. Dreyfus on expertise: The limits of phenomenological analysis. *Continental Philosophy Review* 35: 245–279.

Texas Education Agency. Nd. Skin diseases and disorders. *Health Science Technology Curriculum*. (http://www.texashste.com/html/ap_inte.htm), April 6, 2004.

Thompson, E. 2001. Empathy and consciousness. *Journal of Consciousness Studies* 8 (5–7): 1–32.

Todes, S. 2001. *Body and World*. Massachusetts: MIT Press.

PART III
CONTESTING EXPERTISE

A GROWING NUMBER of theorists and activists have expressed concern that society is endangered by its experts. Unfortunately, it is difficult to generate sound arguments that justify contesting expertise. Undue psychological and political sentiments often compromise epistemic and normative analyses. The essays in this section all confront this issue, and in the cases where theorists unwittingly foster the very problems that they intend to redress, we learn something new about the pitfalls that attend to this process.

Paul Feyerabend made it possible for philosophers of science to write as social critics. Unlike Thomas Kuhn and Imre Lakatos, who both discussed the social dimensions of science in a primarily epistemological context, Feyerabend characterizes his seminal work, *Against Method*, as "humanitarian"; it is a text devoted to "supporting people," not advancing knowledge. Within this normative context, Feyerabend advances the following thesis: modern scientific experts have become ideologues; the more time and energy they devote to advancing a position, the more difficult it becomes for them to be open-minded regarding points of view that call their core beliefs into question. In response to this problem, Feyerabend contends that experts ought to be regarded first and foremost as public servants; duly elected committees of laypeople should regulate all scientific research that can affect the public sphere. In the essay contained in this section, "How to Defend Society Against Science," he suggests that if experts did not distort the value of their achievements, nonexperts would come to realize that they are better suited to contribute to the pursuit of knowledge than they had believed previously. When assessing Feyerabend's position, the reader might want to consider whether his vision of democracy drives him to dichotomize the world into two categories of people: experts and nonexperts. If his account is that restrictive, it might foreclose the possibility that far from being a unified group that shares a common aptitude for regulating and criticizing expert advice, nonexperts are indeed a disparate lot with a variety of background skills.

Perhaps the most radical contemporary outlook on expertise can be reconstructed from Steve Fuller's essay, "The Constitutively Social Character of Expertise." Fuller appears to contend that experts are granted too much social authority, even when it comes to the matters that they specialize in. This position is tied to a social view of expertise, one that purports that everything significant about expertise can be specified when a social field is circumscribed. What Fuller experiments with, therefore, are the normative and epistemological implications that might follow were we to focus our attention on how experts create, maintain, and constantly reinforce an interface in which their claims to cognitive authority are bolstered through networking and rhetorical persuasion. Attention to this detail would reveal instances in which the client's perceived need for an expert's knowledge or skills turns out to be a manufactured desire, which is to say, a goal created by an expert who wants his or her services to be perceived as necessarily beneficial. If Fuller's account of discretionary power is accurate, then the prestige and deference accorded to experts such as physicians, meteorologists, economists, and scientists from every field might need to be tempered.

In "Epistemic Dependence," John Hardwig provides us with reasons to believe that Fuller holds an incorrect view on expertise. Hardwig contends:

> The rational layman will recognize that in matters about which there is good reason to believe that there is expert opinion, he ought (methodologically) not to make up his own mind. His stance on these matters will—if he is rational—usually be rational deference to the epistemic authority of the expert.

Far from granting Hardwig's point, Fuller insists that Hardwig's position concedes too much authority to experts. In *Social Epistemology*, Fuller characterizes the type of argument that Hardwig proffers as the "authoritarian theory of knowledge." According to Fuller, the underlying logic of Hardwig's position can be expressed as follows:

1) The ordinary individual or "layman" holds more beliefs than he could reasonably be expected to have the relevant evidence for.
2) However, in most cases, the layman knows of some other individual, an "expert," who through specialized training has acquired the evidence needed for rationally holding the belief.
3) From (1) and (2) we can conclude either (a) that most of the layman's beliefs are irrationally held, or (b) that it is rational for the layman to hold a belief if someone he recognizes as an expert has evidence for rationally holding it.
4) Since only (3b) saves the intuition that most of our beliefs are rationally held, it follows that the layman is "epistemically dependent" on the authority of experts for all but the beliefs on which he is an expert.
5) Therefore, for most epistemic judgments, it is less rational to "think for oneself" than to defer to the authority of the relevant expert.

We leave it to the reader to decide whether Fuller's reconstruction of Hardwig's argument is reasonable. For example, there might be a significant difference between Hardwig's actual claims about when it is appropriate and rational for a nonexpert to defer to an expert and Fuller's interpretation of the logic supporting Hardwig's account.

The dangers of expertise that Feyerabend and Fuller highlight are subject to further analysis by Edward Said, a thinker who dealt extensively with matters concerning the relation between "professionals" and "amateurs" and "insiders" and "outsiders" throughout his career. In "Opponents, Audiences, Constituencies, and Community," Said attempts to explain precisely how "humanistic culture in general has acted in tacit compliance" with the "antidemocratic view" of the "general public" that is promoted by what he calls the "cult of expertise and professionalism." By linking the legacy of Anglo-American New Criticism to the *ethos* of the modern university, Said indicts those who embrace the "comforts of specialized habits" and cede "objective representation (hence power) of the world to a small coterie of experts and their clients."

Finally, as a means of repositioning critically the views of Feyerabend, Fuller, Hardwig, and Said, we close this section with Don Ihde's "Why Not Science Critics?" Ihde contends that it is a mistake to analyze the problem of expertise from the perspective of an expert-lay divide, as this absolute dichotomy conceals intermediary proficiencies such as the "well-informed amateur." At face value, it might seem problematic to rehabilitate the term "amateur," even by adding a special qualification. For example, celebrity activists who endorse (or challenge) particular scientific platforms can endanger society, particularly when their celebrity status renders them recipients of star-struck adulation. Ihde, however, insists that the well-informed amateur is a "lover" of a particular subject, someone who is neither an "insider," since he or she lacks the official training that credentializes one within expert cultures, nor a total "outsider," since he or she is not completely ignorant of the operational norms governing a particular expert culture. The intuitive problem with such an intermediary position is that it sounds too distanced from expert cultures to warrant careful consideration from them. According to Ihde, such an objection is unwarranted. He contends that because it minimizes bias, distance is a useful—not harmful—feature; it is the central concept to which we can appeal for contesting expertise.

I FIND MYSELF believing all sorts of things for which I do not possess evidence: that smoking cigarettes causes lung cancer, that my car keeps stalling because the carburetor needs to be rebuilt, that mass media threaten democracy, that slums cause emotional disorders, that my irregular heart beat is premature ventricular contraction, that students' grades are not correlated with success in the nonacademic world, that nuclear power plants are not safe (enough) . . . The list of things I believe, though I have no evidence for the truth of them, is, if not infinite, virtually endless. And I am finite. Though I can readily imagine what I would have to do to obtain the evidence that would support any one of my beliefs, I cannot imagine being able to do this for *all* of my beliefs. I believe too much; there is too much relevant evidence (much of it available only after extensive, specialized training); intellect is too small and life too short.

What are we as epistemologists to say about all these beliefs? If I, without the available evidence, nevertheless believe a proposition, are my belief and I in that belief necessarily irrational or nonrational? Is my belief then *mere* belief (Plato's right opinion)? If not, why not? Are there other good reasons for believing propositions, reasons which do not reduce to having evidence for the truth of those propositions? What would these reasons look like?

In this paper I want to consider the idea of intellectual authority, particularly that of experts. I want to explore the "logic" or epistemic structure of an appeal to intellectual authority and the way in which such an appeal constitutes justification for believing and knowing. I have divided the paper into three parts. In the first, I argue that one can have good reasons for believing a proposition if one has good reasons to believe that *others* have good reasons to believe it and that, consequently, there is a kind of good reason for believing which does not constitute evidence for the truth of the proposition. In the second, I urge that because the layman is the epistemic inferior of the expert (in matters in which the expert is expert), rationality sometimes consists in refusing to think for oneself. In the third, I apply the results of these considerations to the concept of knowledge and argue that the expert-layman relationship is essential to the scientific and scholarly pursuit of knowledge.

If I am correct, appeals to epistemic authority are essentially ingredient in much of our knowledge. Appeals to the authority of experts often provide justification for claims to know, as well as grounding rational belief. At the same time, however, the epistemic superiority of the expert to the layman implies rational authority over the layman, undermining the intellectual autonomy of the individual and forcing a reexamination of our notion of rationality. The epistemic individualism implicit in many of our epistemologies is thus called into question, with important implications for how we understand knowledge and the knower, as well as for our conception of rationality.

I.

Restricting ourselves—here and throughout the paper—to belief in and knowledge of propositions for which there is evidence, let us suppose that there are good reasons to believe a proposition—that p. What kinds of things can be good reasons to believe that p? The usual answer to this question is in terms of evidence, "evidence" being defined roughly as anything that counts toward establishing the truth of p (i.e., sound arguments as well as factual information). There is evidence, then, for the truth of p, but it does not follow that everyone has or even can have this evidence.

Suppose that person A has good reasons—evidence—for believing that p, but a second person, B, does not. In *this* sense B has no (or insufficient) reasons to believe that p. However, suppose also that B has good reasons to believe that A has good reasons to believe that p. Does B then, *ipso facto*, have good reasons to believe that p? If so, B's belief is epistemically grounded in an appeal to the authority of A and A's belief. And, if we accept this, we will be able to explain how B's belief can be more than mere belief; how it can, indeed, be rational belief; and how B can be rational in his belief that p. And our problems will be solved . . . or only starting.

Only starting because we are then faced with the prospect, not formerly considered by epistemologists, of a very odd kind of good reason for belief: a reason that does *not* constitute evidence for the truth of p. For B's reasons for believing that p are not evidence for the truth of p. We can see this by noting two things. (1) Although A's evidence counts toward establishing the truth of p, the case for p is not stronger after B discovers that A has this evidence than it was before B found out about A and A's reasons. (2) The chain of appeals to authority must end somewhere, and, if the whole chain of appeals is to be epistemically sound, it must end with someone who possesses the necessary evidence, since truth claims cannot be established by an appeal to authority, nor by investigating what other people believe about them.[1]

But B must have some good reasons to support his belief that p, or that belief would be *mere* belief (Plato's right opinion, again). B does have good reasons, all right; in fact, he has evidence. But his evidence doesn't count toward establishing the truth of p; it counts only toward establishing that A (unlike B himself) "knows what he's talking about" when he says that p. How can B have good reasons to believe that A has good reasons to believe that p when B does not himself have evidence that p? It's easy—B has good reasons to believe that A has conducted the inquiry necessary to have evidence for believing that p.

If the necessary inquiry is simple enough, B's belief that p can be grounded in A's despite the fact that we would not call A an expert. For example, if the service station attendant who checks my oil informs me that it's OK, I would believe him, but I would not call him an expert. However, the more epistemologically interesting cases are those in which expertise is involved—cases in which B has good reason to believe that A is an expert about whether or not p, in consequence of inquiry that has been sustained, prolonged, and systematic.[2]

The layman's appeal to the intellectual authority of the expert, his epistemic dependence on the expert, and his intellectual inferiority to the expert (in matters on which the expert is expert) are all expressed by the formula with which we have been working: B has good reasons to believe that A has good reasons to believe that p. But the layman's epistemic inferiority and dependence can be even more radical—in many such cases, extensive training and special competence may be necessary before B could conduct the necessary inquiry. And, lacking this training and competence, B may not be able to understand A's reasons, or, even if he does understand them, he may not be able to appreciate why they are good reasons.

Michael Polanyi and Harry Prosch[3] put the first part of this point dramatically, taking their examples from the physical sciences:

> The popular conception of science says that science is a collection of observable facts that anybody can verify for himself. We have seen this is not true in the case of expert knowledge, like that needed in diagnosing a disease. Moreover, it is not true in the physical sciences. In the first place, for instance, a layman cannot possibly get hold of the equipment for testing a statement of fact in astronomy or in chemistry. Even supposing that he could somehow get the use of an observatory or a chemical laboratory, he would not know how to use the instruments he found there and might very well damage them beyond repair before he had ever made a single observation; and if he should succeed in carrying out an observation to check up on a statement of science and found a result that contradicted it, he could rightly assume that he had made a mistake, as students do in a laboratory when they are learning to use its equipment.
>
> (184/5)

Moreover, the training and resultant competence to conduct the required inquiry are often accessible only to those with certain talents and abilities. Consequently, B might not *ever* be able to obtain the evidence that supports his belief that *p*. If my own desperate and losing struggle with freshman calculus is a reliable indicator, I might *never* be able to obtain the evidence for my belief that relativity physics is correct, no matter how much time and effort I devoted to the enterprise. I may simply lack the mathematical ability to possess that evidence.

But extensive training and special competence may be necessary before one can assess or even understand the expert's reasons for believing that *p*. Although I might be able to understand studies about the impact of mass media on voters, I am not competent to assess the merits of those studies, unversed as I am in the issues surrounding social science research methods. And, lacking the requisite mathematical training and ability, I cannot even read the books and articles that support my belief that relativity physics is correct.

If, then, layman *B* (1) has not performed the inquiry that would provide the evidence for his belief that *p*, (2) is not competent, and perhaps could not even become competent, to perform that inquiry, (3) is not able to assess the merits of the evidence provided by expert *A*'s inquiry, and (4) may not even be able to understand the evidence and how it supports *A*'s belief that *p*, can *B* nonetheless have good reasons to believe that *A* has good reasons to believe that *p*? I think he can. If so, should we conclude that *B*'s belief that *p* is rationally justified? I think we should, acknowledging that *B*'s belief stands on better epistemic ground than other beliefs which we would call simply irrational or nonrational.

Many epistemologists may be tempted to reject this conclusion because it is so divergent from our received view about the nature of rational belief. But I think we *must* say that *B*'s belief is rationally justified—even if he does not know or understand what *A*'s reasons are—if we do not wish to be forced to conclude that a very large percentage of beliefs in any complex culture are simply and unavoidably irrational or nonrational. For, in such cultures, more is known that is relevant to the truth of one's beliefs than anyone could know by himself. And surely it would be paradoxical for epistemologists to maintain that the more that is known in a culture, the less rational the beliefs of individuals in that culture.

II.

Nevertheless, acceptance of epistemological individualism dies hard. It may well resurface in the guise of a suggestion about the appropriate stance of a responsible and rational layman in relation to the expert. If I am not presently in a position to know what the expert's good reasons for believing that *p* are or to understand why these are good reasons, I am obviously in no position to check the accuracy of what he tells me. What stance should I then take? A plausible and tempting suggestion is that, if I think I have the required ability, I should become informed so that I can assess the reliability of the expert's reports and thus escape my dependence on him and regain my intellectual autonomy.

The idea behind this suggestion lies at the heart of one model of what it means to be an intellectually responsible and rational person, a model which is nicely captured by Kant's statement that one of the three basic rules or maxims for avoiding error in thinking is to "think *for oneself*."[4] This is, I think, an extremely pervasive model of rationality—it underlies Descartes' methodological doubt; it is implicit in most epistemologies; it colors the way we have thought about knowledge. On this view, the very core of rationality consists in preserving and adhering to one's own independent judgment; for how can one be sure one is being informed, not misinformed, if one suspends judgment?

But I submit that this model provides us with a romantic ideal which is thoroughly unrealistic and which, in practice, results in less rational belief and judgment. I could, indeed, escape epistemic dependence on *some* experts; perhaps, if I am talented enough, I could escape reliance on any given expert. I can and do choose where to establish my intellectual autonomy. But if I were to pursue epistemic autonomy across the board, I would succeed only in holding relatively uninformed, unreliable, crude, untested, and therefore *irrational* beliefs. If I would be rational, I can never avoid some epistemic dependence on experts, owing to the fact that I believe more than I can become fully informed about.

Once more, then: if I am not in a position to know what the expert's good reasons for believing that *p* are and why these are good reasons, what stance should I take in relation to the expert? If I do not know these things, I am also in no position to determine whether the person really is an expert. By asking the right questions, I might be able to spot a few quacks, phonies, or incompetents, but only the more obvious ones. For example, I may suspect that my doctor is incompetent, but generally I would have to know what doctors know in order to confirm or dispel my suspicion. Thus, we must face the implications of the fact that laymen do not fully understand what constitutes good reasons in the domain of expert opinion.

Granted, I can check on a given expert and perhaps obtain a ranking of various experts[5]—by relying on other experts. If my doctor says that I should see a cardiologist, I can ask him and other physicians in the community about the local cardiologists. Or if I want to know the effects of mass media on voters, I can go to the political science department and ask who has done the best work in this area and whether there has been significant criticism of it. This checking and ranking of experts can be expressed by extending our formula and its implied chain of authority: *B* has good reasons for believing that *C* has good reasons for believing that *A* has good reasons for believing that *p*. However, by appealing to such a hierarchy of experts I have not regained my epistemic autonomy by avoiding reliance on experts—I have only extended and refined this reliance. Nor could I regain my epistemic autonomy in all cases without believing on the basis of relatively crude and untested reasons.

Granted also, if I do not know and have no way of finding out who the experts are, I will have no way to appeal to the chain of authority. I will then not know who has good reasons to believe that p, to whom to defer, or whose opinion (if any) will give me good reasons for believing that p. This sometimes happens, and, when it does, rational deference becomes impossible. But generally I can find someone whose opinion is more informed than mine and who can refer me to someone who is knowledgeable about whether or not p. And even if a layman, because of his relative inability to discriminate among experts, ends up appealing to a lesser instead of a greater expert, the lesser expert's opinion will still be better than the layman's.[6]

In terms of our formula, then, B could believe that p *either* because B has good reasons to believe that A has good reasons to believe that p, *or* because B has good reasons to believe that C has good reasons to believe that A has good reasons to believe that p. But, in either case, B cannot have sufficiently good reasons not to believe that p or to believe that not p. In other words, the layman cannot rationally refuse to defer to the views of the expert or experts he acknowledges. This does not mean that B can never successfully raise a devastating objection to believing that p or imagine an alternative to believing that p, but it does mean that only someone with A's expertise can make an accurate assessment of the value and validity of the objection or alternative. Under cross-examination by the layman, the expert may admit the cogency of a given point, but *he* (and his fellow experts) must judge whether it is cogent and germane, since they are the only ones who fully understand what is involved in the methods, techniques, premises, and bases of the expert's training and inquiry and how these affect the resultant belief.

The layman can, in other words, propose criticisms and alternatives, but rationally he must allow the expert to dispose of them, for in a conversation with an expert (as opposed to a dialogue among equals),[7] the final court of rational appeal belongs solely to one party, by virtue of that party's greater competence for and commitment to inquiry into the relevant subject matter. The rational layman recognizes that his own judgment, uninformed by training and inquiry as it is, is *rationally inferior* to that of the expert (and the community of experts for whom the expert usually speaks) and consequently can always be *rationally* overruled. Recognizing that the highest court of rational appeal lies outside of himself, the layman may simply have to accept the fact that his objection is not a good one, even though it still seems good to him.

There are, of course, a whole series of ad hominems that permit a layman rationally to refuse to defer to the expert's opinion. The layman can assert that the expert is not a disinterested, neutral witness; that his interest in the outcome of the discussion prejudices his testimony. Or that he is not operating in good faith—that he is lying, for example, or refusing to acknowledge a mistake in his views because to do so would tend to undermine his claim to special competence. Or that he is covering for his peers or knuckling under to social pressure from others in his field, etc., etc. Such ad hominems are not always fallacious, and they sometimes do ground the rational refusal to defer to the statements of experts. But one interesting feature of such ad hominems is that they seem and perhaps are much more admissible, important, and damning in a layman's discussions with experts than they are in dialogues among peers. It doesn't matter so much if one's peers are biased or operating in bad faith; they will be found out. The merits of their arguments can be tested and evaluated rather than just accepted.

With the exception—often an important exception—of such ad hominems, I see no way to avoid the conclusion I have proposed above: that the rational layman will recognize that, in matters about which there is good reason to believe that there is expert opinion, he ought (methodologically) not to make up his own mind. His stance on these matters will—if he is rational—usually be rational deference to the epistemic authority of the expert.

If it is objected that, in cases of divided expert opinion, the layman will have no method for deciding whether or not to believe that p, this is granted.[8] But in such cases the rational layman, recognizing that his own relatively casual and crude inquiry is not competent to resolve issues that even the sustained inquiry of experts cannot resolve, will also recognize that he is confronted with a situation in which he must either suspend belief or—if this is impossible or undesirable—arrive at belief on some admittedly nonrational basis. And if it is objected that layman B can have good reasons to believe that p even when p is false and even when expert A does not have good reasons to believe that p, this is also granted. For B will sometimes be misled by phony or mistaken claims to expertise, despite a careful attempt to ascertain that A is indeed an expert about p; and, moreover, there is simply no guarantee that the views of even the best present experts are coincident with the final truth.

The conclusion that it is sometimes *irrational* to think for oneself—that *rationality* sometimes consists in deferring to epistemic authority and, consequently, in passively and uncritically accepting what we are given to believe— will strike those wedded to epistemic individualism as odd and unacceptable, for it undermines their paradigm of rationality. To others, it may seem too obvious for such belaboring. But in either case, I submit, we should recast our epistemologies and our accounts of rationality to make them congruent with this important fact of modern life.

III.

Although the preceding discussion is obviously relevant to the big word of epistemology—*know*—I have so far astutely avoided using it. But the relevance of the discussion is clear, given the standard analysis of "*A* knows that *p*" in terms of (1) *A* believes that *p*, (2) *A* has good reasons to believe that *p*, and (3) that p is true. The third condition is standardly taken to be the kicker, and it threatens to render the whole analysis inapplicable to knowledge; for A can have good reasons to believe that p even if p is false and B can have good reasons to believe that A has good reasons to believe that p—again, even if p is false. However, this third condition is not my primary concern, for I would argue for a fallibilist conception of knowledge.

I wish, rather, to focus our attention on the second, more neglected condition in the above analysis of "A knows that p." It seems plausible that both A and B must have better or more complete reasons in order to *know* that p than are necessary merely to have good reasons to *believe* that p; for some beliefs, though rational, would not be well founded enough to qualify as knowledge (even on a fallibilist conception of knowledge). Thus it seems reasonable to hold that there is a progression from (1) believing that p (mere belief or right opinion), to (2) having good reasons to believe that p (rational belief), to (3) knowing that p.

What happens, then, if we substitute 'know' for 'has good reasons to believe' in our formula: i.e., B knows that A knows that p? Is it possible to *know* vicariously, as it were, or must *knowers* (as opposed to mere rational believers) stand on their own epistemic feet? I argued above that B can have good reasons to believe that p without having direct reasons or evidence for p. Is the same true for knowing? Or must B know that p *before* he can know that A knows that p, thus precluding an appeal to A's knowledge as the basis and justification of his own claim to know? In other words, recalling the earlier distinction between having evidence that p and another kind of good reason to believe that p, must B possess the evidence for the truth of p in order to *know* that p? Or can knowledge, as well as rational belief, be based on an appeal to epistemic authority?

Suppose someone tells me something that is true without giving me evidence for its truth. Perhaps A tells me that laetrile does not cure cancer without giving me the studies that prove this, much less the concrete data on which those studies are based. But suppose I have good reasons to believe that A is an authority in the field of cancer research and so I believe what he tells me. Do *I* then *know* that laetrile does not cure cancer, or have I achieved something much less than knowledge (perhaps only right opinion or rational belief)? If I then know, it is possible for one to know that p without possessing evidence for the truth of p. But that seems paradoxical or counterintuitive; for, in the cases we are considering, evidence is relevant to establishing knowledge, but we are asking whether it is possible to have this knowledge without the relevant evidence.

Even more paradoxical is the idea that B can know that p even though he doesn't understand that p. Suppose an eminent authority in particle physics tells me that a quark is a fundamental particle, and suppose this is true. But I don't even understand what that means, because I have no notion of what a quark is or what counts as a fundamental particle. However, I check up on the physicist, and, as a result, I *know* that he has unsurpassed credentials. Could I then be said to *know* that a quark is a fundamental particle, though I don't even understand what I know?

To sum up: should we say that B can (1) know that p by knowing that A knows that p, and (2) know this without first knowing that p? Should we say this even if it implies that B can know that p without having evidence for p and perhaps without even understanding p? Instead of attempting to answer these questions directly, I will argue that much of what we want to count as knowledge rests on the epistemic structure expressed by the formula, B knows that A knows that p.[9] I will then offer two conclusions and leave it to the reader to decide which is more epistemologically palatable.

Scientists, researchers, and scholars are, sometimes at least, knowers, and all of these knowers stand on each other's shoulders in the way expressed by the formula: B knows that A knows that p. These knowers could not do their work without presupposing the validity of many other inquiries which they cannot (for reasons of competence as well as time) validate for themselves. Scientists, for example, simply do not repeat the experiments of other scientists unless the experiment is important and something seems fishy about it. It would, moreover, be impossible for anyone to get to the research front in, say, physics or psychology, if he relied only on the results of his own inquiry or insisted on assessing for himself the evidence behind all the beliefs he accepts in his field. Thus, if scientists, researchers, and scholars are knowers, the layman-expert relationship is also present *within* the structure of knowledge, and the expert is an expert partially because he so often takes the role of the layman *within his own field*.

Moreover, research in many fields is increasingly done by teams rather than individuals. For example, it is not uncommon for the title of an article reporting experimental results in particle physics to look like this:

William Bugg, Professor of Physics at the University of Tennessee, Knoxville, and a participant in this experiment, explained how such an experiment is done. This experiment, which recorded charm events and measured the lifespan of the charmed particles, was one of a series of experiments costing perhaps $10 million. After it was funded, about 50 man/years were spent making the needed equipment and the necessary improvements in the Stanford Linear Accelerator. Then approximately 50 physicists worked perhaps 50 man/years collecting the data for the experiment. When the data were in, the experimenters divided into five geographic groups to analyze the data, a process which involved looking at 2 1/2 million pictures, making measurements on 300,000 interesting events, and running the results through computers in order to isolate and measure 47 charm events. The "West Coast group" that analyzed about a third of the data included 40 physicists and technicians who spent about 60 man/years on their analysis.

Charm Photoproduction Cross Section at 20 GeV

K. Abe, T. C. Bacon, J. Ballam, L. Berny, A. V. Bevan, H. H. Bingham, J. E. Brau, K. Braune, D. Brick,
W. M. Bugg, J. Butler, W. Cameron, J. T. Carroll, C. V. Cautis, J. S. Chima, H. O. Cohn, D. C. Colley,
G. T. Condo, S. Dado, R. Diamond, P. J. Dornan, R. Erickson, T. Fieguth, R. C. Field, L. Fortney,
B. Franek, N. Fujiwara, R. Gearhart, T. Glanzman, J. J. Goldberg, G. P. Gopal, A. T. Goshaw,
E. S. Hafen, V. Hagopian, G. Hall, E. R. Hancock, T. Handler, H. J. Hargis, E. L. Hart, P. Haridas,
K. Hasegawa, T. Hayashino, D. Q. Huang, [a] R. I. Hulsizer, S. Isaacson, M. Jobes, G. E. Kalmus,
D. P. Kelsey, J. Kent, T. Kitagaki, J. Lannutti, A. Levy, P. W. Lucas, M. MacDermott, W. A. Mann,
T. Maruyama, R. Merenyi, R. Milburn, C. Milstene, K. C. Moffeit, J. J. Murray, A. Napier,
S. Noguchi, F. Ochiai, S. O'Neale, A. P. T. Palounek, I. A. Pless, M. Rabin, [b] P. Rankin,
W. J. Robertson, A. H. Rogers, E. Ronat, H. Rudnicka, T. Sato, J. Schneps, S. J. Sewell,
J. Shank, A. M. Shapiro, C. K. Sinclair, R. Sugahara, A. Suzuki, K. Takahashi, K. Tamai,
S. Tanaka, S. Tether, H. B. Wald, W. D. Walker, M. Widgoff, C. G. Wilkins, S. Wolbers,
C. A. Woods, Y. Wu, A. Yamaguchi, R. K. Yamamoto, S. Yamashita,
G. Yekutieli, Y. Yoshimura, G. P. Yost, and H. Yuta

*Birmingham University, Birmingham B15 2TT, England, and Brown University, Providence, Rhode Island 02912,
and Duke University, Durham, North Carolina 27706, and Florida State University, Tallahassee, Florida 32306,
and Imperial College, London SW7 2BZ, England, and National Laboratory for High Energy Physics (KEK),
Oho-machi, Tsukuba-gun, Ibaraki 305, Japan, and Oak Ridge National Laboratory, Oak Ridge, Tennessee
37830, and Rutherford Appleton Laboratory, Didcot, Oxon OX11 0QX, England, and Stanford Linear
Accelerator Center, Stanford University, Stanford, California 94305, and Technion–Israel Institute
of Technology, Haifa 32000, Israel, and Tohoku University, Sendai 980, Japan, and Tufts
University, Medford, Massachusetts 02155, and University of California, Berkeley,
California 94720, and University of Tel Aviv, Tel Aviv, Israel, and University of
Tennessee, Knoxville, Tennessee 37916, and Weizmann Institute, Rehovot, Israel*

(Stanford Linear Accelerator Center Hybrid Facility Photon Collaboration)

(Received 2 May 1983)

Forty-seven charm events have been observed in an exposure of the SLAC Hybrid
Facility bubble chamber to a 20-GeV backward-scattered laser beam. Thirty-seven
events survive all the necessary cuts imposed. Based on this number the total charm
cross section is calculated to be $63 ^{+33}_{-26}$ nb.

PACS numbers: 13.60.Le, 13.60.Rj

In this Letter we present results on the charm photoproduction cross section in an experiment using the SLAC Hybrid Facility. Results on lifetimes of charmed particles based on part of the data were published earlier.[1]

The SLAC 1-m hydrogen bubble chamber was exposed to a 20-GeV photon beam produced by Compton scattering of laser light by the 30-GeV electron beam. It was collimated to 3 mm in diameter. The photon beam energy spectrum is shown in Fig. 1. It peaks at 20 GeV with a full width at half maximum of 2 GeV. Most of the data were taken at photon intensities of 20–30 γ / pulse. In order to detect decays of charmed particles, a fourth camera with high-resolution optics having a resolution of 55 μm over a depth of field ± 6 mm was used. The cameras were triggered either on the passage of a charged particle through three multiwire proportional chambers and pointing back to the fiducial volume of

the bubble chamber or on a sufficient energy deposition in an array of lead-glass blocks. Particle identification was provided by ionization measurements in the bubble chamber and light detection in two large-aperture Cherenkov counters. More details of the experimental setup and trigger are given in Ref. 1.

The results presented here are based on 270 000 hadronic interactions found in a restricted fiducial volume. All hadronic events were closely examined for the decays of short-lived particles within 1 cm of the production vertex. When such a decay was found, the following cuts were applied to ensure that the decays which survived were genuine charm decays: (a) Decays with less than two charged products were rejected. (b) Two-prong decays consistent with either photon conversions or strange-particle hypotheses were rejected. To eliminate K^0 decays, the two-body (assumed to be $\pi\pi$) invariant

Obviously, no one person could have done this experiment—in fact, Bugg reports that no one university or national laboratory could have done it—and many of the authors of an article like this will not even know how a given number in the article was arrived at.[10] Furthermore, even if one person could know enough and live long enough to do such an experiment, there would be absolutely no point in his attempting to do so, for his results would have become obsolete long before he completed the experiment. Although Bugg expresses confidence that the team's measurement of the lifespan of charmed particles is a good one, he estimates that within three years some other group will have come up with another technique that will give considerably better results. He consequently expects that within five years the paper will no longer be of general interest.

Finally, Bugg notes that the article's 99 authors represent different specializations with particle physics, but all are experimentalists, so none would be able to undertake the theoretical revisions which might be required as a result of this experiment and which provide a large part of the rationale for doing it. On the other hand, most theoreticians would not be competent to conduct the experiment—and neither the experimentalists nor the theoreticians are competent to design, build, and maintain the equipment without which the experiment could not be run at all.

Obviously, this is an extreme example, though not all that extreme in the realm of particle physics.[11] However, we can see how dependence on other experts pervades any complex field of research when we recognize that most footnotes that cite references are appeals to authority. And when these footnotes are used to establish premises for the study, they involve the author in layman-expert relationships even within his own pursuit of knowledge. Moreover, the horror that sweeps through the scientific community when a fraudulent researcher is uncovered is instructive, for what is at stake is not only public confidence. Rather, each researcher is forced to acknowledge the extent to which his own work rests on the work of others—work which he has not and could not (if only for reasons of time and expense) verify for himself.

Thus in very many cases *within* the pursuit of knowledge, there is clearly a complex network of appeals to the authority of various experts, and the resulting knowledge could not have been achieved by any one person. We then have something like the following:

A knows that m.

B knows that n.

C knows (1) that A knows that m, and (2) that if m, then o.

D knows (1) that B knows that n, (2) that C knows that o, and (3) that if n and o, then p.

E knows that D knows that p.

Suppose that this is the only way to know that *p* and, moreover, that no one who "knows" that *p* knows that *m*, *n*, and *o* except by knowing that others know them. Does *D* or *E* know that *p*? Does anyone know that *p*? Is that *p* known?

Unless we maintain that most of our scientific research and scholarship could *never*, because of the cooperative methodology of the enterprise, result in knowledge, I submit that we must say that *p* is known in cases like this. But if *D* or *E* knows that *p*, we must also say that someone can know "vicariously"—i.e., without possessing the evidence for the truth of what he knows, perhaps without even fully understanding what he knows. And this conclusion would require dramatic changes in our analysis of what knowledge must be.

If the conclusion is unpalatable, another is possible. Perhaps that *p* is known, not by any one person, but by the *community* composed of *A*, *B*, *C*, *D*, and *E*. Perhaps *D* and *E* are not entitled to say, "I know that *p*," but only, "We know that *p*." This community is not reducible to a class of individuals, for no one individual and no one individually knows that *p*. If we take this tack, we could retain the idea that the knower must understand and have evidence for the truth of what he knows, but in doing so we deny that the knower is always an individual or even a class of individuals. This alternative may well point to part of what Peirce may have had in mind when he claimed that the *community* of inquirers is the primary knower and that individual knowledge is derivative.

The latter conclusion may be the more epistemologically palatable; for it enables us to save the old and important idea that *knowing* a proposition requires understanding the proposition and possessing the relevant evidence for its truth. But it will not be very comfortable for those who have a taste for desert landscapes, intellectual autonomy, or epistemic individualism; for it undermines the methodological individualism that is implicit in most epistemology. I believe that it is also deeply disturbing because it reveals the extent to which even our rationality rests on trust and because it threatens some of our most cherished values—individual autonomy and responsibility, equality and democracy. But that is a story for another occasion.

Thus if the arguments of this paper are accepted, some very basic changes in our epistemologies are required. We must recast our conception of what it means for beliefs and persons to be rational. We must also either agree that one can know without possessing the supporting evidence or accept the idea that there is knowledge that is known by the community, not by any individual knower.

NOTES

1. It might still seem that if *B* has good reasons to believe that *A* has good reasons to believe that *p*, then *B* has *evidence* that *p*. The dispute between me and someone inclined to press this objection would turn on delicate epistemological issues involved in clarifying the

concept of evidence. But I would argue that B does *not* have evidence that p, urging, in addition to the arguments presented in the body of this paper, the following. (1) Evidence that p counts against evidence that not p. But consider a case of conflicting experts: A who has evidence that p, and C, who has evidence that not p. In such a case if B believes that p only because he believes that A has good reasons to believe that p, B's reasons do not count against C's; only A's do. (2) It would be possible to construct cases in which B has good reason to believe that A has good reason to believe that p even where we would agree that there is *no* evidence for p. (More on such cases in sec. II below.) But regardless of how this dispute about "evidence" is to be resolved, I would observe that B's reasons are logically dependent on A's. Most of the rest of the points in this paper will follow if this is granted.

2. I assume that we can all agree that there are experts, but I have not attempted in this paper to offer a precise definition of 'expert' or to delineate the range of possible expertise (beyond the introductory proviso that this paper is restricted to belief in and knowledge of propositions for which there is evidence). If the theses of this paper are correct, however, it will become crucial for epistemologists to argue about the definition of 'expert' and the range of actual and possible expertise.

But one observation about my use of 'expert' is in order: it does not presuppose or entail the truth of the expert's views. If one defines 'expert' in terms of the *truth* of his views (as Plato's Gorgias and Thrasymachus do), it is often impossible in principle to say who is an expert—even if one is an expert oneself!—since it is often impossible to say whose view is coincident with the truth. But I submit that it is not similarly impossible to say what constitutes sustained, relevant inquiry and to ascertain who is engaged in it (though there will sometimes be very real problems in making this judgment). And whenever sustained inquiry is both necessary for and efficacious with respect to determining whether or not p, the expert's views are less likely to be mistaken and likely to be less mistaken than an inexpert opinion. Thus, in my use of 'expert', the connection between truth and the views of the expert is not completely severed, though that connection is neither necessary nor simple.

3. *Meaning* (Chicago: University Press, 1977).

4. *Critique of Judgment*, J. H. Bernard, tr. (New York: Hafner, 1951), p. 136; Kant's emphasis. Kant repeats this statement in the *Anthropologie*, p. 118, and in the *Logik*, p. 371, both Cassirer editions (Berlin, 1932).

5. In a series of recent articles, Keith Lehrer has explored the issues concerning the ranking of experts and the opinions of various experts and, consequently, the way to handle the problem of disagreement among experts, all with much more rigor and precision than I can muster here. Cf., e.g., "Social Information," *Monist*, I.x, 4 (October, 1977): 473–487, and also the articles Lehrer refers to in his footnotes to this article.

6. Of course, a more detailed account of the whole issue of identifying relevant experts would have to distinguish among (1) B merely believing that A has good reasons to believe that p, (2) B having some reason to believe that A has good reasons to believe that p, and (3) B having good reasons to believe that A has good reasons to believe that p. And none of this resolves the often excruciating practical problem of identifying who the real or best experts are—e.g., what is the patient faced with conflicting medical opinions to do? But these are logically posterior issues and problems; the argument of this section of the paper is that in any case he should *not* make his own diagnosis, nor even read up some about his problem and then make his own diagnosis.

7. I have attempted to explicate the logic of dialogue among presumed epistemic equals in the area of moral reasoning in my article "The Achievement of Moral Rationality," *Philosophy and Rhetoric*, v I, 3 (Summer 1973): 171–185.

8. If it is possible to rank experts in the ways Lehrer (*op. cit.*) has explored or in some other way, the layman can of course resolve the dilemma posed by divided expert opinion by deferring to the *best* expert opinion. However, there will still be cases in which even the best experts will disagree.

9. This strategy for approaching these issues will mean, of course, that it remains open to a courageous enough epistemologist to avoid my conclusion by embracing the view that the achievements of scientists, researchers, and scholars are not and could not be *knowledge* whenever these achievements are based on cooperative methodologies. This option does not seem very attractive to me, to say the least.

10. Of course, only a few people actually write the article, but it does not follow that these people are masterminds for the whole procedure or that they completely understand the experiment and the analysis of the data. According to Bugg, although a few persons—"the persons most actively involved in working on the data and who therefore understand most about it"—wrote up the experiment (this article is 3 1/2 journal pages long), they really only prepared a draft for revisions and corrections by the other authors. The team then met to argue substantive points about the techniques for analyzing the data and how the article should be presented to best enable other physicists to understand it.

11. Of the 42 articles on elementary particles and fields published by *Physical Review Letters* in the three months from April 25 to July 18, 1983, 11 listed more than 10 authors, 9 listed more than 20 authors, and 5 more than 40 authors. In the same period only 5 articles were by single authors.

12. THE CONSTITUTIVELY SOCIAL CHARACTER OF EXPERTISE

STEVE FULLER

1. A BRIEF HISTORY OF EXPERTISE

"Expert," a contraction of the participle "experienced," first appeared as a noun in French at the start of the Third Republic (about 1870). The general idea was much the same as today, namely, someone whose specialized training enables him [sic!] to speak authoritatively on some matter. However, the original context of use was quite specific. The first experts were called as witnesses in trials to detect handwriting forgeries. These people were experienced in discriminating scripts that appeared indistinguishable to the ordinary observer. Thus, the etymological root of "expertise" in "experience" was carried over as the semantically heightened way in which the expert experienced the relevant environment. In contemporary parlance, the tasks that originally required the service of an expert principally involved "pattern recognition," except that the patterns recognized by the expert were identified in terms of an implicit explanatory framework, one typically fraught with value connotations, as in the case of identifying a script as a "forgery."

When evaluating the likelihood that a script was forged, experts were not expected to publicly exhibit their reasoning. They were not casuists who weighed the relative probability that various general principles applied to the case. Rather, it was on the basis of an expert's previous experience of having successfully identified forgeries that his judgment was now trusted. This is not to say that no one could contest expert judgment, but he would have to be another expert, a colleague. If no colleague came forward to testify against an expert's judgment, then the judgment would stand. The climate of collegiality that harbored the mystique of the expert led journalists of the Third Republic to distinguish experts from the "lay" public, thereby conjuring up a clerical image redolent of the secular religion that Auguste Comte's more zealous followers had been promoting under the rubric of Positivism (Williams, 1983: 129).

Moreover, experts were contrasted not only with the lay public but also with *intellectuals.* This point is important for understanding the source of what might be called the *epistemic power* of expertise. An intellectual takes the entire world as fair game for his judgments, but at the same time he opens himself to scrutiny from all quarters. Indeed, the intellectual's natural habitat is controversy, and often he seems to spend more time on defending and attacking po-

sitions than on developing and applying them. In contrast, the expert's judgments are restricted to his area of training. The credibility of those judgments
are measured in terms of the freedom from contravention that his colleagues
accord him. The mystique of expertise is created by the impression that an expert's colleagues are sufficiently scrupulous that, were it necessary, they would
be able and inclined to redress any misuse or abuse of their expertise. The fact
that they do not means that the expert must be doing something right.

Collegiality enables experts to exert what both Plato and Machiavelli would
have recognized as an ideal form of power. In its ideal form, power thrives on
a counterfactual that never needs to be realized—in less charitable terms, a
persuasive bluff. If a prince's enemies believe that the prince could squash any
uprising, the enemies will lie low, and the prince will seem invincible; however,
if the enemies challenge the prince, and the prince defeats them only with great
difficulty, then the air of princely invincibility will disappear, and the prince
will need to prepare for redoubled efforts by his enemies in the future. Thus,
ideal power is brought down to the level of brute force (Botwinick, 1990:
133–80). Trial attorneys continue to exploit this point whenever they try to undermine the very possibility of expertise in a field by pitting particular experts
against one another. The lawyers do not expect a definitive judgment to emerge
from the crossfire; rather, they expect to show that no such judgment can be
rendered.

Philosophers have traditionally shared the attorney's desire to dissipate the
power of expertise, but without in the process undermining the possibility of
knowledge. Be it embodied in machine or human, philosophers have looked
askance at the epistemological status of expertise. For Karl Popper, the expertise conferred on those trained in the special sciences serves to strategically
contain critical inquiry, as the evaluation of a knowledge claim depends on the
credentials of the claimant. The sociology of knowledge perennially incurs the
wrath of philosophers because it seems to condone this tendency, which cul-
minates in the formation of scientific guilds, or disciplines. These, in turn, divert inquiry away from questions of fundamental principles that go to the
heart of disciplinary identity. Disciplines proliferate explanatory frameworks
(jargons, if you will), while "science," in the philosophically honorific sense,
unifies such frameworks. Not surprisingly, then, Popper regarded Kuhnian
"normal science" as a "danger" to the advance of knowledge, if that advance is
measured in terms of explanatory comprehensiveness, or the Newtonian
virtue of explaining the most by the least (Popper, 1970).

Cognitive scientists display a similar disdain when they contrast the "knowl-
edge-based" or "domain-specific" character of expert systems to more general
purpose problem-solving machines that utilize principles that cut across domains. What is often called the "orthodoxy" in cognitive science holds that
an adequate theory of cognition is not to be found simply by compounding a

number of distinct expert systems, just as an adequate theory of knowledge requires more than simply articulating the research conventions of all the special sciences (Haugeland, 1984). The hope, in both cases, is that whatever principles govern the special cases are not unique to them.

2. SOME SENSES OF THE SOCIAL

To someone whose work is primarily in the design of expert systems, it may seem odd to juxtapose philosophical suspicions about human and machine expertise in the way I have. Perhaps this juxtaposition points to a residual positivist hankering that cognitive science and epistemology (or philosophy of science) share for the "unity of science." Without casting doubts on this diagnosis, I nevertheless want to stress some philosophically disturbing aspects of expertise that have yet to fully grip theorists in this area. As a first pass, these aspects turn on the following truism: *Expertise is a constitutively social phenomenon.* Indeed, as a card-carrying "social epistemologist" (Fuller, 1988, 1993a, 1993b), I am committed to the position that expertise can be exhaustively analyzed as a social phenomenon. But before too many eyebrows are raised, let me begin by listing—in descending order of intuitiveness—four distinct senses in which expertise is constitutively social:

1) The skills associated with an expertise are the product of specialized training. Expertise cannot be picked up casually or as the byproduct of some other form of learning.
2) Both experts and the lay public recognize that expertise is relevant only on certain occasions. No expertise carries universal applicability.
3) The disposition of expertise is dependent on the collegial patterns of the relevant experts. Protracted internecine disputes over fundamentals typically erode expertise.
4) The cognitive significance of an expertise is affected by the availability of expert training and judgment, relative to the need for the expertise. Too many experts or too little need typically devalue the expertise in question.

So far, I have concentrated on sense (3), given its salience in the historical development of the expert as a social role distinct from that of the layperson, the intellectual, and, as we have just seen, even the scientist. However, all four senses echo the twin themes of *boundedness* and *compartmentalization* as essential to the definition of expertise. The cognitive science literature offers several ways of articulating these themes: Simon's heuristics, Minsky's frames, Schank's scripts, Fodor's modules, and, most abstractly, Pylyshyn's cognitive impenetrability. Of course, these terms do not divide up the mind's labor in quite the same way. A similar proviso would have to be attached to sociological markers of expertise, such as "indexicality" and "functional differentiation" (cf. Cicourel, 1968; Knorr-Cetina, 1981). But for our purposes, the most striking comparison may be between, so to speak, the cognitive and political impenetrability of expertise, the so-called autonomy of the professions (cf. Abbott, 1988).

The analogy I wish to draw here is not particularly difficult to grasp, but doing so may alert us to the conceptual baggage that is unwittingly imported in the images we use to characterize expertise. The profession that has most jealously guarded its autonomy—scientists—has often struck a bargain to ensure that professionally produced knowledge remains both cognitively and politically impenetrable. From the charter of the Royal Society in seventeenth-century Britain to the guild right of *Lehrfreiheit* (freedom of inquiry) that German university professors enjoyed under Bismarck, the following two conditions have been met (cf. Proctor, 1991):

(A) The state agrees not to interfere in the internal governance of the profession, on the condition that the profession does not interfere in the governance of the state.

(B) The state agrees to protect the profession from others who might want to interfere with its governance (e.g. other professional, political, or business interests), on the condition that the state is given the first opportunity to appropriate the knowledge produced by the profession, where appropriation includes the right to prevent others from subsequently appropriating the knowledge (e.g. for reasons of national security).

Most theories of the mind in cognitive science are fairly explicit in distinguishing an executive central processor or general problem-solver from domain-specific modules that function in relative autonomy from this unit. Sometimes (as in Simon, 1981; Minsky, 1986) the political imagery of governance is quite strong. However, it is not just any old image of governance, but one that is characteristic of twentieth-century thinking about the state, namely, *democratic pluralism* (cf. Held, 1987: 186–220). The pluralist portrays the state as mediating competing factions in a large democratic society by enabling the factions to flourish without letting any of them override each other or the national interest. The learning process of big democracies consists of these factions settling into interest groups and, ultimately (and ideally), professionally governed associations whose interaction is founded on recognition and respect for each other's work as essential to the business of society. Gradually, then, the state's role as mediator recedes to that of chairman of a corporate board of directors. The only point I wish to make here is that this is not the only, nor necessarily the most desirable, image of democratic governance. It might repay the effort for cognitive scientists to examine alternative forms of governance as a source of new images for arranging the parts of the mind.

3. THERE MAY BE LESS TO EXPERTISE THAN MEETS THE EYE

To explore the social character of expertise implied in senses (1) and (2), consider what might be called a *behaviorist* and a *cognitivist* account of how expertise develops (cf. De Mey 1982: 216):

> *Behaviorist*: Expertise is shaped from repeated encounters with relevant environments, such that increased exposure smooths out the rough edges in the expert's practice until it stabilizes at a normatively acceptable standard, which can then be applied, "off the shelf" as it were, in subsequent encounters.
>
> *Cognitivist*: Expertise consists of a core set of skills that are elaborated in a variety of environments, most of which are unforeseen. These elaborations are stored and themselves elaborated upon in subsequent encounters, all of which serves to confer on expertise a sophistication that is evident to the observer but difficult to articulate.

Thus, whereas the behaviorist sees expertise becoming more *stereotyped* in practice, the cognitivist sees it becoming more *nuanced*. In the ongoing dispute between cognitivists and behaviorists, it is often asserted that the cognitivist won this round. But before the final verdict is delivered, I wish to offer support for the behaviorist by way of revealing the hidden social hand of expertise.

To hear cognitivists (not to mention phenomenologists) emote about the "nuanced" and "craftlike" character of expertise, one would think that an act of professional judgment was tantamount to a magic trick, one in which the audience has attended a little too closely to the magician's gestures and not enough to the circumstances under which the illusion transpires. A professional magician does not perform tricks on demand, say, by adapting his performance to play off the specific gullibilities of his audience. Of course, the magician adapts somewhat to his audience, but before he even agrees to display his expertise, the stage must be set just right, and the audience must already be in the right frame of mind to be receptive to the "magic moment." A magician who is too indiscriminate in his eagerness to please is bound to look bad very quickly. An instructive case in point is the odd magician who submits his performances to the strictures of the experimental method (Collins & Pinch, 1982: on psychokinesis).

It is worth noting that, in these scientific performances, the magician fares no worse than the expert witnesses in medicine and psychiatry whose testimony is routinely heard—and believed—in court (Faust, 1985). The reliability and validity of expert judgment in experimental settings are low all around (Arkes & Hammond, 1986). Perhaps the most celebrated historical case of the hubris of overextended expertise was the rise and fall of the Sophists in 5th Century B.C. Athens. After some fair success as teachers of rhetoric and confidants of the ruling class, the Sophists began to offer their services in competition with more established forms of knowledge in virtually every domain. According to a recent historical study (De Romilly, 1992), the Sophists were perceived as opportunistic colonizers of conventional practices who failed to cultivate the trust required for their own practices to succeed. Consequently, the Sophists were soon ridiculed by the people we now regard as the founders of classical philosophy and drama—people with a greater surface respect for tradition than the Sophists had.

Proponents of the operant conditioning paradigm in behaviorism would recognize what is going on here. Successful experts realize that the secret to their success lies in noting the sorts of situations in which clients come away most satisfied with expert judgment (and hence reward the expert appropriately) and then maximizing the opportunities for situations of that sort to arise in the future. In short, the smart expert controls the contingencies by which her behavior is reinforced (cf. Haddock & Houts, 1992). She does not easily submit to an experimental test of her expertise, over which she exerts little control. The strategy for achieving the relevant sense of control involves several tactics:

(i) pre-selecting clients before officially engaging in treatment, typically through an interview process, an important aspect of which is to gain the confidence of the prospective client;

(ii) learning to refuse certain clients, no matter how lucrative their business might be, if it appears as though they will resist treatment or, in some way, be likely to make the expert look bad in the long run;

(iii) persuading the prospective client that her avowed problem is really one that the expert has seen many times before; often this is done by first showing the client that she has not conceptualized her problem correctly; once the problem receives its proper formulation—which neatly coincides with what the expert is in the best position to treat—then treatment can begin;

(iv) obscuring any discussion of the exact method of treatment by recasting the client's problem in the jargon of the expertise, which will presumably lead the client to infer that anything that must be described in such esoteric ways must be subject to a treatment that would be equally difficult to convey.

The most obvious result of these four tactics is to shape the prospective client's behavior so that her problem falls into one of the stereotyped patterns with which the expert is familiar. A less obvious but equally important result is that the expert can now exert *spin control* over how the client understands her situation after the treatment. If the client's problem is solved, the expert can confidently claim credit for the solution. No suspicions of "spontaneous remission" are likely to be raised, especially if the client has paid dearly for treatment. But perhaps more important, if the client's problem remains unsolved after the treatment, then the expert can claim, with only slightly less confidence, that other unforeseen factors intervened, including the client's own recalcitrance. Whatever happens, therefore, reinforces the expert's competence and integrity. On the basis of these considerations, I conclude that the key to understanding the distinct character of expertise may lie less in its associated skills than in the discretionary control that the expert has in deploying those skills. (Philosophers of science and cognitive scientists who are uncomfortable with brute talk of "control" may substitute this piece of genteel intellectualism: Experts have heightened metaknowledge of the *ceteris paribus* clause, or relevance conditions, for applying their expertise.) Thus, if the above pre-selection strategy fails to mold the client into shape, the expert can then tell the client that the problem lies outside her expertise, which will probably cause the client to believe that her problem remains unsolved, not because the expert was incompetent, but because the client has yet to locate the right expert, which may itself reflect the client's own failure to understand the nature of her problem. The element of trust crucial for the maintenance of expertise may be seen in the willingness with which the client holds herself responsible for an expert's inability to come to grips with her problem (cf. Gambetta, 1988). From what I have said so far, it may seem that my thinking about the social dimension of expertise has been strongly based on psychiatric encounters, which have often been subject to unflattering depictions as confidence games. However, the same observations apply, perhaps with greater import, to experts operating in the arena of public policy, especially those trained in medicine, engineering, or economics. For, the biggest single problem facing the future of democracy may be *cognitive authoritarianism*, the tendency to cede an ever larger share of the realm of participatory politics to expert rule (Fuller, 1988: 277–88). The conversion is accomplished as government officials become convinced that the public has ill-formed conceptions of its own needs, needs that are best shaped and addressed by the relevant experts. When government fails to act speedily on this conversion, and hence does not clear the political environment to enable the expert's stereotyped knowledge to take effect, the expert will often appear as a moral censor, appealing to his special knowledge—which would supposedly be efficacious if politicians secured the relevant background conditions—as a norm against which the state of society is criticized.

But can all this talk about the strategically discretionary character of expertise be applied to computerized expert systems? I do not see why not. Suppose a knowledge engineer has been asked to design an expert system that will offer advice on playing the stock market. After some time, the knowledge engineer returns with a product that she announces was constructed from in-depth interviews with four of the best stock analysts on Wall Street, making sure that respected spokespersons for all the relevant market perspectives—fundamentalists, chartists, insiders, and traders—were canvassed (cf. Smith, 1981). The effect of this pedigree on the client will be similar to that of the diploma and license that hang on the wall of the human expert's office and invariably engage the client's peripheral vision during a consultation. If the knowledge engineer designs the interface with protocols that make interaction between client and expert appear stilted, then the client will probably interpret that to mean that the expert is concerned with getting to the heart of the client's problem without dragging in superfluous information. Likewise, if the expert seems to give the client advice that causes her to lose money in the market, then the client may wonder whether a human expert could have really done any better, or that perhaps she did not input all the information that was relevant for the expert system to provide better advice. Moreover, the client's inclination to assume responsibility for the bad advice increases with the amount of money that she had to originally spend to purchase the system. That an AI pioneer, Joseph Weizenbaum (1976), should appeal to *moral*, rather than technical, grounds for restricting the use of expert systems reflects the propensity of clients to invest the same level of trust in computers as in human beings.

4. GLOBAL CONSTRUCTIVISM AND
THE POLITICAL ECONOMY OF EXPERTISE

The senses (1)–(3) in which expertise is constitutively social are consistent with a *constructivist* sociological orientation (cf. Knorr-Cetina, 1981). Constructivists typically minimize the attribution of intrinsic properties to cognitive agents. Instead, they unpack such so-called intrinsic properties into relational ones, in which the relata are two mutually interpretive agents who jointly negotiate who will be credited with which properties. As we have seen, an adept expert can shift the burden of responsibility onto the client for the unpleasant consequences of following expert advice. But for the constructivist, there is no "fact of the matter" about whether the expert's incompetence or the client's recalcitrance is to blame, until the transaction has actually taken place. In discussions of cognitive science that acknowledge that society is more than a metaphor for the mind, constructivism is often presented as *the* sociological perspective. Empirically speaking, the dominance of constructivism cannot be denied, but I will ultimately appeal to sense (4) of expertise's sociality in order to introduce a different, and more comprehensive, sociological perspective.

Constructivism is a more heterogeneous doctrine than it may first appear. Weizenbaum and others whose opposition to the cognitive authority of either computer or human experts is mainly moral presuppose that such experts can, indeed, exercise all manner of authority, if they are not limited by convention. However, a milder species of constructivism is represented by Daniel Dennett's (e.g., 1987) instrumentalist approach to intentionality, which makes one's cognitive status dependent on another's interpretive stance. Dennett's constructivism is "asymmetrical," in that he does not grant the computer the same degree of agency as most human beings in constructing their respective identities. By contrast, a committed sense of symmetry is the hallmark of the more radical constructivists, the ethnographers who can be increasingly found on the sites where knowledge engineering occurs (cf. Greenbaum & Kyng, 1991). Here knowledge engineers, human and computer experts, clients, and other people and artifacts are portrayed as engaged in a mutually reinforcing cooperative venture. I will now address the limitations of this version of constructivism.

Drawing on the work of cultural anthropologists, especially Clifford Geertz (1983), ethnographic constructivism makes much of the "local" character of expert knowledge, which is brought out very clearly in the design of expert systems. The idiosyncrasy of locales is brought out in the knowledge engineer's interviews with both client and expert, followed by the process of adapting the expert system to the client's specific needs and abilities. These ethnographic accounts are meant to stand in striking contrast to the accounts typically given by the designers of AI systems in basic research settings, who often influence the way in which the applied researchers, the knowledge engineers, conceptualize their activities. Specifically, workers in AI tend to attribute properties to their programs (typically once embodied in a machine) that the ethnographers would prefer to regard as "boundary objects" in terms of which the identities (or cognitive capacities) of a variety of agents are negotiated. For example, the degree of satisfaction that the client gets from using an expert system has implications for how much of human expertise was successfully transferred to the computer program. While I wholeheartedly endorse this reinterpretation of expertise as a corrective to the accounts given by AI researchers, it nevertheless shares the fatal flaw of its opponents. Ironically, the flaw is the tendency to universalize from a single case. Let me explain.

It is one thing to say that all knowledge is local. That is rather boring. It is another to say that all locales are different from one another. That is more exciting. Ethnographers infer the exciting proposition from the boring one all the time. But what licenses this inference? Certainly, there have not been enough ethnographies of knowledge engineering to license the exciting proposition as an empirical generalization. In fact, the few ethnographies available do not depict locales so radically different from one another. My point here is not to decide the issue by philosophical argument but to observe that the ethnographic appeal to locality presupposes a *conceptual cartography* whereby one imagines that the spatio-temporal distance between locales represents a conceptual distance as well. In that sense, "the local" presupposes "the global," an image of the whole. Admittedly, sometimes this image turns out to be right; but other times it doesn't. However, that is a matter for empirical inquiry, and it is not clear that the ethnographic brand of constructivism encourages the appropriate sort of empirical inquiry. For, to learn about the global properties of knowledge engineering, one needs to discover the pattern by which expertise is *distributed* across a representative sample of locales and the *aggregated* consequences of such a distribution for the knowledge system as a whole. Here I start to speak the language of *political economy*, and to signal the quest for statistical correlations among variables that are hypothesized to be salient for understanding how expertise works.

What sorts of people are the producers and consumers of expert knowledge? Provided with a serviceable answer, we can study the distribution of expertise by focusing on a cognitively relevant locus of scarcity: A client has only so much time and money to spend consulting an expert, be it human or computer. Under what circumstances does the client feel she has gotten what she has paid for, and, when she does not feel that way, how is the blame apportioned: Who receives the lion's share of incompetence—the client or the expert? A key to understanding the distribution of expertise is to see how each side tries to convert its own sense of frustration into a perception of the other's liabilities. It would be fair to suppose that expert computers today receive far more attributions of incompetence than expert humans. A constructivist would diagnose this difference in terms of the client's lack of time, imagination, or interest in interpreting the computer as performing intelligently—perhaps because the client feels either that she has better things to do at this point, and the computer is in no position to prevent her from doing them, or that she would have to end up interpreting the computer as doing something other than she would have wanted (cf. Fuller, 1993b: 179–85). However, as people become more accustomed to dealing with expert computers, this difference in attribution is likely to disappear. But before concluding that we are projecting a future in which experts of all sorts are engaged in mutually satisfying relationships with their clients, we need to consider the aggregated consequences of people increasingly turning to computers for advice.

As we have seen, the history of expertise teaches that the expert is not a universalizable social role. There are no "experts" in areas that are regarded as commonsense or part of general education or easy to acquire without specialized training. Consequently, knowledge engineers are in the curious position of potentially destroying expertise as they diligently codify it and make it available to more people in user-friendly packages. While this consequence is bound to elude any on-site description of the knowledge engineer's work, it is nevertheless felt by professional associations that believe that knowledge engineers are indirectly deskilling their members. For, even as the human expert retains discretionary control over when, where, and how she uses her expertise, she may be losing discretionary control at the meta-level, namely, over who— or what—else counts as an expert in her field. Librarians have so far been most vocal in their concerns (cf. Pfaffenberger, 1990), but attempts by many doctors and lawyers to limit the scope of the interviews they give to knowledge engineers reflect similar worries.

At first glance, it may seem that the proliferation of expert systems is the ideal vehicle for democratizing expertise, as it would seem to put expert knowledge within the reach of more people. Just because the knowledge engineer can extract elements of expertise from her interviews with experts, it does not follow that the expertise remains intact once it is programmed into a computer for a client. After all, if expertise is indeed constitutively social, then altering the context in which expert knowledge is deployed should alter the character of the knowledge itself. Such change may be witnessed in the course of designing the interface that enables the client to interact with the expert system. Here the tendency has been to "go ergonomic" by designing interfaces that require the client to change his ordinary patterns of thought and behavior as little as possible (Downes, 1987). Less charitably put, the ergonomic approach reinforces the client's cognitive biases and thereby minimizes the learning experience that he might derive from engaging with the expertise as a form of knowledge. A potentially "dialectical" exchange is thus rendered merely "instrumental" (cf. Adorno & Horkheimer, 1972). The result is a spurious sense of autonomy, whereby the client's powers appear to be extended only because the environment in which he acts has been changed to his advantage (Fuller, 1986).

Thus, while the expert humans may lose some of their power as their clients increasingly rely on computerized systems, the clients themselves may not, in turn, become epistemically empowered. *Experts are deskilled without clients being reskilled.* Where has the original power of expertise gone? That power would seem to have dissipated somewhere in the knowledge engineering process, specifically when expertise was converted into a *tool* that exerted few of its own demands on the user (cf. Fields, 1987).

The utopian vision of democratized expertise is foiled by the simple fact that expertise, and perhaps knowledge more generally, is what economists call a *positional good* (Hirsch, 1977). A positional good is one whose value is directly tied to others not having it. Economists have generally refused to count knowledge as a positional good, preferring instead the classical philosophical position that knowledge is an "ethereal good," one whose value is not determined by the laws of supply and demand (Fuller, 1992). However, professional associations realize all too well the positional character of expertise.

The existence of positional goods is the dark secret of the welfare state. According to welfare economics, capitalism can avert a Marxist revolution because lingering inequalities of wealth will be resolved once a level of productivity is reached that enables everyone to be supported at a minimally acceptable standard. At most, stabilizing this situation will require a modest redistribution of income through progressive taxation. Overlooked in this scenario is that, as more goods are made more generally available, the perceived value of the goods may decline as they no longer serve to discriminate people in socially relevant ways (Bourdieu, 1984). Knowledge-intensive goods display such positionality effects. Higher education is perhaps the most obvious case in point: As it becomes easier for people to complete college, more postgraduate degrees are needed to acquire the same credentials. Should it become impossible either to stop the production of degree-holders or to set up additional barriers in the credentialing process, higher education will then no longer be seen as imparting an especially valued form of knowledge. Instead, it will take the place of bare literacy and the high school diploma as the minimum threshold for entry into the job market. Can anything be done reverse such positionality effects, or is the value of knowledge-intensive goods doomed to continual deflation?

The political scientist Yaron Ezrahi (1990) has argued that the scientific enterprise has come to consume so many resources and to produce so many questionable consequences that we may be reaching the end of the period (which began with the Enlightenment) when knowledge is presumed to be a public good. Ezrahi envisages that scientific forms of knowledge will gradually acquire the social character of artforms: their support will be privatized and their products customized to client tastes, which are presumed *not* to be universalizable. The expansion of intellectual property law to cover more instances of "basic research" suggests that Ezrahi's prognosis is already taking shape (Fuller, 1991). Given their own interest in customizing expertise for user demand, knowledge engineers clearly contribute to this overall trend toward privatization. Indeed, human experts may soon find the need to seek legal protection for their expertise, if only to earn royalties from the expert systems designed on the basis of it. In that way, knowledge engineers would not benefit too much from the "Japan Effect" of learning how to manufacture expertise more efficiently than the original experts themselves (cf. Weil & Snapper, 1990). This form of legal protection would, in turn, require a new category of intellectual property beyond the usual three of patent, copyright, and trademark.

But even if expertise were to become entirely market-driven, the skills surrounding the expertise would still attract human practitioners for the same reasons as art continues to do. The skills would be detached from the fame, fortune, or power that had been previously tied to them. Most of the perverse consequences of positional goods rest on such coupling (Crouch, 1983). For example, higher education is populated by a few people who are interested in the education process itself and many more who view it as a credentialing process, the surest route to a job. Decoupling those two groups would presumably help restore the integrity of higher education.

Knowledge engineers have a crucial role to play in the future disposition of expertise and knowledge more generally. Customized expert systems will hasten the demise of expertise and turn Ezrahi's image into a reality. However, the proliferation of such systems may also limit the client's potential for cognitive growth. A page from the history of manufacturing may prove instructive in resolving this dilemma.

Once the demand for manufactured products grew to a critical level, customization yielded to mass production (Beniger, 1986: 291–343). This transition was accompanied by the design of quality control standards for the mass produced goods. In the process of defining the minimum level of acceptability for a particular good, manufacturers effectively forced potential customers to adapt their behavior to the set dimensions of the good. Typically, these adaptations were dictated by the manufacturer's desire to cut costs, but knowledge engineers could collectively set guidelines for the design of expert systems, the successful use of which required clients to expand their cognitive repertoire. The sort of behavioral changes I envision here may be quite subtle. For example, an online library search system may discourage disciplinary provincialism by requiring the client to initiate searches by using protocols that are tied less to the jargon of specific disciplines and more to the exact topic or problem that client wishes to tackle. The system's database would, in turn, draw on the literatures of several disciplines so as not simply to confirm the course of inquiry that the client would be naturally inclined to follow (Cronin & Davenport, 1988: 316–27).

5. BUT IS EXPERTISE REALLY KNOWLEDGE?

It remains an open question whether the epistemic power of science is tied more to its sheer practice or to its status as a positional good. Those keen on retaining the ethereal quality of knowledge may no doubt want to make a strong distinction between "genuine knowledge" and "mere expertise" (cf. Ford & Agnew, 1992). They will object to my apparent conflation of these two concepts. Whereas expertise may ultimately reduce to matters of status and trust, the objector may argue, the test of knowledge is precisely that it does not lose its force as its availability increases. In response, let me grant the objector's distinction as the basis for an empirical hypothesis. If there are indeed types of "information" or "skills" whose power does not diminish with their increased availability, then I will gladly call them "genuine knowledge." However, my guess is that there are limits to the optimal distribution of these cognitive products. Consider the following homely observations:

(a) Everyone in a town may know the location of a particular store and the time it opens for business. However, if many of these people decide to act on that knowledge at roughly the same time to purchase the same goods, then a larger percentage of them will probably return home empty-handed than if fewer of them knew about the store in the first place. *Here knowledge lacked efficacy because the knowledgeable got in each other's way.* (A more realistic version of this situation is one in which everyone decides to take the same expert's advice on which stock to purchase.)

(b) It is often assumed that information freely exchanged among a large network of peers breeds the sort of critical inquiry that is necessary for genuine epistemic progress: the larger and freer the network, the more critical the inquiry. Unfortunately, this assumption presumes, contrary to fact, that inquirers have an inexhaustible ability and inclination to attend to each other's work. Yet, by the time the network of inquiry attains the dimensions of "Big Science," inquirers become more concerned with finding allies than opponents, and hence are likely to simply ignore work that cannot be immediately used for one's own purposes (cf. Fuller, 1994). *Here knowledge lacked efficacy because more of it was available than could be assimilated.*

Knowledge was undermined in (a) because too many people possessed the same information, whereas in (b) it was because each person possessed too much information. In neither case did these skewed distributions actually convert a truth into a falsehood, but from a pragmatic standpoint, they might as well have. In other words, attention to the socially distributed character of knowledge may help explain the intuitions that have traditionally led philosophers to posit a conception of knowledge that "transcends" the constitutively social character of expertise.

REFERENCES

Abbott, A. (1988). *The System of Professions.* Chicago: Chicago.
Adorno, T. & Horkheimer, M. (1972). *The Dialectic of Enlightenment.* New York: Continuum.
Arkes, H. & Hammond, K., Eds. (1986). *Judgment and Decision Making.* Cambridge: Cambridge.
Beniger, J. (1986). *The Control Revolution.* Cambridge: Harvard.
Botwinick, A. (1990). *Skepticism and Political Participation.* Philadelphia: Temple.
Bourdieu, P. (1984). *Distinction.* Cambridge: Harvard.
Cicourel, A. (1968). *Cognitive Sociology.* Harmondsworth: Penguin.

Collins, H. & Pinch, T. (1982). *Frames of Meaning*. London: Routledge & Kegan Paul.

Cronin, B. & Davenport, E. (1988). *Post-Professionalism: Transforming the Information Heartland*. London: Taylor Graham.

Crouch, C. (1983). Market failure. In A. Ellis & K. Kumar (Eds.), *Dilemmas of Liberal Democracies*. London: Tavistock.

De Mey, M. (1982). *The Cognitive Paradigm*. Dordrecht: Reidel.

De Romilly, J. (1992). *The Great Sophists in Periclean Athens*. Oxford: Clarendon.

Dennett, D. (1987). *The Intentional Stance*. Cambridge: MIT.

Downes, S. (1987). A philosophical ethnography of human-computer interaction research. *Social Epistemology*, 1, 27–36.

Ezrahi, Y. (1990). *The Descent of Icarus*. Cambridge: Harvard.

Faust, D. (1985). *The Limits of Scientific Reasoning*. Minneapolis: Minnesota.

Fields, C. (1987). The computer as tool. *Social Epistemology*, 1, 5–26.

Ford, K. & Agnew, N. (1992). Expertise: socially situated, personally constructed, and "reality" relevant. Paper presented at the AAAI Spring Symposium on the Cognitive Aspects of Knowledge Acquisition. Stanford.

Fuller, S. (1986). User-friendliness: friend or foe? *Logos*, 7, 93–98.

Fuller, S. (1988). *Social Epistemology*. Bloomington: Indiana.

Fuller, S. (1991). Studying the proprietary grounds of knowledge. *Journal of Social Behavior and Personality*, 6 (6), 105–28.

Fuller, S. (1992). Knowledge as product and property. In N. Stehr & R. Ericson (Eds.), *The Culture and Power of Knowledge* (pp. 157–90). Berlin: Walter de Gruyter.

Fuller, S. (1993a). *Philosophy of Science and Its Discontents*. 2nd edn. (Orig. 1989) New York: Guilford.

Fuller, S. (1993b). *Philosophy, Rhetoric, and the End of Knowledge: The Coming of Science & Technology Studies*. Madison: University of Wisconsin Press.

Fuller, S. (1994). The social psychology of scientific knowledge: another strong programme. In W. Shadish & S. Fuller (Eds.), *Social Psychology of Science*. New York: Guilford: 162–80.

Gambetta, D., Ed. (1988). *Trust*. Oxford: Blackwell.

Geertz, C. (1983). *Local Knowledge*. New York: Basic.

Greenbaum, J. & Kyng, M., Eds. (1991). *Design at Work*. Hillsdale NJ: Lawrence Erlbaum.

Haddock, K. & Houts, A. (1992). Answers to philosophical and sociological uses of psychologism in science studies. In R. Giere (Ed.), *Cognitive Models of Science* (pp. 367–99). Minneapolis: Minnesota.

Haugeland, J. (1984). *Artificial Intelligence: The Very Idea*. Cambridge: MIT Press.

Held, D. (1987). *Models of Democracy*. Oxford: Polity.

Hirsch, F. (1977). *Social Limits to Growth*. London: Routledge & Kegan Paul.

Knorr-Cetina, K. (1981). *The Manufacture of Knowledge*. Oxford: Pergamon.

Minsky, M. (1986). *The Society of Mind*. New York: Simon & Schuster.

Pfaffenberger, B. (1990). *Democratizing Information*. Boston: G. K. Hall.

Popper, K. (1970). Normal science and its dangers. In I. Lakatos & A. Musgrave (Eds.), *Criticism and the Growth of Knowledge* (pp. 51–58). Cambridge: Cambridge.

Proctor, R. (1991). *Value-Neutral Science?* Cambridge: Harvard.

Simon, H. (1981). *The Sciences of the Artificial*. Cambridge: MIT.

Smith, C. (1981). *The Mind of the Market.* Totowa NJ: Rowman & Littlefield.

Weil, V. & Snapper, J., Eds. (1990). *Owning Scientific and Technical Information.* New Brunswick NJ: Rutgers.

Weizenbaum, J. (1976). *Computer Power and Human Reason.* San Francisco. W. H. Freeman.

Williams, R. (1983). *Keywords.* Oxford: Oxford.

13. HOW TO DEFEND SOCIETY AGAINST SCIENCE

PAUL FEYERABEND

PRACTITIONERS OF A strange trails, friends, enemies, ladies and gentlemen: Before starting with my talk, let me explain to you how it came into existence.

About a year ago I was short of funds. So I accepted an invitation to contribute to a book dealing with the relation between science and religion. To make the book sell I thought I should make my contribution a provocative one and that most provocative statement one can make about the relation between science and religion is that science is a religion. Having made the statement the core of my article I discovered that lots of reasons, lots of excellent reasons, could be found for it. I enumerated the reasons, finished my article, and got paid. That was stage one.

Next I was invited to a Conference for the Defence of Culture. I accepted the invitation because it paid for my flight to Europe. I also must admit that I was rather curious. When I arrived in Nice I had no idea what I would say. Then while the conference was taking its course I discovered that everyone thought very highly of science and that everyone was very serious. So I decided to explain how one could defend culture from science. All the reasons collected in my article would apply here as well and there was no need to invent new things. I gave my talk, was rewarded with an outcry about my "dangerous and ill considered ideas," collected my ticket and went on to Vienna. That was stage number two.

Now I am supposed to address you. I have a hunch that in some respect you are very different from my audience in Nice. For one, you look much younger. My audience in Nice was full of professors, businessmen, television executives, and the average age was about 58 1/2. Then I am quite sure that most of you are considerably to the left of most of the people in Nice. As a matter of fact, speaking somewhat superficially I might say that you are a leftist audience while my audience in Nice was a rightist audience. Yet despite all these differences you have some things in common. Both of you, I assume, respect science and knowledge. Science, of course, must be reformed and must be made less authoritarian. But once the reforms are carried out, it is a valuable source of knowledge that must not be contaminated by ideologies of a different kind. Secondly, both of you are serious people. Knowledge is a serious matter, for the

The following article is a revised version of a talk given to the Philosophy Society at Sussex University in November 1974.

Right as well as for the Left, and it must be pursued in a serious spirit. Frivolity is out, dedication and earnest application to the task at hand is in. These similarities are all I need for repeating my Nice talk to you with hardly any change. So, here it is.

FAIRYTALES

I want to defend society and its inhabitants from all ideologies, science included. All ideologies must be seen in perspective. One must not take them too seriously. One must read them like fairytales which have lots of interesting things to say but which also contain wicked lies, or like ethical prescriptions which may be useful rules of thumb but which are deadly when followed to the letter.

Now, is this not a strange and ridiculous attitude? Science, surely, was always in the forefront of the fight against authoritarianism and superstition. It is to science that we owe our increased intellectual freedom vis-à-vis religious beliefs; it is to science that we owe the liberation of mankind from ancient and rigid forms of thought. Today these forms of thought are nothing but bad dreams—and this we learned from science. Science and enlightenment are one and the same thing—even the most radical critics of society believe this. Kropotkin wants to overthrow all traditional institutions and forms of belief, with the exception of science. Ibsen criticises the most intimate ramifications of nineteenth-century bourgeois ideology, but he leaves science untouched. Levi-Strauss has made us realise that Western Thought is not the lonely peak of human achievement it was once believed to be, but he excludes science from his relativization of ideologies. Marx and Engels were convinced that science would aid the workers in their quest for mental and social liberation. Are all these people deceived? Are they all mistaken about the role of science? Are they all the victims of a chimaera?

To these questions my answer is a firm *Yes and No.*

Now, let me explain my answer.

My explanation consists of two parts, one more general, one more specific.

The general explanation is simple. Any ideology that breaks the hold a comprehensive system of thought has on the minds of men contributes to the liberation of man. Any ideology that makes man question inherited beliefs is an aid to enlightenment. A truth that reigns without checks and balances is a tyrant who must be overthrown and any falsehood that can aid us in the overthrow of this tyrant is to be welcomed. It follows that seventeenth- and eighteenth-century science indeed *was* an instrument of liberation and enlightenment. It does not follow that science is bound to *remain* such an instrument. There is nothing inherent in science or in any other ideology that makes it *essentially* liberating. Ideologies can deteriorate and become stupid religions. Look at Marxism. And that the science of today is very different from the science of 1650 is evident at the most superficial glance.

For example, consider the role science now plays in education. Scientific "facts" are taught at a very early age and in the very same manner in which religious "facts" were taught only a century ago. There is no attempt to waken the critical abilities of the pupil so that he may be able to see things in perspective. At the universities the situation is even worse, for indoctrination is here carried out in a much more systematic manner. Criticism is not entirely absent. Society, for example, and its institutions, are criticised most severely and often most unfairly, and this already at the elementary school level. But science is excepted from the criticism. In society at large the judgement of the scientist is received with the same reverence as the judgement of bishops and cardinals was accepted not too long ago. The move towards "demythologization," for example, is largely motivated by the wish to avoid any clash between Christianity and scientific ideas. If such a clash occurs, then science is certainly right and Christianity wrong. Pursue this investigation further and you will see that science has now become as oppressive as the ideologies it had once to fight. Do not be misled by the fact that today hardly anyone gets killed for joining a scientific heresy. This has nothing to do with science. It has something to do with the general quality of our civilization. Heretics in science are still made to suffer from the *most severe* sanctions this relatively tolerant civilization has to offer.

But—is this description not utterly unfair? Have I not presented the matter in a very distorted light by using tendentious and distorting terminology? Must we not describe the situation in a very different way? I have said that science has become rigid, that it has ceased to be an instrument of *change* and *liberation*, without adding that it has found the *truth*, or a large part thereof. Considering this additional fact we realise, so the objection goes, that the rigidity of science is not due to human wilfulness. It lies in the nature of things. For once we have discovered the truth—what else can we do but follow it?

This trite reply is anything but original. It is used whenever an ideology wants to reinforce the faith of its followers. "Truth" is such a nicely neutral word. Nobody would deny that it is commendable to speak the truth and wicked to tell lies. Nobody would deny that—and yet nobody knows what such an attitude amounts to. So it is easy to twist matters and to change allegiance to truth in one's everyday affairs into allegiance to the Truth of an ideology which is nothing but the dogmatic defense of that ideology. And it is of course *not* true that we *have* to follow the truth. Human life is guided by many ideas. Truth is one of them. Freedom and mental independence are others. If Truth, as conceived by some ideologists, conflicts with freedom, then we have a choice. We may abandon freedom. But we may also abandon Truth. (Alternatively, we may adopt a more sophisticated idea of truth that no longer contra-dicts freedom; that was Hegel's solution.) My criticism of modern science is that it inhibits freedom of thought. If the reason is that it has found the truth and now follows it then I would say that there are better things than first find-ing, and then following such a monster.

This finishes the general part of my explanation.

There exists a more specific argument to defend the exceptional position science has in society today. Put in a nutshell the argument says (1) that science has finally found the correct method for achieving results and (2) that there are many results to prove the excellence of the method. The argument is mis-taken—but most attempts to show this lead into a dead end. Methodology has by now become so crowded with empty sophistication that it is extremely dif-ficult to perceive the simple errors at the basis. It is like fighting the hydra—cut off one ugly head, and eight formalizations take its place. In this situation the only answer is superficiality: when sophistication loses content then the only way of keeping in touch with reality is to be crude and superficial. This is what I intend to be.

Against Method

There is a method, says part (1) of the argument. What is it? How does it work? One answer which is no longer as popular as it used to be is that science works by collecting facts and inferring theories from them. The answer is unsatisfac-tory as theories never follow from facts in the strict logical sense. To say that they may yet be supported by facts assumes a notion of support that (a) does not show this defect and is (b) sufficiently sophisticated to permit us to say to what extent, say, the theory of relativity is supported by the facts. No such no-tion exists today nor is it likely that it will ever be found (one of the problems is that we need a notion of support in which grey ravens can be said to support "All ravens are black"). This was realised by conventionalists and transcenden-tal idealists who pointed out that theories shape and order facts and can there-fore be retained come what may. They can be retained because the human mind either consciously or unconsciously carries out its ordering function. The trouble with these views is that they assume for the mind what they want to explain for the world, viz., that it works in a regular fashion. There is only one view which overcomes all these difficulties. It was invented twice in the nineteenth century, by Mill, in his immortal essay *On Liberty*, and by some Darwinists who extended Darwinism to the battle of ideas. This view takes the bull by the horns: theories cannot be justified and their excellence cannot be shown without reference to other theories. We may explain the *success* of a the-ory by reference to a more comprehensive theory (we may explain the success of Newton's theory by using the general theory of relativity); and we may explain our preference for it by comparing it with other theories.

Such a comparison does not establish the intrinsic excellence of the theory we have chosen. As a matter of fact, the theory we have chosen may be pretty lousy. It may contain contradictions, it may conflict with well-known facts, it may be cumbersome, unclear, *ad hoc* in decisive places, and so on. But it may still be better than any other theory that is available at the time. It may in fact be the best lousy theory there is. Nor are the standards of judgement chosen in an absolute manner. Our sophistication increases with every choice we make, and so do our standards. Standards compete just as theories compete and we choose the standards most appropriate to the historical situation in which the choice occurs. The rejected alternatives (theories; standards; "facts") are not eliminated. They serve as correctives (after all, we may have made the wrong choice) and they also explain the content of the preferred views (we understand relativity better when we understand the structure of its competitors; we know the full meaning of freedom only when we have an idea of life in a totalitarian state, of its advantages—and there are many advantages—as well as of its disadvantages). Knowledge so conceived is an ocean of alternatives channelled and subdivided by an ocean of standards. It forces our mind to make imaginative choices and thus makes it grow. It makes our mind capable of choosing, imagining, criticising.

Today this view is often connected with the name of Karl Popper. But there are some very decisive differences between Popper and Mill. To start with, Popper developed his view to solve a special problem of epistemology—he wanted to solve "Hume's problem." Mill, on the other hand, is interested in conditions favourable to human growth. His epistemology is the result of a certain theory of man, and not the other way around. Also Popper, being influenced by the Vienna Circle, improves on the logical form of a theory before discussing it while Mill uses every theory in the form in which it occurs in science. Thirdly, Popper's standards of comparison are rigid and fixed, while Mill's standards are permitted to change with the historical situation. Finally, Popper's standards eliminate competitors once and for all: theories that are either not falsifiable or falsifiable and falsified have no place in science. Popper's criteria are clear, unambiguous, precisely formulated; Mill's criteria are not. This would be an advantage if science itself were clear, unambiguous, and precisely formulated. Fortunately, it is not.

To start with, no new and revolutionary scientific theory is ever formulated in a manner that permits us to say under what circumstances we must regard it as endangered: many revolutionary theories are unfalsifiable. Falsifiable versions do exist, but they are hardly ever in agreement with accepted basic statements: every moderately interesting theory is falsified. Moreover, theories have formal flaws, many of them contain contradictions, *ad hoc* adjustments, and so on and so forth. Applied resolutely, Popperian criteria would eliminate science without replacing it by anything comparable. They are useless as an aid to science. In the past decade this has been realised by various thinkers, Kuhn and Lakatos among them. Kuhn's ideas are interesting but, alas, they are much too vague to give rise to anything but lots of hot air. If you don't believe me, look at the literature. Never before has the literature on the philosophy of science been invaded by so many creeps and incompetents. Kuhn encourages people who have no idea why a stone falls to the ground to talk with assurance about scientific method. Now I have no objection to incompetence but I do object when incompetence is accompanied by boredom and self-righteousness. And this is exactly what happens. We do not get interesting false ideas, we get boring ideas or words connected with no ideas at all. Secondly, wherever one tries to make Kuhn's ideas more definite, one finds that they are *false*. Was there ever a period of normal science in the history of thought? No—and I challenge anyone to prove the contrary.

Lakatos is immeasurably more sophisticated than Kuhn. Instead of theories he considers research programmes which are sequences of theories connected by methods of modification, so-called heuristics. Each theory in the sequence may be full of faults. It may be beset by anomalies, contradictions, ambiguities. What counts is not the shape of the single theories, but the tendency exhibited by the sequence. We judge historical developments and achievements over a period of time, rather than the situation at a particular time. History and methodology are combined into a single enterprise. A research programme is said to progress if the sequence of theories leads to novel predictions. It is said to degenerate if it is reduced to absorbing facts that have been discovered without its help. A decisive feature of Lakatos' methodology is that such evaluations are no longer tied to methodological rules which tell the scientist either to retain or to abandon a research programme. Scientists may stick to a degenerating programme; they may even succeed in making the programme overtake its rivals and they therefore proceed rationally whatever they are doing (provided they continue calling degenerating programmes degenerating and progressive programmes progressive). This means that Lakatos offers *words* which *sound* like the elements of a methodology; he does not offer a methodology. There is no method according to the most advanced and sophisticated methodology in existence today. This finishes my reply to part (1) of the specific argument.

AGAINST RESULTS

According to part (2), science deserves a special position because it has produced *results*. This is an argument only if it can be taken for granted that nothing else has ever produced results. Now it may be admitted that almost everyone who discusses the matter makes such an assumption. It may also be admitted that it is not easy to show that the assumption is false. Forms of life different from science either have disappeared or have degenerated to an extent that makes a fair comparison impossible. Still, the situation is not as hopeless as it was only a decade ago. We have become acquainted with methods of medical diagnosis and therapy which are effective (and perhaps even more effective than the corresponding parts of Western medicine) and which are yet based on an ideology that is radically different from the ideology of Western science. We have learned that there are phenomena such as telepathy and telekinesis which are obliterated by a scientific approach and which could be used to do research in an entirely novel way (earlier thinkers such as Agrippa of Nettesheim, John Dee, and even Bacon were aware of these phenomena). And then—is it not the case that the Church saved souls while science often does the very opposite? Of course, nobody now believes in the ontology that underlies this judgement. Why? Because of ideological pressures identical with those which today make us listen to science to the exclusion of everything else. It is also true that phenomena such as telekinesis and acupuncture may eventually be absorbed into the body of science and may therefore be called "scientific." But note that this happens only after a long period of resistance during which a science *not yet* containing the phenomena wants to get the upper hand over forms of life that contain them. And this leads to a further objection against part (2) of the specific argument. The fact that science has results counts in its favour only if these results were achieved by science alone, and without any outside help. A look at history shows that science hardly ever gets its results in this way. When Copernicus introduced a new view of the universe, he did not consult *scientific* predecessors, he consulted a crazy Pythagorean such as Philolaos. He adopted his ideas and he maintained them in the face of all sound rules of scientific method. Mechanics and optics owe a lot to artisans, medicine to midwives and witches. And in *our* own day we have seen how the interference of the state can advance science: when the Chinese communists refused to be intimidated by the judgement of experts and ordered traditional medicine back into universities and hospitals there was an outcry all over the world that science would now be ruined by China. The very opposite occurred: Chinese science advanced and Western science learned from it. Wherever we look we see that great scientific advances are due to outside interference which is made to prevail in the face of the most basic and most "rational" methodological rules. The lesson is plain: there does not exist a single argument that could be used to support the exceptional role which science today plays in society. Science has done many things, but so have other ideologies. Science often proceeds systematically, but so do other ideologies (just consult the records of the many doctrinal debates that took place in the Church) and, besides, there are no overriding rules which are adhered to under any circumstances; there is no "scientific methodology" that can be used to separate science from the rest. *Science is just one of the many ideologies that propel society and it should be treated as such* (this statement applies even to the most progressive and most dialectical sections of science). What consequences can we draw from this result?

The most important consequence is that there must be a *formal separation between state and science* just as there is now a formal separation between state and church. Science may influence society but only to the extent to which any political or other pressure group is permitted to influence society. Scientists may be consulted on important projects but the final judgement must be left to the democratically elected consulting bodies. These bodies will consist mainly of laymen. Will the laymen be able to come to a correct judgement? Most certainly, for the competence, the complications and the successes of science are vastly exaggerated. One of the most exhilarating experiences is to see how a lawyer, who is a layman, can find holes in the testimony, the technical testimony, of the most advanced expert and thus prepare the jury for its verdict. Science is not a closed book that is understood only after years of training. It is an intellectual discipline that can be examined and criticized by anyone who is interested and that looks difficult and profound only because of a systematic campaign of obfuscation carried out by many scientists (though, I am happy to say, not by all). Organs of the state should never hesitate to reject the judgement of scientists when they have reason for doing so. Such rejection will educate the general public, will make it more confident, and it may even lead to improvement. Considering the sizeable chauvinism of the scientific establishment we can say: the more Lysenko affairs, the better (it is not the *interference* of the state that is objectionable in the case of Lysenko, but the *totalitarian* interference which kills the opponent rather than just neglecting his advice). Three cheers to the fundamentalists in California who succeeded in having a dogmatic formulation of the theory of evolution removed from the textbooks and an account of Genesis included. (But I know that they would become as chauvinistic and totalitarian as scientists are today when given the chance to run society all by themselves. Ideologies are marvelous when used in the companies of other ideologies. They become boring and doctrinaire as soon as their merits lead to the removal of their opponents.) The most important change, however, will have to occur in the field of education.

EDUCATION AND MYTH

The purpose of education, so one would think, is to introduce the young into life, and that means: into the *society* where they are born and into the *physical universe* that surrounds the society. The method of education often consists in the teaching of some *basic myth*. The myth is available in various versions.

More advanced versions may be taught by initiation rites which firmly implant them into the mind. Knowing the myth, the grown-up can explain almost everything (or else he can turn to experts for more detailed information). He is the master of Nature and of Society. He understands them both and he knows how to interact with them. However, *he is not the master of the myth that guides his understanding.*

Such further mastery was aimed at, and was partly achieved, by the Presocratics. The Presocratics not only tried to understand the *world.* They also tried to understand, and thus to become the masters of, the *means of understanding the world.* Instead of being content with a single myth they developed many and so diminished the power which a well-told story has over the minds of men. The sophists introduced still further methods of reducing the debilitating effect of interesting, coherent, "empirically adequate" etc. etc. tales. The achievements of these thinkers were not appreciated and they certainly are not understood today. When teaching a myth we want to increase the chance that it will be understood (i.e. no puzzlement about any feature of the myth), believed, *and accepted.* This does not do any harm when the myth is counterbalanced by other myths: even the most dedicated (i.e. totalitarian) instructor in a certain version of Christianity cannot prevent his pupils from getting in touch with Buddhists, Jews and other disreputable people. It is very different in the case of science, or of rationalism where the field is almost completely dominated by the believers. In this case it is of paramount importance to strengthen the minds of the young, and "strengthening the minds of the young" means strengthening them *against* any easy acceptance of comprehensive views. What we need here is an education that makes people *contrary, counter-suggestive*, without making them incapable of devoting themselves to the elaboration of any single view. How can this aim be achieved?

It can be achieved by protecting the tremendous imagination which children possess and by developing to the full the spirit of contradiction that exists in them. On the whole children are much more intelligent than their teachers. They succumb, and give up their intelligence because they are bullied, or because their teachers get the better of them by emotional means. Children can learn, understand, and keep separate two to three different languages ("children" and by this I mean three to five year olds, *not* eight year olds who were experimented upon quite recently and did not come out too well; why? because they were already loused up by incompetent teaching at an earlier age). Of course, the languages must be introduced in a more interesting way than is usually done. There are marvelous writers in all languages who have told marvelous stories—let us begin our language teaching with *them* and not with "der Hund hat einen Schwanz" and similar inanities. Using stories we may of course also introduce "scientific" accounts, say, of the origin of the world and thus make the children acquainted with science as well. But science must not be given any special position except for pointing out that there are lots of people who believe in it. Later on the stories which have been told will be supplemented with "reasons," where by reasons I mean further accounts of the kind found in the tradition to which the story belongs. And, of course, there will also be contrary reasons. Both reasons and contrary reasons will be told by the experts in the fields and so the young generation becomes acquainted with all kinds of sermons and all types of wayfarers. It becomes acquainted with them, it becomes acquainted with their stories, and every individual can make up his mind which way to go. By now everyone knows that you can earn a lot of money and respect and perhaps even a Nobel Prize by becoming a scientist, so many will become scientists. They will *become* scientists *without having been taken in by the ideology of science,* they will *be* scientists *because they have made a free choice.* But has not much time been wasted on unscientific subjects and will this not detract from their competence once they have become scientists? Not at all! The progress of science, of good science depends on novel ideas and on intellectual freedom: science has very often been advanced by outsiders (remember that Bohr and Einstein regarded themselves as outsiders). Will not many people make the wrong choice and end up in a dead end? Well, that depends on what you mean by a "dead end." Most scientists today are devoid of ideas, full of fear, intent on producing some paltry result so that they can add to the flood of inane papers that now constitutes "scientific progress" in many areas. And, besides, what is more important? To lead a life which one has chosen with open eyes, or to spend one's time in the nervous attempt of avoiding what some not so intelligent people call "dead ends"? Will not the number of scientists decrease so that in the end there is nobody to run our precious laboratories? I do not think so. Given a choice many people may choose science, for a science that is run by free agents looks much more attractive than the science of today which is run by slaves, slaves of institutions and slaves of "reason." And if there is a temporary shortage of scientists the situation may always be remedied by various kinds of incentives. Of course, scientists will not play any predominant role in the society I envisage. They will be more than balanced by magicians, or priests, or astrologers. Such a situation is unbearable for many people, old and young, right and left. Almost all of you have the firm belief that at least *some* kind of truth has been found, that it must be preserved, and that the method of teaching I advocate and the form of society I defend will dilute it and make it finally disappear. You have this firm belief; many of you may even have reasons. But what *you have to consider is that the absence of good contrary reasons is due to a historical accident;* it does *not* lie in the nature of things. Build up the kind of society I recommend and the views you now despise (without knowing them, to be sure) will return in such splendour that you will have to work hard to maintain your own position and will perhaps be entirely unable to do so. You do not believe me? Then look at history. Scientific astronomy was firmly founded on Ptolemy and Aristotle, two of the greatest minds in the history of Western Thought. Who upset their well-argued, empirically adequate and precisely formulated system? Philolaos the mad and antediluvian Pythagorean. How was it that Philolaos could stage such a comeback? Because he found an able defender: Copernicus. Of course, you may follow your intuitions as I am following mine. But remember that your intuitions are the result of your "scientific" training where by science I also mean the science of Karl Marx. My training, or, rather, my non-training, is that of a journalist who is interested in strange and bizarre events. Finally, is it not utterly irresponsible, in the present world situation, with millions of people starving, others enslaved, downtrodden, in abject misery of body and mind, to think luxurious thoughts such as these? Is not freedom of choice a luxury under such circumstances? Is not the flippancy and the humour I want to see combined with the freedom of choice a luxury under such circumstances? Must we not give up all self indulgence and *act*? Join together, and *act*? That is the most important objection which today is raised against an approach such as the one recommended by me. It has tremendous appeal, it has the appeal of unselfish dedication. Unselfish dedication—to what? Let us see!

We are supposed to give up our selfish inclinations and dedicate ourselves to the liberation of the oppressed. And selfish inclinations are what? They are our wish for maximum liberty of thought in the society in which we live *now*, maximum liberty not only of an abstract kind, but expressed in appropriate institutions and methods of teaching. This wish for concrete intellectual and physical liberty in our own surroundings is to be put aside for the time being. This assumes, first, that we do not need this liberty for our task. It assumes that we can carry out our task with a mind that is firmly closed to some alternatives. It assumes that the correct way of liberating others *has always been found* and that all that is needed is to carry it out. I am sorry, I cannot accept such doctrinaire self-assurance in such extremely important matters. Does this mean that we cannot act at all? It does not. But it means that *while acting we have to try to realise as much of the freedom I have recommended so that our actions may be corrected in the light of the ideas we get while increasing our freedom.* This will slow us down, no doubt, but are we supposed to charge ahead simply because some people tell us that they have found an explanation for all the misery and an excellent way out of it? Also we want to liberate people not to make them succumb to a new kind of slavery, *but to make them realise their own wishes*, however different these wishes may be from our own. Self-righteous and narrow-minded liberators cannot do this. As a rule they soon impose a slavery that is worse, because more systematic, than the very sloppy slavery they have removed. And as regards humour and flippancy the answer should be obvious. Why would anyone want to liberate anyone else? Surely not because of some *abstract* advantage of liberty but because liberty is the best way to free development *and thus to hap-*

piness. We want to liberate people so that *they can smile.* Shall we be able to do this if we ourselves have forgotten how to smile and are frowning on those who still remember? Shall we then not spread another disease, comparable to the one we want to remove, the disease of puritanical self-righteousness? Do not object that dedication and humour do not go together—Socrates is an excellent example to the contrary. *The hardest task needs the lightest hand or else its completion will not lead to freedom but to a tyranny much worse than the one it replaces.*

14. OPPONENTS, AUDIENCES, CONSTITUENCIES, AND COMMUNITY

EDWARD W. SAID

WHO WRITES? For whom is the writing being done? In what circumstances? These, it seems to me, are the questions whose answers provide us with the ingredients making for a politics of interpretation. But if one does not wish to ask and answer the questions in a dishonest and abstract way, some attempt must be made to show why they are questions of some relevance to the present time. What needs to be said at the beginning is that the single most impressive aspect of the present time—at least for the "humanist," a description for which I have contradictory feelings of affection and revulsion—is that it is manifestly the Age of Ronald Reagan. And it is in this age as a context and setting that the politics of interpretation and the politics of culture are enacted.

I do not want to be misunderstood as saying that the cultural situation I describe here caused Reagan, or that it typifies Reaganism, or that everything about it can be ascribed or referred back to the personality of Ronald Reagan. What I argue is that a particular situation within the field we call "criticism" is not merely related to but is an integral part of the currents of thought and practice that play a role within the Reagan era. Moreover, I think, "criticism" and the traditional academic humanities have gone through a series of developments over time whose beneficiary and culmination is Reaganism. Those are the gross claims that I make for my argument.

A number of miscellaneous points need to be made here. I am fully aware that any effort to characterize the present cultural moment is very likely to seem quixotic at best, unprofessional at worst. But that, I submit, is an aspect of the present cultural moment, in which the social and historical setting of critical activity is a totality felt to be benign (free, apolitical, serious), uncharacterizable as a whole (it is too complex to be described in general and tendentious terms), and somehow outside history. Thus it seems to me that one thing to be tried—out of sheer critical obstinacy—is precisely *that* kind of generalization, *that* kind of political portrayal, *that* kind of overview condemned by the present dominant culture to appear inappropriate and doomed from the start.

It is my conviction that culture works very effectively to make invisible and even "impossible" the actual *affiliations* that exist between the world of ideas and scholarship, on the one hand, and the world of brute politics, corporate and state power, and military force, on the other. The cult of expertise and professionalism, for example, has so restricted our scope of vision that a positive (as opposed to an implicit or passive) doctrine of noninterference among fields has set in. This doctrine has it that the general public is best left ignorant, and the most crucial policy questions affecting human existence are best left to "experts," specialists who talk about their specialty only, and—to use the word first given wide social approbation by Walter Lippmann in *Public Opinion* and *The Phantom Public*—"insiders," people (usually men) who are endowed with the special privilege of knowing how things really work and, more important, of being close to power.[1]

Humanistic culture in general has acted in tacit compliance with this anti-democratic view, the more regrettably since, both in their formulation and in the politics they have given rise to, so-called policy issues can hardly be said to enhance human community. In a world of increasing interdependence and political consciousness, it seems both violent and wasteful to accept the notion, for example, that countries ought to be classified simply as pro-Soviet or pro-American. Yet this classification—and with it the reappearance of a whole range of cold war motifs and symptoms (discussed by Noam Chomsky in *Towards a New Cold War*)—dominates thinking about foreign policy. There is little in humanistic culture that is an effective antidote to it, just as it is true that few humanists have very much to say about the problems starkly dramatized by the 1980 Report of the Independent Commission on International Development Issues, *North-South: A Programme for Survival.* Our political discourse is now choked with enormous, thought-stopping abstractions, from terrorism, Communism, Islamic fundamentalism, and instability, to moderation, freedom, stability, and strategic alliances, all of them as unclear as they are both potent and unrefined in their appeal. It is next to impossible to think about human society either in a global way (as Richard Falk eloquently does in *A Global Approach to National Policy* [1975]) or at the level of everyday life. As Philip Green shows in *The Pursuit of Inequality*, notions like equality and welfare have simply been chased off the intellectual landscape. Instead a brutal Darwinian picture of self-help and self-promotion is proposed by Reaganism, both domestically and internationally, as an image of the world ruled by what is being called "productivity" or "free enterprise."

Add to this the fact that liberalism and the Left are in a state of intellectual disarray and fairly dismal perspectives emerge. The challenge posed by these perspectives is not how to cultivate one's garden despite them but how to understand cultural work occurring within them. What I propose here, then, is a rudimentary attempt to do just that, notwithstanding a good deal of inevitable incompleteness, overstatement, generalization, and crude characterization. Finally, I will very quickly propose an alternative way of undertaking cultural work, although anything like a fully worked-out program can only be done collectively and in a separate study.

My use of "constituency," "audience," "opponents," and "community" serves as a reminder that no one writes simply for oneself. There is always an Other; and this Other willy-nilly turns interpretation into a social activity, albeit with unforeseen consequences, audiences, constituencies, and so on. And, I would add, interpretation is the work of intellectuals, a class badly in need today of moral rehabilitation and social redefinition. The one issue that urgently requires study is, for the humanist no less than for the social scientist, the status of *information* as a component of knowledge: its sociopolitical status, its contemporary fate, its economy (a subject treated recently by Herbert Schiller in *Who Knows: Information in the Age of the Fortune 500*). We all think we know what it means, for example, to *have* information and to write and interpret texts containing information. Yet we live in an age which places unprecedented emphasis on the production of knowledge and information, as Fritz Machlup's *Production and Distribution of Knowledge in the United States* dramatizes clearly. What happens to information and knowledge, then, when IBM and AT&T—two of the world's largest corporations—claim that what they do is to put "knowledge" to work "for the people"? What is the role of humanistic knowledge and information if they are not to be unknowing (many ironies there) partners in commodity production and marketing, so much so that what humanists do may in the end turn out to be a quasi-religious concealment of this peculiarly unhumanistic process? A true secular politics of interpretation sidesteps this question at its peril.

1

At a recent MLA convention, I stopped by the exhibit of a major university press and remarked to the amiable sales representative on duty that there seemed to be no limit to the number of highly specialized books of advanced literary criticism his press put out. "Who reads these books?" I asked, implying of course that however brilliant and important most of them were they were difficult to read and therefore could not have a wide audience—or at least an audience wide enough to justify regular publication during a time of economic crisis. The answer I received made sense, assuming I was told the truth. People who write specialized, advanced (i.e., New New) criticism faithfully read each other's books. Thus each such book could be assured of, but wasn't necessarily always getting, sales of around three thousand copies, "all other things being equal." The last qualification struck me as ambiguous at best, but it needn't detain us here. The point was that a nice little audience had been built and could be routinely mined by this press; certainly, on a much larger scale, publishers of cookbooks and exercise manuals apply a related principle as they churn out what may seem like a very long series of unnecessary books, even if an expanding crowd of avid food and exercise aficionados is not quite the same thing as a steadily attentive and earnest crowd of three thousand critics reading each other.

What I find peculiarly interesting about the real or mythical three thousand is that whether they derive ultimately from the Anglo-American New Criticism (as formulated by I. A. Richards, William Empson, John Crowe Ransom, Cleanth Brooks, Allen Tate, and company, beginning in the 1920s and continuing for several decades thereafter) or from the so-called New New Criticism (Roland Barthes, Jacques Derrida, et al., during the 1960s), they vindicate, rather than undermine, the notion that intellectual labor ought to be divided into progressively narrower niches. Consider very quickly the irony of this. New Criticism claimed to view the verbal object as in itself it really was, free from the distractions of biography, social message, even paraphrase. Matthew Arnold's critical program was thereby to be advanced not by jumping directly from the text to the whole of culture but by using a highly concentrated verbal analysis to comprehend cultural values available only through a finely wrought literary structure finely understood.

Charges made against the American New Criticism that its ethos was clubby, gentlemanly, or Episcopalian are, I think, correct only if it is added that in practice New Criticism, for all its elitism, was strangely populist in intention. The idea behind the pedagogy, and of course the preaching, of Brooks and Robert Penn Warren was that everyone properly instructed could feel, perhaps even act, like an educated gentleman: In its sheer projection this was by no means a trivial ambition. No amount of snide mocking at their quaint gentility can conceal the fact that, in order to accomplish the conversion, the New Critics aimed at nothing less than the removal of *all* of what they considered the specialized rubbish—put there, they presumed, by professors of literature—standing between the reader of a poem and the poem. Leaving aside the questionable value of the New Criticism's ultimate social and moral message, we must concede that the school deliberately and perhaps incongruously tried to create a wide community of responsive readers out of a very large, potentially unlimited, constituency of students and teachers of literature.

In its early days, the French *nouvelle critique*, with Barthes as its chief apologist, attempted the same kind of thing. Once again the guild of professional literary scholars was characterized as impeding responsiveness to literature. Once again the antidote was what seemed to be a specialized reading technique based on a near jargon of linguistic, psychoanalytic, and Marxist terms, all of which proposed a new freedom for writers and literate readers alike. The philosophy of *écriture* promised wider horizons and a less restricted community, once an initial (and as it turned out painless) surrender to structuralist activity had been made. For despite structuralist prose, there was no impulse among the principal structuralists to exclude readers; quite the contrary, as Barthes' often abusive attacks on Raymond Picard show, the main purpose of critical reading was to create new readers of the classics who might otherwise have been frightened off by their lack of professional literary accreditation.

For about four decades, then, in both France and the United States, the schools of "new" critics were committed to prying literature and writing loose from confining institutions. However much it was to depend upon carefully learned technical skills, reading was in very large measure to become an act of public depossession. Texts were to be unlocked or decoded, then handed on to anyone who was interested. The resources of symbolic language were placed at the disposal of readers who it was assumed suffered the debilitations of either irrelevant "professional" information or the accumulated habits of lazy inattention.

Thus French and American New Criticism were, I believe, competitors for authority within mass culture, not other-worldly alternatives to it. Because of what became of them, we have tended to forget the original missionary aims the two schools set for themselves. They belong to precisely the same moment that produced Jean-Paul Sartre's ideas about an engaged literature and a committed writer. Literature was about the world, readers were in the world; the question was not *whether* to be but *how* to be, and this was best answered by carefully analyzing language's symbolic enactments of the various existential possibilities available to human beings. What the Franco-American critics shared was the notion that verbal discipline could be self-sufficient once you learned to think pertinently about language stripped of unnecessary scaffolding; in other words, you did not need to be a professor to benefit from Donne's metaphors or Saussure's liberating distinction between *langue* and *parole*. And so the New Criticism's precious and cliquish aspect was mitigated by its radically anti-institutional bias, which manifested itself in the enthusiastic therapeutic optimism to be observed in both France and the United States. Join humankind against the schools: this was a message a great many people could appreciate.

How strangely perverse, then, that the legacy of both types of New Criticism is the private-clique consciousness embodied in a kind of critical writing that has virtually abandoned any attempt at reaching a large, if not a mass, audience. My belief is that both in the United States and in France the tendency toward formalism in New Criticism was accentuated by the academy. For the fact is that a disciplined attention to language can only thrive in the rarefied atmosphere of the classroom. Linguistics and literary analysis are features of the modern school, not of the marketplace. Purifying the language of the tribe— whether as a project subsumed within modernism or as a hope kept alive by embattled New Criticisms surrounded by mass culture—always moved further from the really big existing tribes and closer toward emerging new ones, comprised of the acolytes of a reforming or even revolutionary creed who in the end seemed to care more about turning the new creed into an intensely separatist orthodoxy than about forming a large community of readers.

To its unending credit, the university protects such wishes and shelters them under the umbrella of academic freedom. Yet advocacy of *close reading* or of *écriture* can quite naturally entail hostility to outsiders who fail to grasp the salutary powers of verbal analysis; moreover, persuasion too often has turned out to be less important than purity of intention and execution. In time the guild adversarial sense grew as the elaborate techniques multiplied, and an interest in expanding the constituency lost out to a wish for abstract correctness and methodological rigor within a quasi-monastic order. Critics read each other and cared about little else.

The parallels between the fate of a New Criticism reduced to abandoning universal literacy entirely and that of the school of F. R. Leavis are sobering. As Francis Mulhern reminds us in *The Moment of Scrutiny*, Leavis was not a formalist himself and began his career in the context of generally Left politics. Leavis argued that great literature was fundamentally opposed to a class society and to the dictates of a coterie. In his view, English studies ought to become the cornerstone of a new, fundamentally democratic outlook. But largely because the Leavisites concentrated their work both in and for the university, what began as a healthy oppositional participation in modern industrial society changed into a shrill withdrawal from it. English studies became narrower and narrower, in my opinion, and critical reading degenerated into decisions about what should or should not be allowed into the great tradition.

I do not want to be misunderstood as saying that there is something inherently pernicious about the modern university that produces the changes I have been describing. Certainly there is a great deal to be said in favor of a university manifestly not influenced or controlled by coarse partisan politics. But one thing in particular about the university—and here I speak about the modern university without distinguishing between European, American, or Third World and socialist universities—does appear to exercise an almost totally unrestrained influence: the principle that knowledge ought to exist, be sought after, and disseminated in a very divided form. Whatever the social, political, economic, and ideological reasons underlying this principle, it has not long gone without its challengers. Indeed, it may not be too much of an exaggeration to say that one of the most interesting motifs in modern world culture has been the debate between proponents of the belief that knowledge can exist in a synthetic universal form and, on the other hand, those who believe that knowledge is inevitably produced and nurtured in specialized compartments. Georg Lukács' attack on reification and his advocacy of "totality," in my opinion, very tantalizingly resemble the wide-ranging discussions that have been taking place in the Islamic world since the late nineteenth century on the need for mediating between the claims of a totalizing Islamic vision and modern specialized science. These epistemological controversies are therefore centrally important to the workplace of knowledge production, the university, in which *what* knowledge is and how it ought to be discovered are the very lifeblood of its being.

The most impressive recent work concerning the history, circumstances, and constitution of modern knowledge has stressed the role of social convention. Thomas Kuhn's "paradigm of research," for example, shifts attention away from the individual creator to the communal restraints upon personal initiative. Galileos and Einsteins are infrequent figures not just because genius is a rare thing but because scientists are borne along by agreed-upon ways to do research, and this consensus encourages uniformity rather than bold enterprise. Over time this uniformity acquires the status of a discipline, while its subject matter becomes a field or territory. Along with these goes a whole apparatus of techniques, one of whose functions is, as Michel Foucault has tried to show in *The Archaeology of Knowledge*, to protect the coherence, the territorial integrity, the social identity of the field, its adherents and its institutional presence. You cannot simply choose to be a sociologist or a psychoanalyst; you cannot simply make statements that have the status of knowledge in anthropology; you cannot merely suppose that what you say as a historian (however well it may have been researched) enters historical discourse. You have to pass through certain rules of accreditation, you must learn the rules, you must speak the language, you must master the idioms, and you must accept the authorities of the field—determined in many of the same ways—to which you want to contribute.

In this view of things, expertise is partially determined by how well an individual learns the rules of the game, so to speak. Yet it is difficult to determine in absolute terms whether expertise is *mainly* constituted by the social conventions governing the intellectual manners of scientists or, on the other hand, mainly by the putative exigencies of the subject matter itself. Certainly convention, tradition, and habit create ways of looking at a subject that transform it completely; and just as certainly there are generic differences between the subjects of history, literature, and philology that require different (albeit related) techniques of analysis, disciplinary attitudes, and commonly held views. Elsewhere I have taken the admittedly aggressive position that Orientalists, area-studies experts, journalists, and foreign-policy specialists are not always sensitive to the dangers of self-quotation, endless repetition, and received ideas that their fields encourage, for reasons that have more to do with politics and ideology than with any "outside" reality. Hayden White has shown in his work that historians are subject not just to narrative conventions but also to the virtually closed space imposed on the interpreter of events by verbal retrospection, which is very far from being an objective mirror of reality. Yet even these views, although they are understandably repugnant to many people, do not go as far as saying that everything about a "field" can be reduced either to an interpretive convention or to political interest.

Let us grant, therefore, that it would be a long and potentially impossible task to prove empirically that, on the one hand, there could be objectivity so far as knowledge about human society is concerned or, on the other, that all knowledge is esoteric and subjective. Much ink has been spilled on both sides of the debate, not all of it useful, as Wayne Booth has shown in his discussion of scientism and modernism, *Modern Dogma and the Rhetoric of Assent*. An instructive opening out of the impasse—to which I want to return a bit later—has been the body of techniques developed by the school of reader-response critics: Wolfgang Iser, Norman Holland, Stanley Fish, and Michael Riffaterre, among others. These critics argue that since texts without readers are no less incomplete than readers without texts, we should focus attention on what happens when both components of the interpretive situation interact. Yet with the exception of Fish, reader-response critics tend to regard interpretation as an essentially private, interiorized happening, thereby inflating the role of solitary decoding at the expense of its just as important social context. In his latest book, *Is There a Text in This Class?*, Fish accentuates the role of what he calls interpretive communities, groups as well as institutions (principal among them the classroom and pedagogues) whose presence, much more than any unchanging objective standard or correlative of absolute truth, controls what we consider to be knowledge. If, as he says, "interpretation is the only game in town," then it must follow that interpreters who work mainly by persuasion and not scientific demonstration are the only players.

I am on Fish's side there. Unfortunately, though, he does not go very far in showing why, or even how, some interpretations are more persuasive than others. Once again we are back to the quandary suggested by the three thousand advanced critics reading each other to everyone else's unconcern. Is it the inevitable conclusion to the formation of an interpretive community that its constituency, its specialized language, and its concerns tend to get tighter, more airtight, more self-enclosed as its own self-confirming authority acquires more power, the solid status of orthodoxy, and a stable constituency? What is the acceptable humanistic antidote to what one discovers, say, among sociologists, philosophers, and so-called policy scientists who speak only to and for each other in a language oblivious to everything but a well-guarded, constantly shrinking fiefdom forbidden to the uninitiated?

For all sorts of reasons, large answers to these questions do not strike me as attractive or convincing. For one, the universalizing habit by which a system of thought is believed to account for everything too quickly slides into a quasi-religious synthesis. This, it seems to me, is the sobering lesson offered by John Fekete in *The Critical Twilight*, an account of how New Criticism led directly to Marshall McLuhan's "technocratic-religious eschatology." In fact, interpretation and its demands add up to a rough game, once we allow our-selves to step out of the shelter offered by specialized fields and by fancy all-embracing mythologies. The trouble with visions, reductive answers, and sys-tems is that they homogenize evidence very easily. Criticism as such is crowded out and disallowed from the start, hence impossible; and in the end one learns to manipulate bits of the system like so many parts of a machine. Far from tak-ing in a great deal, the universal system as a universal type of explanation ei-ther screens out everything it cannot directly absorb or it repetitively churns out the same sort of thing all the time. In this way it becomes a kind of con-spiracy theory. Indeed, it has always seemed to me that the supreme irony of what Derrida has called logocentrism is that its critique, deconstruction, is as insistent, as monotonous, and as inadvertently systematizing as logocentrism itself. We may applaud the wish to break out of departmental divisions, there-fore, without at the same time accepting the notion that one single method for doing so exists. The unheeding insistence of Rene Girard's "interdisciplinary" studies of mimetic desire and scapegoat effects is that they want to convert all human activity, all disciplines, to one thing. How can we assume this one thing covers everything that is essential, as Girard keeps suggesting?

This is only a relative skepticism, for one can prefer foxes to hedgehogs without also saying that all foxes are equal. Let us venture a couple of crucial distinctions. To the ideas of Kuhn, Foucault, and Fish we can usefully add those of Giovanni Battista Vico and Antonio Gramsci. Here is what we come up with. Discourses, interpretive communities, and paradigms of research are produced by intellectuals, Gramsci says, who can either be religious or secular. Now Gramsci's implicit contrast of secular with religious intellectuals is less fa-miliar than his celebrated division between organic and traditional intellectu-als. Yet it is no less important for that matter. In a letter of 17 August 1931, Gramsci writes about an old teacher from his Cagliari days, Umberto Cosmo:

> It seemed to me that I and Cosmo, and many other intellectuals at this time (say the first fifteen years of the century) occupied a certain common ground: we were all to some degree part of the movement of moral and in-tellectual reform which in Italy stemmed from Benedetto Croce, and whose first premise was that modern man can and should live without the help of religion . . . positivist religion, mythological religion, or whatever brand one cares to name. . . . [2] This point appears to me even today to be the major contribution made to international culture by modern Italian intellectuals, and it seems to me a civil conquest that must not be lost.[3]

Benedetto Croce of course was Vico's greatest modern student, and it was one of Croce's intentions in writing about Vico to reveal explicitly the strong secu-lar bases of his thought and also to argue in favor of a secure and dominant civil culture (hence Gramsci's use of the phrase "civil conquest"). "Conquest" has perhaps a strange inappropriateness to it, but it serves to dramatize Gramsci's contention—also implicit in Vico—that the modern European state is possible not only because there is a political apparatus (army, police force, bureaucracy) but because there is a civil, secular, and nonecclesiastical society making the state possible, providing the state with something to rule, filling the state with its humanly generated economic, cultural, social, and intellectual production.

Gramsci was unwilling to let the Vichian-Crocean achievement of civil society's secular working go in the direction of what he called "immanentist thought." Like Arnold before him, Gramsci understood that if nothing in the social world is natural, not even nature, then it must also be true that things exist not only because they come into being and are created by human agency (*nascimento*) but also because by coming into being they displace something else that is already there: this is the combative and emergent aspect of social change as it applies to the world of culture linked to social history. To adapt from a statement Gramsci makes in *The Modern Prince*, "reality (and hence cultural reality) is a product of the application of human will to the society of things" and since also "everything is political, even philosophy and philosophies," we are to understand that in the realm of culture and of thought each production exists not only to earn a place for itself but to displace, win out over, others.[4] All ideas, philosophies, views, and texts aspire to the consent of their consumers, and here Gramsci is more percipient than most in recognizing that there is a set of characteristics unique to civil society in which texts— embodying ideas, philosophies, and so forth—acquire power through what Gramsci describes as diffusion, dissemination into and hegemony over the world of "common sense." Thus ideas aspire to the condition of acceptance, which is to say that one can interpret the meaning of a text by virtue of what in its mode of social presence enables its consent by either a small or a wide group of people.

The secular intellectuals are implicitly present at the center of these considerations. Social and intellectual authority for them does not derive directly from the divine but from an analyzable history made by human beings. Here Vico's counterposing of the sacred with what he calls the gentile realm is essential. Created by God, the sacred is a realm accessible only through revelation: it is ahistorical because complete and divinely untouchable. But whereas Vico has little interest in the divine, the gentile world obsesses him. "Gentile" derives from *gens*, the family group whose exfoliation in time generates history. But "gentile" is also a secular expanse because the web of filiations and affiliations that composes human history—law, politics, literature, power, science, emotion—is informed by *ingegno*, human ingenuity and spirit. This, and not a divine *fons et origo*, is accessible to Vico's new science.

But here a very particular kind of secular interpretation and, even more interestingly, a very particular conception of the interpretive situation is entailed. A direct index of this is the confusing organization of Vico's book, which seems to move sideways and backward as often as it moves forward. Because in a very precise sense God has been excluded from Vico's secular history, that history, as well as everything within it, presents its interpreter with a vast horizontal expanse, across which are to be seen many interrelated structures. The verb "to look" is therefore frequently employed by Vico to suggest what historical interpreters need to do. What one cannot see or look at—the past, for example— is to be divined; Vico's irony is too clear to miss, since what he argues is that only by putting oneself in the position of the maker (or divinity) can one grasp how the past has shaped the present. This involves speculation, supposition, imagination, sympathy; but in no instance can it be allowed that something other than human agency caused history. To be sure, there are historical laws of development, just as there is something that Vico calls divine Providence mysteriously at work inside history. The fundamental thing is that history and human society are made up of numerous efforts crisscrossing each other, frequently at odds with each other, always untidy in the way they involve each other. Vico's writing directly reflects this crowded spectacle.

One last observation needs to be made. For Gramsci and Vico, interpretation must take account of this secular horizontal space only by means appropriate to what is present there. I understand this to imply that no single explanation sending one back immediately to a single origin is adequate. And just as there are no simple dynastic answers, there are no simple discrete historical formations or social processes. A heterogeneity of human involvement is therefore equivalent to a heterogeneity of results, as well as of interpretive skills and techniques. There is no center, no inertly given and accepted authority, no fixed barriers ordering human history, even though authority, order, and distinction exist. The secular intellectual works to show the absence of divine originality and, on the other side, the complex presence of historical actuality. The conversion of the absence of religion into the presence of actuality is secular interpretation.

2

Having rejected global and falsely systematic answers, one had better speak in a limited and concrete way about the contemporary actuality, which so far as our discussion here is concerned is Reagan's America, or, rather, the America inherited and now ruled over by Reaganism. Take literature and politics, for example. It is not too much of an exaggeration to say that an implicit consensus has been building for the past decade in which the study of literature is considered to be profoundly, even constitutively nonpolitical. When you discuss Keats or Shakespeare or Dickens, you may touch on political subjects, of course, but it is assumed that the skills traditionally associated with modern literary criticism (what is now called rhetoric, reading, textuality, tropology, or deconstruction) are there to be applied to *literary* texts, not, for instance, to a government document, a sociological or ethnological report, or a newspaper. This separation of fields, objects, disciplines, and foci constitutes an amazingly *rigid* structure which, to my knowledge, is almost never discussed by literary scholars. There seems to be an unconsciously held norm guaranteeing the simple essence of "fields," a word which in turn has acquired the intellectual authority of a natural objective fact. Separation, simplicity, silent norms of pertinence: this is one depoliticizing strain of considerable force, since it is capitalized on by professions, institutions, discourses, and a massively reinforced consistency of specialized fields. One corollary of this is the proliferating orthodoxy of separate fields. "I'm sorry I can't understand this—I'm a literary critic, not a sociologist."

The intellectual toll this has taken in the work of the most explicitly political of recent critics—Marxists, in the instance I shall discuss here—is very high. Fredric Jameson has recently produced what is by any standard a major work of intellectual criticism, *The Political Unconscious.* What it discusses, it discusses with a rare brilliance and learning: I have no reservations at all about that. He argues that priority ought to be given to the political interpretation of literary texts and that Marxism, as an interpretive act as opposed to other methods, is "that 'untranscendable horizon' subsumes such apparently antagonistic or incommensurable critical operations [as the other varieties of interpretive act] assigning them an undoubted sectoral validity within itself, and thus at once cancelling and preserving them."[5] Thus Jameson avails himself of all the most powerful and contradictory of contemporary methodologies, enfolding them in a series of original readings of modern novels, producing in the end a working through of three "semantic horizons" of which the third "phase" is the Marxist: hence, from *explication de texte,* through the ideological discourses of social classes, to the ideology of form itself, perceived against the ultimate horizon of human history.

It cannot be emphasized too strongly that Jameson's book presents a remarkably complex and deeply attractive argument to which I cannot do justice here. This argument reaches its climax in Jameson's conclusion, in which the Utopian element in all cultural production is shown to play an underanalyzed and liberating role in human society; additionally, in a much too brief and suggestive passage, Jameson touches on three political discussions (involving the state, law, and nationalism) for which the Marxist hermeneutic he has outlined, fully a negative as well as a positive hermeneutic, can be particularly useful.

We are still left, however, with a number of nagging difficulties. Beneath the surface of the book lies an unadmitted dichotomy between two kinds of "Politics": (1), the politics defined by political theory from Hegel to Louis Althusser and Ernst Bloch; (2), the politics of struggle and power in the everyday world, which in the United States at least has been won, so to speak, by Reagan. As to why this distinction should exist at all, Jameson says very little. This is even more troubling when we realize that Politics 2 is only discussed once, in the course of a long footnote. There he speaks in a general way about "ethnic groups, neighborhood movements, . . . rank-and-file labor groups," and so on and quite perspicaciously enters a plea for alliance politics in the United States as distinguished from France, where the totalizing global politics imposed on nearly every constituency has either inhibited or repressed their local development (p. 54). He is absolutely right of course (and would have been more so had he extended his arguments to a United States dominated by only two parties). Yet the irony is that in criticizing the global perspective and admitting its radical discontinuity with local alliance politics, Jameson is also advocating a strong hermeneutic globalism which will have the effect of subsuming the local in the synchronic. This is almost like saying: Don't worry; Reagan is merely a passing phenomenon: the cunning of history will get him too. Yet except for what suspiciously resembles a religious confidence in the teleological efficacy of the Marxist vision, there is no way, to my mind, by which the local is necessarily going to be subsumed, cancelled, preserved, and resolved by the synchronic. Moreover, Jameson leaves it entirely up to the reader to guess what the connection is between the synchrony and theory of Politics 1 and the molecular struggles of Politics 2. Is there continuity or discontinuity between one realm and the other? How do quotidian politics and the struggle for power enter into the hermeneutic, if not by simple instruction from above or by passive osmosis?

These are unanswered questions precisely because, I think, Jameson's assumed constituency is an audience of cultural-literary critics. And this constituency in contemporary America is premised on and made possible by the separation of disciplines I spoke about earlier. This further aggravates the discursive separation of Politics 1 from Politics 2, creating the obvious impression that Jameson is dealing with autonomous realms of human effort. And this has a still more paradoxical result. In his concluding chapter, Jameson suggests allusively that the components of class consciousness—such things as group solidarity against outside threats—are at bottom utopian "insofar as all such (class-based) collectivities are *figures* for the ultimate concrete collective life of an achieved Utopian or classless society." Right at the heart of this thesis we find the notion that "ideological commitment is not first and foremost a matter of moral choice but of the taking of sides in a struggle between embattled groups" (pp. 291, 290). The difficulty here is that whereas moral choice is a cat-

egory to be rigorously de-Platonized and historicized, there is no inevitability—logical or otherwise—for reducing it completely to "the taking of sides in
a struggle between embattled groups." On the molecular level of an individual
peasant family thrown off its land, who is to say whether the desire for restitu-
tion is exclusively a matter of taking sides or of making the moral choice to resist dispossession. I cannot be sure. But what is so indicative of Jameson's position is that from the global, synchronic hermeneutic overview, moral choice
plays no role and, what is more, the matter is not investigated empirically or
historically (as Barrington Moore has tried to do in *Injustice: The Social Basis
of Obedience and Revolt*).

Jameson has certainly earned the right to be one of the preeminent spokes-
men for what is best in American cultural Marxism. He is discussed this way by
a well-known English Marxist, Terry Eagleton, in a recent article, "The Idealism
of American Criticism." Eagleton's discussion contrasts Jameson and Frank
Lentricchia with the main currents of contemporary American theory which,
according to Eagleton, "develops by way of inventing new idealist devices for
the repression of history."[6] Nevertheless, Eagleton's admiration for Jameson
and Lentricchia does not prevent him from seeing the limitations of their
work, their political "unclarity," their lingering pragmatism, eclecticism, the relationship of their hermeneutic criticism to Reagan's ascendancy, and—in
Jameson's case specially—their nostalgic Hegelianism. This is not to say, however, that Eagleton expects either of them to toe the current ultra-Left line,
which alleges that "the production of Marxist readings of classical texts is class-
collaborationism." But he is right to say that "the question irresistibly raised for
the Marxist reader of Jameson is simply this: How is a Marxist-structuralist
analysis of a minor novel of Balzac to help shake the foundations of capitalism?"
Clearly the answer to this question is that such readings won't; but what does
Eagleton propose as an alternative? Here we come to the disabling cost of rigidly
enforced intellectual and disciplinary divisions, which also affects Marxism.

For we may as well acknowledge that Eagleton writes about Jameson as a fellow Marxist. This is intellectual solidarity, yes, but within a "field" defined
principally as an intellectual discourse existing solely within an academy that
has left the extra-academic outside world to the new Right and to Reagan. It
follows with a kind of natural inevitability that if one such confinement is acceptable, others can be acceptable: Eagleton faults Jameson for the practical in-
effectiveness of his Marxist-structuralism but, on the other hand, meekly takes
for granted that he and Jameson inhabit the small world of literary studies,
speak its language, deal only with its problematics. Why this should be so is
hinted at obscurely by Eagleton when he avers that "the ruling class" determines what uses are made of literature for the purpose of "ideological reproduction" and that as revolutionaries "we" cannot select "the literary terrain
on which the battle is to be engaged." It does not seem to have occurred to

Eagleton that what he finds weakest in Jameson and Lentricchia, their marginality and vestigial idealism, is what also makes him bewail their rarefied discourse at the same time that he somehow accepts it as his own. The very same specialized ethos has been attenuated a little more now: Eagleton, Jameson, and Lentricchia are literary Marxists who write for literary Marxists, who are in cloistral seclusion from the inhospitable world of real politics. Both "literature" and "Marxism" are thereby confirmed in their apolitical content and methodology: literary criticism is still "only" literary criticism, Marxism only Marxism, and politics is mainly what the literary critic talks about longingly and hopelessly.

This rather long digression on the consequences of the separation of "fields" brings me directly to a second aspect of the politics of interpretation viewed from a secular perspective rigorously responsive to the Age of Reagan. It is patently true that, even within the atomized order of disciplines and fields, methodological investigations can and indeed do occur. But the prevailing mode of intellectual discourse is militantly antimethodological, if by methodological we mean a questioning of the structure of fields and discourses themselves. A principle of silent exclusion operates within and at the boundaries of discourse; this has now become so internalized that fields, disciplines, and their discourses have taken on the status of immutable durability. Licensed members of the field, which has all the trappings of a social institution, are identifiable as belonging to a guild, and for them words like "expert" and "objective" have an important resonance. To acquire a position of authority within the field is, however, to be involved internally in the formation of a canon, which usually turns out to be a blocking device for methodological and disciplinary self-questioning. When J. Hillis Miller says, "I believe in the established canon of English and American Literature and the validity of the concept of privileged texts," he is saying something that has moment by virtue neither of its logical truth nor of its demonstrable clarity.[7] Its power derives from his social authority as a well-known professor of English, a man of deservedly great reputation, a teacher of well-placed students. And what he says more or less eliminates the possibility of asking whether canons (and the imprimatur placed upon canons by a literary critic) are more methodologically necessary to the order of dominance within a guild than they are to the secular study of human history.

If I single out literary and humanistic scholars in what I am saying, it is because, for better or worse, I am dealing with texts, and texts are the very point of departure and culmination for literary scholars. Literary scholars read and they write, both of which are activities having more to do with wit, flexibility, and questioning than they do with solidifying ideas into institutions or with bludgeoning readers into unquestioning submission. Above all it seems to me that it goes directly against the grain of reading and writing to erect barriers between texts or to create monuments out of texts—unless, of course, literary scholars believe themselves to be servants of some outside power requiring this duty from them. The curricula of most literature departments in the university today are constructed almost entirely out of monuments, canonized into rigid dynastic formation, serviced and reserviced monotonously by a shrinking guild of humble servitors. The irony is that this is usually done in the name of historical research and traditional humanism, and yet such canons often have very little historical accuracy to them. To take one small example, Robert Darnton has shown that

> much of what passes today as 18th century French literature wasn't much read by Frenchmen in the 18th century. . . . We suffer from an arbitrary notion of literary history as a canon of classics, one which was developed by professors of literature in the 19th and 20th centuries—while in fact what people of the 18th century were reading was very different. By studying the publisher's accounts and papers at [the Société Typographique de] Neufchatel I've been able to construct a kind of bestseller list of pre-revolutionary France, and it doesn't look anything like the reading lists passed out in classrooms today.[8]

Hidden beneath the pieties surrounding the canonical monuments is a guild solidarity that dangerously resembles a religious consciousness. It is worth recalling Michael Bakunin in *Dieu et l'état*: "In their existing organization, monopolizing science and remaining thus outside social life, the *savants* form a separate caste, in many respects analogous to the priesthood. Scientific abstraction is their God, living and real individuals are their victims, and they are the consecrated and licensed sacrificers."[9] The current interest in producing enormous biographies of consecrated great authors is one aspect of this priestifying. By isolating and elevating the subject beyond his or her time and society, an exaggerated respect for single individuals is produced along with, naturally enough, awe for the biographer's craft. There are similar distortions in the emphasis placed on autobiographical literature whose modish name is "self-fashioning."

All this, then, atomizes, privatizes, and reifies the untidy realm of secular history and creates a peculiar configuration of constituencies and interpretive communities: this is the third major aspect of a contemporary politics of interpretation. An almost invariable rule of order is that very little of the *circumstances* making interpretive activity possible is allowed to seep into the interpretive circle itself. This is peculiarly (not to say distressingly) in evidence when humanists are called in to dignify discussions of major public issues. I shall say nothing here about the egregious lapses (mostly concerning the relationship between the government-corporate policymakers and humanists on questions of national and foreign policy) to be found in the Rockefeller Foundation-funded report *The Humanities in American Life*. More crudely dramatic for my purposes is another Rockefeller enterprise, a conference on "The Reporting of Religion in the Media," held in August 1980. In addressing his opening remarks to the assembled collection of clerics, philosophers, and other humanists, Martin Marty evidently felt it would be elevating the discussion somewhat if he brought Admiral Stansfield Turner, head of the CIA, to his assistance: he therefore "quoted Admiral Turner's assertion that United States intelligence agencies had overlooked the importance of religion in Iran, 'because everyone knew it had so little place and power in the modern world.'" No one seemed to notice the natural affinity assumed by Marty between the CIA and scholars. It was all part of the mentality decreeing that humanists were humanists and experts experts no matter who sponsored their work, usurped their freedom of judgment and independence of research, or assimilated them unquestioningly to state service, even as they protested again and again that they were objective and nonpolitical.

Let me cite one small personal anecdote at the risk of overstating the point. Shortly before my book *Covering Islam* appeared, a private foundation convened a seminar on the book to be attended by journalists, scholars, and diplomats, all of whom had professional interests in how the Islamic world was being reported and represented in the West generally. I was to answer questions. One Pulitzer Prize–winning journalist who is now the foreign news editor of a leading Eastern newspaper, was asked to lead the discussion, which he did by summarizing my argument briefly and on the whole not very accurately. He concluded his remarks by a question meant to initiate discussion: "Since you say that Islam is badly reported [actually my argument in the book is that "Islam" isn't something to be reported or nonreported: it is an ideological abstraction], could you tell us how we should report the Islamic world in order to help clarify the U.S.'s strategic interests there?" When I objected to the question, on the grounds that journalism was supposed to be either reporting or analyzing the news and not serving as an adjunct to the National Security Council, no attention was paid to what in everyone's eyes was an irrelevant naiveté on my part. Thus have the security interests of the state been absorbed silently into journalistic interpretation: expertise is therefore supposed to be unaffected by its institutional affiliations with power, although of course it is exactly those affiliations—hidden but assumed unquestioningly—that make the expertise possible and imperative.

Given this context, then, a constituency is principally a clientele people who use (and perhaps buy) your services because you and others belonging to your guild are certified experts. For the relatively unmarketable humanists whose wares are "soft" and whose expertise is almost by definition marginal, their constituency is a fixed one composed of other humanists, students, government and corporate executives, and media employees, who use the humanist to assure a harmless place for "the humanities" or culture or literature in the society. I hasten to recall, however, that this is the role voluntarily accepted by humanists whose notion of what they do is neutralized, specialized, and non-political in the extreme. To an alarming degree, the present continuation of the humanities depends, I think, on the sustained self-purification of humanists for whom the ethic of specialization has become equivalent to minimizing the content of their work and increasing the composite wall of guild conscious-ness, social authority, and exclusionary discipline around themselves. Oppo-nents are therefore not people in disagreement with the constituency but peo-ple to be kept out, nonexperts and nonspecialists, for the most part.

Whether all this makes an interpretive *community*, in the secular and non-commercial, noncoercive sense of the word, is very seriously to be doubted. If a community is based principally on keeping people out and on defending a tiny fiefdom (in perfect complicity with the defenders of other fiefdoms) on the basis of a mysteriously pure subject's inviolable integrity, then it is a reli-gious community. The secular realm I have presupposed requires a more open sense of community as something to be won and of audiences as human be-ings to be addressed. How, then, can we understand the present setting in such a way as to see in it the possibility of change? How can interpretation be inter-preted as having a secular, political force in an age determined to deny inter-pretation anything but a role as mystification?

3

I shall organize my remarks around the notion of *representation*, which, for lit-erary scholars at least, has a primordial importance. From Aristotle to Auerbach and after, mimesis is inevitably to be found in discussions of literary texts. Yet as even Auerbach himself showed in his monographic stylistic studies, tech-niques of representation in literary work have always been related to, and in some measure have depended on social formations. The phrase "la cour et la ville," for example, makes primarily *literary* sense in a text by Nicolas Boileau, and although the text itself gives the phrase a peculiarly refined local meaning, it nevertheless presupposed both an audience that knew he referred to what Auerbach calls "his social environment" and the social environment itself, which made references to it possible. This is not *simply* a matter of reference, since, from a verbal point of view, referents can be said to be equal and equally verbal. Even in very minute analyses, Auerbach's view does, however, have to do with the *coexistence* of realms—the literary, the social, the personal—and the way in which they make use of, affiliate with, and represent each other.

With very few exceptions, contemporary literary theories assume the rela-tive independence and even autonomy of literary representation over (and not just from) all others. Novelistic verisimilitude, poetic tropes, and dramatic metaphors (Lukács, Harold Bloom, Francis Ferguson) are representations to and for themselves of the novel, the poem, the drama: this, I think, accurately sums up the assumptions underlying the three influential (and, in their own way, typical) theories I have referred to. Moreover, the organized study of literature—*en soi* and *pour soi*—is premised on the constitutively primary act of literary (that is artistic) representation, which in turn absorbs and incorporates other realms, other representations, secondary to it. But all this institutional weight has precluded a sustained, systematic examination of the coexistence of and the interrelationship between the literary and the social, which is where representation—from journalism, to political struggle, to economic production and power—plays an extraordinarily important role. Confined to the study of one representational complex, literary critics accept and paradoxically ignore the lines drawn around what they do.

This is depoliticization with a vengeance, and it must, I think, be understood as an integral part of the historical moment presided over by Reaganism. The division of intellectual labor I spoke of earlier can now be seen as assuming a *thematic* importance in the contemporary culture as a whole. For if the study of literature is "only" about literary representation, then it must be the case that literary representations and literary activities (writing, reading, producing the "humanities," and arts and letters) are essentially ornamental, possessing at most secondary ideological characteristics. The consequence is that to deal with literature as well as the broadly defined "humanities" is to deal with the nonpolitical, although quite evidently the political realm is presumed to lie just beyond (and beyond the reach of) literary, and hence *literate*, concern.

A perfect recent embodiment of this state of affairs is the 30 September 1981 issue of *The New Republic*. The lead editorial analyzes the United States' policy toward South Africa and ends up supporting this policy, which even the most "moderate" of Black African states interpret (correctly, as even the United States explicitly confesses) as a policy supporting the South African settler-colonial regime. The last article of the issue includes a mean personal attack on me as "an intellectual in the thrall of Soviet totalitarianism," a claim that is as disgustingly McCarthyite as it is intellectually fraudulent. Now at the very center of this issue of the magazine—a fairly typical issue by the way—is a long and decently earnest book review by Christopher Hill, a leading Marxist historian. What boggles the mind is not the mere coincidence of apologies for apartheid rubbing shoulders with good Marxist sense but how the one antipode includes (without any reference at all) what the other, the Marxist pole, performs unknowingly.

There are two very impressive points of reference for this discussion of what can be called the national culture as a nexus of relationships between "fields," many of them employing representation as their technique of distribution and production. (It will be obvious here that I exclude the creative arts and the natural sciences.) One is Perry Anderson's "Components of the National Culture" (1969);[10] the other is Regis Debray's study of the French intelligentsia, *Teachers, Writers, Celebrities* (1980). Anderson's argument is that an absent intellectual center in traditional British thought about society was vulnerable to a "white" (antirevolutionary, conservative) immigration into Britain from Europe. This in turn produced a blockage of sociology, a technicalization of philosophy, an idea-free empiricism in history, and an idealist aesthetics. Together these and other disciplines form "something like a closed system," in which subversive discourses like Marxism and psychoanalysis were for a time quarantined; now, however, they too have been incorporated. The French case, according to Debray, exhibits a series of three hegemonic conquests in time. First there was the era of the secular universities, which ended with World War I. That was succeeded by the era of the publishing houses, a time between the wars when Galimard-NRF—agglomerates of gifted writers and essayists that included Jacques Rivière, André Gide, Marcel Proust, and Paul Valéry—replaced the social and intellectual authority of the somewhat overproductive, mass-populated universities. Finally, during the 1960s, intellectual life was absorbed into the structure of the mass media: worth, merit, attention, and visibility slipped from the pages of books to be estimated by frequency of appearance on the television screen. At this point, then, a new hierarchy, what Debray calls a mediocracy, emerges, and it rules the schools and the book industry.

There are certain similarities between Debray's France and Anderson's England, on the one hand, and Reagan's America, on the other. They are interesting, but I cannot spend time talking about them. The differences are, however, more instructive. Unlike France, high culture in America is assumed to be above politics as a matter of unanimous convention. And unlike England, the intellectual center here is filled not by European imports (although they play a considerable role) but by an unquestioned ethic of objectivity and realism, based essentially on an epistemology of separation and difference. Thus each field is separate from the others because the subject matter is separate. Each separation corresponds immediately to a separation in function, institution, history, and purpose. Each discourse "represents" the field, which in turn is supported by its own constituency and the specialized audience to which it appeals. The mark of true professionalism is accuracy of representation of society, vindicated in the case of sociology, for instance, by a direct correlation between representation of society and corporate and/or governmental interests, a role in social policymaking, access to political authority. Literary studies, conversely, are realistically *not* about society but about masterpieces in need of periodic adulation and appreciation. Such correlations make possible the use of words like "objectivity," "realism," and "moderation" when used in sociology or in literary criticism. And these notions in turn assure their own confirmation by careful selectivity of evidence, the incorporation and subsequent neutralization of dissent (also known as pluralism), and networks of insiders, experts whose presence is due to their conformity, not to any rigorous judgment of their past performance (the good team player always turns up).

But I must press on, even though there are numerous qualifications and refinements to be added at this point (e.g., the organized relationship between clearly affiliated fields such as political science and sociology versus the use by one field of another unrelated one for the purposes of national policy issues; the network of patronage and the insider/outsider dichotomy; the strange cultural encouragement of theories stressing such "components" of the structure of power as chance, morality, American innocence, decentralized egos, etc.). The particular mission of the humanities is, in the aggregate, to represent *noninterference* in the affairs of the everyday world. As we have seen, there has been a historical erosion in the role of letters since the New Criticism, and I have suggested that the conjuncture of a narrowly based university environment for technical language and literature studies with the self-policing, self-purifying communities erected even by Marxist, as well as other disciplinary, discourses, produced a very small but definite function for the humanities: to represent humane marginality, which is also to preserve and if possible to conceal the hierarchy of powers that occupy the center, define the social terrain, and fix the limits of use functions, fields, marginality, and so on. Some of the corollaries of this role for the humanities generally and literary criticism in particular are that the institutional presence of humanities guarantees a space for the deployment of free-floating abstractions (scholarship, taste, tact, humanism) that are defined in advance as indefinable; that when it is not easily domesticated, "theory" is employable as a discourse of occultation and legitimation; that self-regulation is the ethos behind which the institutional humanities allow and in a sense encourage the unrestrained operation of market forces that were traditionally thought of as subject to ethical and philosophical review.

Very broadly stated, then, noninterference for the humanist means laissez-faire: "they" can run the country, we will explicate Wordsworth and Schlegel. It does not stretch things greatly to note that noninterference and rigid specialization in the academy are directly related to what has been called a counterattack by "highly mobilized business elites" in reaction to the immediately preceding period during which national needs were thought of as fulfilled by resources allocated collectively and democratically. However, working through foundations, think tanks, sectors of the academy, and the government, corporate elites according to David Dickson and David Noble "proclaimed a new age of reason while remystifying reality." This involved a set of "interrelated epistemological and ideological imperatives, which are an extrapolation from the noninterference I spoke about earlier. Each of these imperatives is in congruence with the way intellectual and academic "fields" view themselves internally and across the dividing lines:

1) The rediscovery of the self-regulating market, the wonders of free enterprise, and the classical liberal attack on government regulation of the economy, all in the name of liberty.

2) The reinvention of the idea of progress, now cast in terms of "innovation" and "industrialization," and the limitation of expectations and social welfare in the quest for productivity.

3) The attack on democracy, in the name of "efficiency," "manageability," "governability," "rationality," and "competence."

4) The remystification of science through the promotion of formalized decision methodologies, the restoration of the authority of expertise, and the renewed use of science as legitimation for social policy through deepening industry ties to universities and other "free" institutions of policy analysis and recommendation.[11]

In other words, (1) says that literary criticism minds its own business and is "free" to do what it wishes with no community responsibility whatever. Hence at one end of the scale, for instance, is the recent successful attack on the NEH for funding too many socially determined programs and, at the other end, the proliferation of private critical languages with an absurdist bent presided over paradoxically by "big name professors," who also extoll the virtues of humanism, pluralism, and humane scholarship. Retranslated, (2) has meant that the number of jobs for young graduates has shrunk dramatically as the "inevitable" result of market forces, which in turn prove the marginality of scholarship that is premised on its own harmless social obsolescence. This has created a demand for sheer innovation and indiscriminate publication (e.g., the sudden increase in advanced critical journals; the departmental need for experts and courses in theory and structuralism), and it has virtually destroyed the career trajectory and social horizons of young people within the system. Imperatives (3) and (4) have meant the recrudescence of strict professionalism for sale to any client, deliberately oblivious of the complicity between the academy, the government, and the corporations, decorously silent on the large questions of social, economic, and foreign policy.

Very well: if what I have been saying has any validity, then the politics of interpretation demands a dialectical response from a critical consciousness worthy of its name. Instead of noninterference and specialization, there must be *interference*, a crossing of borders and obstacles, a determined attempt to generalize exactly at those points where generalizations seem impossible to make. One of the first interferences to be ventured, then, is a crossing from literature, which is supposed to be subjective and powerless, into those exactly parallel realms, now covered by journalism and the production of information, that employ representation but are supposed to be objective and powerful. Here we have a superb guide in John Berger, in whose most recent work there is the basis of a major critique of modern representation. Berger suggests that if we regard photography as coeval in its origins with sociology and positivism (and I would add the classic realistic novel), we see that

> what they shared was the hope that observable quantifiable facts recorded by experts, would constitute the proven truth that humanity required. Precision would replace metaphysics; planning would resolve conflicts. What happened, instead, was that the way was opened to a view of the world in which everything and everybody could be reduced to a factor in a calculation, and the calculation was profit.[12]

Much of the world today is represented in this way: as the McBride Commission Report has it, a tiny handful of large and powerful oligarchies control about ninety percent of the world's information and communication flows. This domain, staffed by experts and media executives is, as Herbert Schiller and others have shown, affiliated to an ever smaller number of governments, at the very same time that the rhetoric of objectivity, balance, realism, and freedom covers what is being done. And for the most part, such consumer items as "the news"—a euphemism for ideological images of the world that determine political reality for a vast majority of the world's population—hold forth, untouched by interfering secular and critical minds, who for all sorts of obvious reasons are not hooked into the systems of power.

This is not the place, nor is there time, to advance a fully articulated program of interference. I can only suggest in conclusion that we need to think about breaking out of the disciplinary ghettos in which as intellectuals we have been confined, to reopen the blocked social processes ceding objective representation (hence power) of the world to a small coterie of experts and their clients, to consider that the audience for literacy is not a closed circle of three thousand professional critics but the community of human beings living in society, and to regard social reality in a secular rather than a mystical mode, despite all the protestations about realism and objectivity.

Two concrete tasks—again adumbrated by Berger—strike me as particularly useful. One is to use the visual faculty (which also happens to be dominated by visual media such as television, news photography, and commercial film, all of them fundamentally immediate, "objective," and ahistorical) to restore the nonsequential energy of lived historical memory and subjectivity as fundamental components of meaning in representation. Berger calls this an alternative use of photography: using photomontage to tell other stories than the official sequential or ideological ones produced by institutions of power. (Superb examples are Sarah Graham-Brown's photo-essay *The Palestinians and Their Society* and Susan Meisalas' *Nicaragua*.) Second is opening the culture to experiences of the Other which have remained "outside" (and have been repressed or framed in a context of confrontational hostility) the norms manufactured by "insiders." An excellent example is Malek Alloula's *Le Harem colonial*, a study of early twentieth-century postcards and photographs of Algerian harem women. The pictorial capture of colonized people by colonizer, which signifies power, is reenacted by a young Algerian sociologist, Alloula, who sees his own fragmented history in the pictures, then reinscribes this history in his text as the result of understanding and making that intimate experience intelligible for an audience of modern European readers.

In both instances, finally, we have the recovery of a history hitherto either misrepresented or rendered invisible. Stereotypes of the Other have always been connected to political actualities of one sort or another, just as the truth of lived communal (or personal) experience has often been totally sublimated in official narratives, institutions, and ideologies. But in having attempted— and perhaps even successfully accomplishing—this recovery, there is the crucial next phase: connecting these more politically vigilant forms of interpretation to an ongoing political and social praxis. Short of making that connection, even the best intentioned and the cleverest interpretive activity is bound to sink back into the murmur of mere prose. For to move from interpretation to its politics is in large measure to go from undoing to doing, and this, given the currently accepted divisions between criticism and art, is risking all the discomfort of a great unsettlement in ways of seeing and doing. One must refuse to believe, however, that the comforts of specialized habits can be so seductive as to keep us all in our assigned places.

NOTES

1. See Ronald Steel, *Walter Lippmann and the American Century* (Boston, 1980), pp. 180–85 and 212–16.

2. Antonio Gramsci to Tatiana Schucht, in Giuseppe Fiori, *Antonio Gramsci: Life of a Revolutionary*, trans. Tom Nairn (London, 1970), p. 74.

3. Gramsci to Schucht, *Lettere dal Carcere* (Turin, 1975), p. 466; my translation.

4. Gramsci, *Selections from the Prison Notebooks*, trans. Quintin Hoare and Geoffrey Nowell Smith (New York, 1971), p. 171.

5. Fredric Jameson, *The Political Unconscious* (Ithaca, N.Y., 1981), p. 10; all further references to this work will be included in the text. Perhaps not incidentally, what Jameson claims for Marxism here is made the central feature of nineteenth-century British fiction by Deirdre David, *Fictions of Resolutions in Three Victorian Novels* (New York, 1980).

6. Terry Eagleton, "The Idealism of American Criticism," *New Left Review* 127 (May–June 1981): 59.

7. J. Hillis Miller, "The Function of Rhetorical Study at the Present Time," *ADE Bulletin* 62 (September 1979): 12.

8. Robert Darnton, "A Journeyman's Life under the Old Regime: Work and Culture in an Eighteenth-Century Printing Shop," *Princeton Alumni Weekly*, 7 September 1981, p. 12.

9. Michael Bakunin, *Selected Writings*, ed. and trans. Arthur Lehning (London, 1973), p. 160.

10. See Perry Anderson, "Components of the National Culture," in *Student Power*, ed. Alexander Cockburn and Robin Blackburn (London, 1969).

11. David Dickson and David Noble, "By Force of Reason: The Politics of Science and Policy," in *The Hidden Election*, ed. Thomas Ferguson and Joel Rogers (New York, 1981), p. 267.

12. John Berger, "Another Way of Telling," *Journal of Social Reconstruction* (January–March 1980): 64.

15. WHY NOT SCIENCE CRITICS?

DON IHDE

THE IDEA FOR my title was suggested quite a few years ago by Langdon Winner. Langdon had sent me a copy of a collection of his essays to read and respond to which eventually became *The Whale and the Reactor*. And, although his topic was philosophy of technology and his experience was what many of us felt in technology studies at the time, the point applies equally well to science, or even better, to what is now often called technoscience. Here is Langdon's point:

> [This] project . . . is a work of criticism, a fact that some readers will find troubling. If, in contrast, this were literary criticism, everyone would immediately understand that the underlying aim is positive. A critic of literature examines a text, analyzing its features, evaluating its qualities, seeking a deeper appreciation that might be useful to other readers of the same text. In a similar way, critics of music, theater and the arts have a valuable, well-established role, serving as a bridge between artists and audiences. Alas, the criticism of [technoscience] is not welcomed in the same manner. Writers who venture beyond the most ordinary conceptions of tools and uses, writers who investigate ways in which technical forms are implicated in the basic patterns and problems of our culture are met with the charge that they are merely "antitechnology" [or "antiscience"] or "blaming [technoscience]". All who have stepped forward as critics in this field— Lewis Mumford, Paul Goodman, Jacques Ellul, Ivan Illich, and others— have been tarred with the same brush, an expression of a desire to stop the dialogue rather than expand it.
>
> (Winner, *Paths of Technopolis*, p. 3)

The contrast between art and literary criticism and what I shall call 'technoscience criticism' is marked. Few would call art or literary critics "anti-art" or "anti-literature" in the working out, however critically, of their products. And while it may indeed be true that given works of art or given texts are excoriated, demeaned, or severely dealt with, one does not usually think of the critic as generically "anti-art" or "anti-literature." Rather, it is precisely because the critic is passionate about his or her subject matter that he or she becomes a 'critic.' That is simply not the case with science or technoscience criticism.

In part this is because art and literary criticism is institutionalized. It is so much a part of the artistic and literary tradition that critics meet, publish, and talk in the same contexts as the artists and writers. And, contrarily, there simply is no such forum within science or technology. The critic—as I shall show below—is either regarded as an outsider, or if the criticism arises from the inside, is soon made to be a quasi-outsider. Why is this the case?

We are now at the juncture where I may announce the theses which I wish to argue for: First, I am obviously holding that I think something like 'technoscience criticism' ought to be part of the social discourse concerning technoscience and that this role ought to be a recognized and legitimated role. And, in a related fashion, while I am holding that the technoscience community resists precisely this role and bears the self-serving primary responsibility for closing off the critical dialogue, I am also implicitly holding that 'technoscience criticism' while often occurring, has not taken the place it could occupy in the places where we should expect it to occur. This contention comes from my own experience in the founding of a growth of North American philosophy of technology—which does frequently offer technology criticism and frequently gets for its efforts an 'anti-technology' label—and for my more recent experience and work in the philosophy of science, which until recently has almost strenuously avoided anything which could be called 'science criticism' except in the narrowest of conceptual senses. I shall later examine a few instances of technoscience criticism which have and do occur, but which are, at best, limitedly successful.

I. BARRIERS TO TECHNOSCIENCE CRITICISM

The most obvious barrier to the formation of an institutionalized technoscience criticism lies in the role of late modern technoscience itself. Technoscience, as institution, began in early modernity by casting itself as the 'other' of religion. Its mythologies, drawn from Classical pre-Christian and often materialist (Democritean/Epicurean) sources; its anti-authoritarianism, including the Galilean claim to have exceeded the Scriptures and Church Father's insights by replacing these with the new sighting possible through his telescope; and the much stronger later anti-religiosity of the Enlightenment which cast religion as 'superstition' and science as 'rationality,' all led to the Modernist substitution of what I am calling technoscience for religion.

In the process, science—whether advertently or inadvertently—itself took on a quasi-theological characteristic. To be critical of the new 'true faith' was to be, in effect, 'heretical,' now called 'irrational.' Functionally speaking, this resistance to criticism serves to keep the critics externally located, as 'others.' And while none of this is news, it maintains itself within the institutional characteristics of technoscience's own belief structure.

The success of this science/religion inversion is instanced in the child's text-book version of Columbus's voyage. We grew up believing that while Columbus knew or at least believed the world to be spherical (hence he was a rational, scientifically informed navigator), his crew believed the world to be flat (hence they were religious and superstitious) and if he went too far they would fall of the end of the earthocean. This myth about the fifteenth century, as Valerie Flint showed in her *The Imaginative Landscape of Christopher Columbus* (Princeton, 1992), was itself an early twentieth century invention. Indeed, the time of invention, now incorporated into our pre-late twentieth century de-construction of Columbian history, was when the Scopes trial was underway. What Flint showed was that even the most moderately informed individual of the fifteenth century believed the earth to be spherical, with the exception of a very small, obscure group of "flat-earth" sectarians. The early twentieth century inventors of the rational-scientific vs. superstitious-religious binary simply elevated the texts of the flat-earthers beyond proportion and claimed this was a widespread belief. But it was an apparently successful polemic which did get institutionalized into our science-dominated education insofar as many children still believe in the invented story. And this is but one example of a dominant mythology which still functions.

However, the science/religion inversion is too general to account for the resistance to an institutionalized technoscience criticism. Instead, I wish to turn our attention to two features of technoscience which are both more deeply embedded in technoscience praxis and which relate more closely to the art-literature criticism analogy. The first relates to science texts:

Bruno Latour's *Science in Action* argues that science-as-institution has successfully created a social form which contains its own form of critique into carefully constructed modes of contestation. Science-as-institution incorporates structured trials of strength through which controversies are settled without damage to the basic institution itself.

In a strategy derived both from phenomenology and deconstruction, Latour deliberately inverts what we usually take to be the scientific self-interpretation—the interpretation which usually begins with an inquiry into Nature and ends with a well formulated 'law' or theorem concerning a natural phenomenon finally promulgated in a text—by beginning with results, a 'text' or scientific article, and working reconstructively backwards to appeals to Nature.

Typically, scientific 'texts' or literature appear as articles in scientific journals. I cannot here trace out the extreme complexity of the construction of such articles, which in Latour's interpretation are carefully crafted results of trials of strength, but everyone is familiar with the fact that virtually all such articles are (a) multi-authored, (b) written in a deliberately anonymous or authorless style, and (c) couched in both quantitative and visualized chart forms. Here, already, is a very 'unliterary' form.

Latour asks: who reads such texts? and how are such texts—for my purposes here—'criticized' or challenged? First, most people do not read such 'texts' at all! Rather, the readers are already usually members of a select community. Indeed, the technical opacity of such texts is part of the form of the text itself. The text "puts off" any ordinary reading. The text does not invite one in unless the reader is already 'an expert' in that style of reading.

But, then assuming that one knows how to 'read' such a 'text,' what are the possible outcomes? Latour cites three: first, one simply "goes along" with the text. One accepts, believes, and if in the field, quickly incorporates the findings which then become part of the larger system of science-as-institution. Or, if you wish to challenge or criticize the 'text,' you find that it can't be (often) done through textual criticism per se, but have to go to an entirely different level— you have to go to the laboratory which produces the conditions for the text. Were there to be a counterpart requirement for literature, it would be one which required us to return to the Agora to refute Plato!

> Here is Latour's version of what happens: The peculiarity of the scientific literature is now clear: the only three possible readings all lead to the demise of the text. If you give up, the text does not count and might as well not have been written at all. If you go along, you believe it so much that it is quickly abstracted, abridged, stylized and sinks into tacit practice. Lastly, if you work through the authors' trials, you quit the text and enter the laboratory. Thus the scientific text is chasing its readers away whether or not it is successful. Made for attack and defense, it is no more a place for a leisurely stay than a bastion or bunker. This makes it quite different from the reading of the Bible, Stendhal or the poems of T. S. Eliot.
>
> (Latour, *Science in Action*, p. 61)

I would like to go on into the context of the laboratory and on into the appeal to Nature which is where the leaving of the text leads in technoscience, but I cannot. I can only say, that to challenge the text at the level of the laboratory ultimately calls for one to construct a counter-laboratory if the challenge is to be carried through. This is a thoroughly technologically embodied process which implies money, gangs of operators, and even an educational process. But it also leaves the potential critic in a very unusual and uncomfortable position.

This points to the second inherent problem for the development of technoscience criticism. And that is the dimension of knowledge-power, or better put, knowledge-expertise which functions within late modern technoscience. In this context I shall turn to Raphael Sassower for illumination from his *Knowledge Without Expertise*. In this book, Sassower turns to an interesting early modern example which occurred in the British Association for the Advancement of Science. The debate internal to the BAAS was whether to keep or expel Section F from its ranks and thereby effectively 'defrock' economics from the status of being a 'science.' This historical example excellently shows how knowledge relates to power, but in the particular form of 'expertise' whereby only 'experts' are empowered to make decisions. Generalizing on this modern form of technocracy, Sassower notes:

> If accepted, the myth [of expertise] has an immediate pragmatic consequence since it suggests that only experts can and should make decisions about their speciality, and that only experts in the same field may judge each other's decisions. What about the non-experts? They seem unqualified to be external reviewers of the decisions of experts, for they do not possess the specialized knowledge that qualifies experts to make certainty claims. In this sense, then the myth of expertise guarantees, . . . that experts judge other experts and that experts are shielded and even insulated from public reproach.
>
> (Sassower, *Knowledge Without Expertise*, p. 65)

I probably need not remind many here how the myth of expertise operates as a two-edged sword in so many academic contexts: in philosophy, for example, shall the dominant philosophical traditions (by number still analytic philosophers) be the sole arbiters of what counts as philosophy? or, does the counter-ploy of counter-expertise, only Continental philosophers should judge Continental results, come into play? But in the realm of technoscience criticism the usual role expertise plays relates to the claim on the part of science-as-institution is that only the scientifically informed may be certified as critics.

I want here to enter two examples of criticism in action, to show how the critic is initially 'other,' or made 'other,' external to institutionalized technoscience. The first instance is autobiographical and as a critic (a philosopher) I was already identified as an external critic:

The occasion is an interdisciplinary panel of scientists, convened to debate the issue then facing Long Islanders about the Shoreham nuclear plant. The plant had just gone low-level operational, prior to the approval of an evacuation plan (imagine evacuating Long Island!) in the case of a nuclear disaster. One of the panelists was Max Dresden, an acerbic and outspoken physicist who in his past had also been associated with the Manhattan Project. His presentation turned out to be a defense of expertise and a diatribe against even allowing public discussion of expert conclusions. He contended (this was before Chernobyl) that nuclear energy was the cleanest, safest, and ultimately the cheapest source of electric power and that it was 'irrational' to oppose—out of ignorance—the opening of the Shoreham plant.

During the discussion, I entered the fray, at first provoking Max to reiterate in even stronger terms his defense of expertise—he now claimed that no one should be allowed to vote on issues of such technical complexity. So, I asked him whether the Shoreham debate was a 'scientific' or a 'political' debate and he eagerly admitted, ruefully, that it was 'political.' I then asked him if he were expert at politics, and he huffed and said no, to which I replied, then, according to your expertise argument you ought to have nothing to say about politics, but merely leave it up to the political process to have its day. (Of course, everyone knows, the Shoreham plant has been decommissioned.) The next day at the Faculty Club, Max came over and as loudly as possible attacked me, saying he wished the entire Philosophy Department could be dismantled given its 'antiscientific' tendencies! I probably need not further explicate how this illustrates the externalization of criticism from within an institutionalized 'myth of expertise.'

The second instance is one which begins with the critic as in insider, a "whistle blower" example: I suspect everyone here remembers the news coverage of the 1991 "Gulf War." It was a trial run of one of our "Star Wars" developments, the anti-missile missile, the "Patriot." The newsbroadcasts showed over and over again the presumed 'interceptions' and claimed hits up to 95% effectiveness. If, then, you followed the more critical analyses to follow, you will probably recall that there was an admission that effectiveness or 'hits' declined to about 24%. Part of this admission was due to the early-on analysis performed by Theodore Postal, a ballistics expert and MIT scientist who took news videotapes used to make the hit claims and subjected them to magnified, enhanced, and computer image techniques which on closer inspection showed that claimed hits were not hits at all. Eventually, he concluded that there may not have been a single, verifiable hit which had been made by a Patriot! Needless to say, this claim was not appreciated by Raytheon, the manufacturer of the missile, nor by his colleague, Shaoul Ezekiel, who had advised Raytheon, and eventually not even by MIT itself which got caught in the cross-fire of claims and anti-claims.

The battle turned nasty: Raytheon implied that Postal had actually doctored the tapes, but later reduced this to the claim, suggested by Ezekiel, that the grain structure and imaging of video tapes was simply too gross to draw the conclusions drawn. The battle continues to this day, particularly between Postal and Ezekiel concerning ethical conduct, with MIT trying to shy away due to the large amounts it gets annually from Raytheon. (see *Science*, 23 February 1996, pp. 1050–1052).

Nor is this some isolated instance. In a study of the "costs of whistle blowing" *Science* (5 January 1996, p. 35) reports that more than two thirds of whistle blowers (within science as an institution) experience negative effects ranging from 'ostracism' through 'pressure to drop allegations,' to the actual nonrenewals or losses of jobs. The long drawn out 'David Baltimore' case is another of these scenarios, in which the whistle blower—not the offender who faked the notebooks—was fired. The insider critic is isolated and, if possible, often separated and thus made into an outsider or 'other.'

While the above scenario would not be much different for business corporations, neither would we be surprised about this ostracization from the corporate sector within business, but for the popular image of science as being more like a Church in the claims about critical concern for truth, this may come as a surprise, although not for those of us close enough to realize that science-as-institution is today much more like the corporate world than it is a church!

II. Science Criticism

The implicit trajectory clearly shows that there is a role for science criticism. This is not to say there is none, quite to the contrary, the examples cited shows that there is both external and internal criticism which does take place, regardless of costs. But, equally, the role of criticism is not one which parallels the role of the art or literary critic, nor is science criticism validated in the same way. Rarely, unless the criticism is so extreme as to provoke public outrage, is an art or literary critic fired, ostracized, or threatened. Moreover, the sector from which one would expect such an institutionalized criticism to originate, namely the philosophy of science, has also not performed this task adequately.

I do not have time here to trace out the reasons for this lacuna, other than to suggest that the heretofore dominant traditions of the philosophy of science (derived from Positivist and analytic traditions, more recently from pragmatic analytic traditions) did indeed take the passionate view of their subject matter which the presumed literary critic takes towards literature, but the result was not criticism so much as an attempt to justify and even to imitate science, in short, to make philosophy more 'like' science.

And when philosophy of science did become normative, it did so in the name of an idealized rationalistic conceptualism. Early positivist attempts to isolate the pure logical form of science and then normatively judge science resulted in the laughable results which proclaimed such sciences as geology "unscientific" or to relegate most of the biological sciences into a kind of 'softness' akin to sociology.

Two other areas of philosophy came a little closer to establishing science criticism: I refer to the various types of 'applied ethics' domains which arose most prominently in the medical contexts, and aspects of the philosophy of technology, in which case the degree of tellingness of critique remains marked by the accolade "anti-technology," as noted by Winner above. But each is at best a partial success. Applied medical ethics has learned and has become partially institutionalized inside medical schools and hospitals, and it does perform evaluative and reflective exercises. Philosophers of technology, on the other hand, have been prone to be far too generic in criticism, both by reifying technologies as Technology with the capital "T," and by making too sweeping claims about 'alienation,' the subsumption of 'Nature' to 'Technology,' etc. And, throughout, both external and internal critics remain to be taken as 'others.' So, do we end with failure? with the impossibility of science criticism in anything like an analogy to art and literary criticism?

III. WHAT WOULD A SCIENCE CRITIC LOOK LIKE?

Given this state of affairs, it now behooves me to make some projections. If I am calling for science critics, what would they look like? What would they do? And, where would they be?

Continuing to pursue the art-literature analogy, I would say the science critic would have to be a well-informed, indeed much better than simply a well-informed amateur, in its sense as a 'lover' of the subject matter, and yet not the total insider. Increasingly some philosophers of science have caught this: Ian Hacking, in his work on science instruments, particularly the microscope, has called for philosophers to "go native" in some degree. And, I also agree with him that this is more a matter of learning science practice with its forms of tacit and operational know-how than it is a matter of conceptual analysis.

Secondly, again like the art or literary critic, it is probably equally important that the amateur not be a fully practicing artist or literary writer. Just as we are probably worst at our own self-criticism, that move just away from self-identity is needed to position the critical stance. Something broader, something more interdisciplinary, something more 'distant' is needed for criticism.

So far, so much like art and literary criticism. But I think technoscience is in many ways a special case which calls for more than occurs in art and literary criticism. For one thing, art and literary criticism still follows the 'authorship paradigms' of its subject matters. Critics remain, like writers and artists, individuals who both do and like to sign their own names, to be personally responsible, and thus bear both the praise and blame for results. Science is not like that: its style is anonymous, impersonal, and above all, corporate or intersubjective. But underneath, even the process of discovery is an intersubjective and increasingly multiple perspectived process. I should like to suggest that the critic, the philosopher, must more thoroughly enter into this process.

To do this, one aspect is collaboration. At Stony Brook, we have one example of such a collaboration in the results from our Logic Lab and the co-authored work of Patrick Grim and Gary Mar on computer modeled philosophical problems. Their research—although not an example of science criticism in any direct sense—has reached as high as notice from *Scientific American* (April, 1993) and is noted for the innovations in fuzzy logics and game theory based on cooperative models. We simply must 'gang up' and produce through the works of cooperative and intersubjective work, our critical results.

Another aspect, related both to the 'going native' and the collaborative result, is that the science critic must get in on the origins rather than the results of the technoscientific process. I have long argued that one flaw in applied ethics fields is that they are like the ambulance corps which attends the battlefield—they fix up the wounded but do not either prevent the battle or ameliorate its consequences. Only when the critic is, in this metaphor, present at the strategy planning of the generals can the critic hope to affect the outcomes.

Here examples of critical participation at foundational stages are even rarer and harder to broach since the exclusionary forces within science-as-institution mitigate against such presence. But I can cite examples (I cite a lately discovered such example in my *Philosophy of Technology* [Paragon, 1993]), mostly some I have discovered in the last several years in trips to northern European technical universities.

In Scandinavian and Dutch technical universities, philosophers have found themselves within research teams and while sometimes assigned the evaluation and consideration of ethical and social outcomes in assessment contexts, sometimes other skills are called for. And, increasingly, I have found myself drawn into these contexts by being asked to review and respond to research design.

I will end with one autobiographical example: In Denmark there is a team of researchers who are dealing with a certain problem of medical crises which occur in operating rooms. Picture the patient, unconscious and anesthetized, but hooked up to an array of machines around the room which give readings of vital signs. Dials, audible alarms, oscilloscopes, all are part of the hermeneutic display to be 'read' by the practitioners. In turn, each device is programmed to go off at a pre-set level. During the operation, however, most alarms that go off, experienced physicians have learned, are 'false alarms' and here one reaches a certain critical juncture. If one ignores the alarm and it is 'genuine,' clearly there is a danger to the patient; yet, on the other hand, if the alarm is 'false' and one stops to fix the situation, other dangers occur.

The dilemma discovered was that the most experienced physicians were more likely to ignore the alarms—sometimes with disastrous results—than less experienced physicians. So, the problem became one of how does one determine a 'real' from a 'false' crisis? Moreover, this was determined to be an explicitly 'hermeneutic' problem, a matter of right reading. But, like science, the result is not the fictive world, but the 'nature' or patient beyond the instrumental texts. What does the critic do, or what can the critic do? It is here that I place him or her to enact 'science criticism.'

JULIA ANNAS is Regents Professor in the Department of Philosophy at the University of Arizona. She has published widely in ancient philosophy, particularly epistemology and ethics, and more recently on modern virtue ethics. Her most recent book is *Platonic Ethics Old and New* and she is working on a book on virtue ethics.

SCOTT BREWER joined the Harvard Law School faculty in 1991, after having taught undergraduate philosophy courses at Harvard, Yale, and Dartmouth, and after clerking for Judge Harry T. Edwards (U.S. Circuit Court for the District of Columbia) and Justice Thurgood Marshall (U.S. Supreme Court). He received a J.D. degree and M.A. in philosophy from Yale and a Ph.D. in philosophy from Harvard. His teaching and scholarship focus on jurisprudence, evidence, and contracts.

ROBERT P. CREASE is a Professor in the Department of Philosophy at Stony Brook University and a historian at Brookhaven National Laboratory. His most recent books are *The Prism and the Pendulum: The Ten Most Beautiful Experiments in Science*, *Making Physics: A Biography of Brookhaven National Laboratory*, and *The Play of Nature: Experimentation as Performance*.

H. M. COLLINS is Distinguished Research Professor of Sociology and Director of the Centre for the Study of Knowledge Expertise and Science at Cardiff University. In 2004, the University of Chicago Press published his 860 page *Gravity's Shadow: The Search for Gravitational Waves*; in 2005, they will publish his coauthored *Dr. Golem: How To Think About Medicine*. He is working on a book on expertise.

HUBERT DREYFUS is Professor of Philosophy in the Graduate School at the University of California at Berkeley. His publications include: *What Computers (Still) Can't Do*, *Being-in-the-World: A Commentary on Division I of Heidegger's Being and Time*, *Mind over Machine: The Power of Human Intuition and Expertise in the Era of the Computer* (with Stuart Dreyfus), and *On the Internet*.

ROBERT EVANS is a Senior Lecturer in Sociology at the Cardiff School of Social Sciences. His research draws on the sociology of scientific knowledge to examine the role of citizens and experts in decisions involving science. These concerns are reflected in the special issue of *Science Technology and Human Values* that he edited, with "Demarcation Socialised" as its theme.

PAUL FEYERABEND taught at the University of California in Berkeley. His most famous publications are *Against Method, Science in a Free Society, Science as an Art, Three Dialogues on Knowledge,* and *Beyond Reason.*

STEVE FULLER is Professor of Sociology at the University of Warwick, England. He is best known for the research program of social epistemology. His chapter in this book is elaborated in *Knowledge Management Foundations.*

ALVIN I. GOLDMAN is Board of Governors Professor at Rutgers University, with appointments in Philosophy and the Center for Cognitive Science. His most recent books are *Knowledge in a Social World* and *Pathways to Knowledge: Private and Public,* both published by Oxford University Press. He is a former President of the American Philosophical Association (Pacific Division).

JOHN HARDWIG is a Professor and head of the Department of Philosophy at the University of Tennessee. He now works primarily in bioethics. A collection of his essays in bioethics was recently published as *Is There a Duty to Die? and Other Essays in Bioethics.*

DON IHDE is Distinguished Professor of Philosophy and Director of the Technoscience Research Group at Stony Brook University. Recent books include *Chasing Technoscience: Matrix for Materiality* (edited with Evan Selinger), *Bodies in Technology,* and *Expanding Hermeneutics: Visualism in Science.*

HÉLÈNE MIALET is a Visiting Assistant Professor at the University of California, Berkeley. She has held positions at Cornell and Oxford University and postdoctoral fellowships at the Max Planck Institute and Cambridge University. Her book, *The Subject of Invention,* will be published this winter by the Presses Universitaires de Grenoble. She is currently finishing a book on Stephen Hawking entitled *Hawking Incorporated,* which is under contract with the University of Chicago Press.

JOHN MIX, a consultant to nonprofit organizations in Manhattan, is an independent researcher with interests in technoscience, particularly environmental sciences, ecology, and evolution.

EVAN SELINGER is an Assistant Professor in the Department of Philosophy at the Rochester Institute of Technology. His most recent books are *Chasing Technoscience: Matrix for Materiality* (edited with Don Ihde) and *Expanding (Post) Phenomenology: A Critical Companion to Ihde.*

EDWARD W. SAID was a Professor of English and Comparative Literature at Columbia University. A prolific writer, his books included *Orientalism; The World, the Text and the Critic; Blaming the Victims; Culture and Imperialism; Peace and Its Discontents: Essays on Palestine in the Middle East Peace Process; End of the Peace Process: Oslo and After;* and *Power, Politics, and Culture.*

PETER SINGER is Ira W. DeCamp Professor of Bioethics at Princeton University, and also holds an appointment as Laureate Professor at the University of Melbourne. His books include *Animal Liberation, Practical Ethics, How Are We to Live?, Rethinking Life and Death, Writings on an Ethical Life, One World,* and *Pushing Time Away.* Together with Renata Singer, he has coedited *The Moral of the Story: An Anthology of Ethics Through Literature.* He is also the author of the major article on ethics for the *Encyclopedia Britannica.*

STEPHEN TURNER is Graduate Research Professor, Department of Philosophy, University of South Florida. He has written extensively on the relation of politics and science and on the social and political implications of expertise. His books in this area include a disciplinary history, *The Impossible Science: An Institutional Analysis of American Sociology* (with Jonathan Turner), *Sociology Responds to Fascism* (edited with Dirk Käsler), and most recently, *Liberal Democracy 3.0: Civil Society in an Age of Experts.*

ACKNOWLEDGMENTS

Chapter 1. Reproduced with permission from *Philosophy and Phenomenological Research* 63:1 (2001): 85–109.

Chapter 2. Reproduced with permission of Harry Collins and Robert Evans from *Social Studies of Science* 32:2 (2002): 235–296. Copyright (© Sage Publications 2005) by permission of Sage Publications Ltd.

Chapter 3. Reproduced with permission of The Yale Law Journal Company and William S. Hein Company from *The Yale Law Journal*, vol. 107, 1535–1681.

Chapter 4. Reproduced with permission of Steven Turner from *Social Studies of Science* 31:1 (2001): 125–149. Copyright (© Sage Publications 2005) by permission of Sage Publications Ltd.

Chapter 5. Reproduced with permission of Peter Singer from *Analysis* 32 (1972): 115–117.

Chapter 6. Reproduced with permission of Hubert Dreyfus and Routledge Publishing from *On the Internet*.

Chapter 7. Reproduced with permission of Springer Science and Business Media, from the Kluwer Academic Publishers journal *Continental Philosophy Review* 35 (2005): 245–279.

Chapter 8. Reproduced with permission of Hélène Mialet from *Social Studies of Science* 29:4 (1999): 551–582. Copyright (© Sage Publications 2005) by permission of Sage Publications Ltd.

Chapter 9. Reproduced with permission of Julia Annas and Cambridge University Press from *Social Philosophy & Policy* 18:2 (2001): 236–256. Copyright © Social Philosophy & Policy Foundation.

Chapter 10. Reproduced with permission of Springer Science and Business Media, from the Kluwer Academic Publishers journal *Phenomenology and the Cognitive Sciences* 3:2 (2004): 145–163.

Chapter 11. Reproduced with permission of John Hardwig and *The Journal of Philosophy* 82 (1985): 335–349.

Chapter 12. Reproduced with permission of Steven Fuller and the *International Journal of Expert Systems* 7:1 (1994): 51–64.

Chapter 13. Reproduced with permission from *Radical Philosophy* 11 (1975): 3–8.

Chapter 14. Reproduced with permission from *Critical Inquiry* 9:1 (1981): 236–256, published by The University of Chicago Press.

Chapter 15. Reproduced with permission from Don Ihde and *International Studies in Philosophy* 29 (1997): 45–54.

Printed in the USA
CPSIA information can be obtained
at www.ICGtesting.com
JSHW021321221024
72173JS00012B/1634/J

9 780231 136440